T0256636

Bayesian Models

Bayesian Models

A Statistical Primer for Ecologists

N. Thompson Hobbs and
Mevin B. Hooten

PRINCETON UNIVERSITY PRESS
PRINCETON AND OXFORD

Published by Princeton University Press, 41 William Street, Princeton, New Jersey 08540
In the United Kingdom: Princeton University Press, 6 Oxford Street, Woodstock,
Oxfordshire OX20 1TW

press.princeton.edu

Library of Congress Cataloging-in-Publication Data

Hobbs, N. Thompson.
Bayesian models : a statistical primer for ecologists / N. Thompson Hobbs and Mevin B. Hooten.
pages cm
Includes bibliographical references and index.
ISBN 978-0-691-15928-7 (hardcover) 1. Ecology–Statistical methods.
2. Bayesian statistical decision theory. I. Hooten, Mevin B., 1976– II. Title.
QH541.15.S72H63 2015
577.01′5195–dc23
2015000021

British Library Cataloging-in-Publication Data is available

This book has been composed in Times LT Std and Futura Std

Typeset by S R Nova Pvt Ltd, Bangalore, India

1 3 5 7 9 10 8 6 4 2

Contents

Preface

Why This Book?

This book is about the process of gaining new knowledge about ecology using models and data. We wrote it for several reasons. An overarching motivation is that our satisfaction with scientific work has been palpably enhanced by understanding this process from start to finish, with no gaps where faith must fill in for understanding. We write this book because what we write about is easily the most intellectually satisfying material we have learned in our careers. It has made the work we do every day more enjoyable. We are confident that your research in ecology will be accomplished with greater satisfaction and reward if you master the concepts we describe here.

There are more specific motivations as well. We teach graduate students ecological modeling using Bayesian methods. Colleagues whose students have taken our classes often ask us what they should read to get a big picture understanding of the Bayesian approach. They seek a book explaining statistical principles of Bayesian modeling written in language accessible to nonstatisticians. They may never write a line of computer code implementing Bayesian methods, but they realize the importance of understanding them. Our colleagues want to be able to appreciate contemporary scientific literature, to review papers and proposals, and to mentor students who *do* know how to write code. We decided to write the book they asked for.

Of course, there are now many excellent texts on Bayesian modeling in ecology (e.g., Clark, 2007; McCarthy, 2007; Royle and Dorazio, 2008; Link and Barker, 2010; Kéry, 2010; Kéry and Schaub, 2012), texts that we use all the time. For the most part, these books emphasize computational methods over concise explanation of basic principles. Many of these are difficult to appreciate without a background in mathematical statistics—training that most ecologists lack. Our book will complement the existing crop by providing the basic understanding of mathematical and statistical principles needed to use more advanced texts effectively.

A third motivation also comes from our personal experience. Students and colleagues often come to us with a modeling problem they are working on, a problem that uses Bayesian methods. They can write code in one of the popular implementations of Gibbs samplers, WinBUGS (Lunn et al., 2000) or JAGS (Plummer, 2003), and often they bring a thick stack of code with them. Although they can create computer programs that give them answers, they cannot write the mathematical expression for the model that underpins their work. They are unsure about their starting point and hence are not entirely confident about where they have ended up. They have difficulty writing a mathematical expression that clearly communicates their analysis in manuscripts and proposals. As you will see as this book unfolds, we believe that reliable analysis must begin with a model written in mathematical symbols and operators, not in a computer language. Writing models is the foundation of good science.

The ability to write models leads to our next motivation. There is a diminishing set of important questions in ecology that can be answered by a single investigator working alone, even a very good one. Instead, what remains are problems solvable only by the application of intersecting sets of talents, skills, and knowledge. This book offers hold-in-your-hand evidence of the value of collaboration between a statistician and an ecologist, but there are many other examples (Gross et al., 2002, 2005; Clark, 2003b, 2005; Latimer et al., 2006; Farnsworth et al., 2006; Cressie et al., 2009; Rotella et al., 2009; Webb et al., 2010; Eaton and Link, 2011; Wilson et al., 2011; Fiechter et al., 2013; Peterson et al., 2013). These collaborations require a mutual understanding of basic principles and a shared vocabulary.

Our final reason for writing this book relates to the first one. Most ecologists working today, save perhaps some of the youngest ones, received training in statistics emphasizing procedures over principles. We emerged from this training understanding how to do "data analysis" using a suite of recipes—t-tests, analysis of variance, regression, general linear models, and so on. The diligent and ambitious among us might have added work in sampling and multivariate statistics. Mathematical statistics was reserved for statistics majors.

The outcome of this approach to training is seen in a revealing way in the frontispiece of a once widely used text (Sokal and Rohlf, 1995), which displays a table of analyses resembling a dichotomous key in taxonomy. If our data are like x, then we should use analysis y; otherwise, we should use analysis z. Those of us trained this way had precious little understanding of *why* we should use analysis y over z beyond the authority provided by that table. Moreover, there was a limited range of kinds of data for which this taxonomic approach would serve, and we were stymied if the observations we worked hard to obtain were not found somewhere in the table. Sometimes we would heroically bend those observations to make them fit in one of the table's narrowly defined cells. A narrow range of approaches to analysis constrains the questions that ecologists are willing to ask—we are uncomfortable posing questions for which we see no analytical route to insight. If the only tool in our locker is analysis of variance, then the world we study must be composed of randomized plots. Some of the current Bayesian books are organized in a similar way—around procedures rather than principles.

All research problems that ecologists seek to solve have aspects in common and aspects that are unique. The unique features of these problems argue for a principled approach to insight. A lean set of modeling principles can substitute for volumes of statistical facts. Understanding these principles enables us to design routes to insight uniquely suited to each of the diverse problems we confront over the course of a research career. Providing this understanding is the main reason for writing this book.

Goals

The overarching goal of this book is to train ecologists in the basic statistical principles needed to use and interpret Bayesian models. An essential part of meeting that goal is to teach how to write out accurate mathematical expressions for Bayesian models linking observations to ideas about how ecological systems work. These expressions form the foundation for inference. It is our ultimate aim to increase the intellectual satisfaction of ecologists with their teaching, research, and peer review by providing a solid, intuitive understanding of how we learn from data and models in the Bayesian approach. Finally, we aim to enhance the quality of collaboration between ecologists and statisticians by training ecologists in the mathematical concepts and language of statistics.

Approach

Organization

The book is organized to provide the understanding needed to support a general process for model building in ecology, a process that applies to virtually all research problems. We outline a flow of tasks in building models that we have found helpful explaining the process (figure 0.0.1). The sequence here is not immutable, but it offers a useful schematic of the steps needed to build a revealing Bayesian model. We will return to this diagram throughout the book to show how specific topics fit into the larger task of model building.

We have organized the book in three parts. The aim of part I is to provide a basic understanding of the principles underpinning Bayesian analysis (fig. 0.0.1). We begin part I with a preview of the entire book to motivate what will follow. We then spend some time encouraging thinking about deterministic models and how they have been traditionally used in ecology (fig. 0.0.1 A). Next, we cover basic principles of probability and probability distributions. Think of this as a crash course in mathematical statistics, a prospect that might not be thrilling on the face of it, but we urge you to read this material carefully if it is not familiar to you. Gaining familiarity will allow you to understand the powerful ideas that follow and, ultimately, to understand what you are doing when applying Bayesian methods to your own research.

Knowledge of statistical distributions provides the foundation for understanding maximum likelihood approaches to parameter estimation. Likelihood is a central element of Bayesian models. We explain the link between likelihood and Bayesian approaches to inference by explaining the theory underpinning the Bayesian approach. We develop the concept of Bayes as "likelihood reweighted." We then dissect Bayes' theorem piece by piece to explain its components: the posterior distribution, the likelihood, the prior, and the marginal distribution of the data. We introduce some uniquely Bayesian concepts likely to be unfamiliar to most ecologists, for example conjugate relationships between likelihoods and priors, which turn out to be critical in the second part of the book. We finish part I by applying the basic statistical concepts we have developed to specific problems in ecological research, illustrating how mathematical expressions are built to link models and data (fig. 0.0.1 B). We show how models of complex phenomena, models with many parameters and latent quantities, can be broken into manageable chunks using hierarchical modeling.

Part II lays out the nuts and bolts of how we use the principles developed in part I to learn about parameters, unobservable states, and derived quantities. We clearly explain Markov chain Monte Carlo and Gibbs sampling, the numerical algorithms that have revolutionized the ability to gain insight from hierarchical models (fig. 0.0.1 C). Part II closes by describing how we check models to assure their fidelity to statistical assumptions and how we make inference from a single model or from multiple models (fig. 0.0.1 D).

Part III includes a series of problems and worked solutions drawn from several subdisciplines of ecology. These problems require application of the concepts and principles we developed in part I and II, emphasizing model specification (fig. 0.0.1 B), a skill that we believe is not emphasized in other texts. Our intention in part III is to encourage active model building and to show how the same approach to model specification can be fruitfully applied to a broad range of problems in ecology.

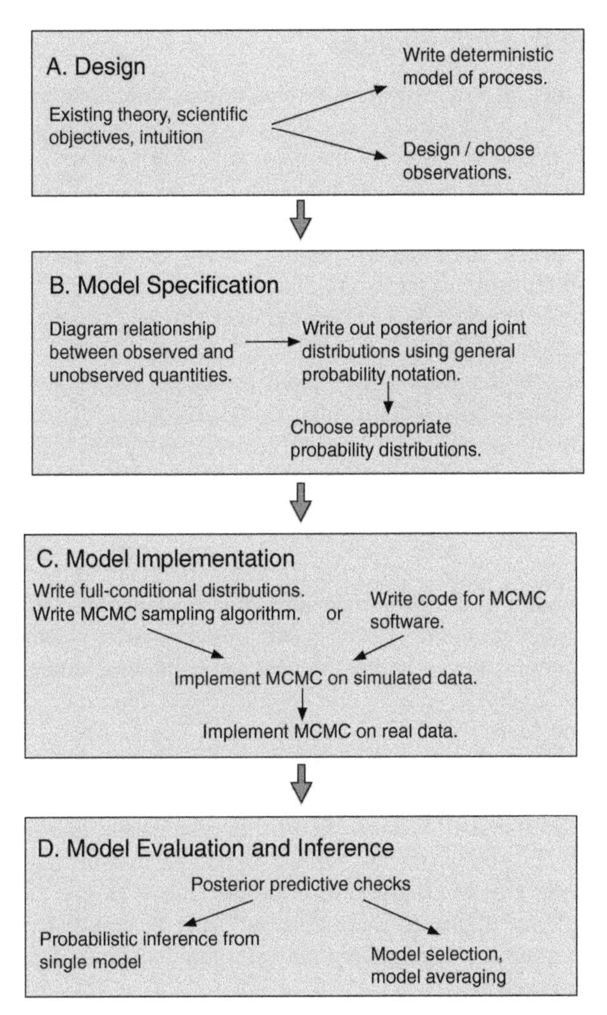

Figure 0.0.1. Gaining insight from Bayesian models involves the same sequence of steps for virtually all research problems, steps that fall into four broad groups. The sequence of steps is indicated by the long and short arrows. This book is organized to explain these steps in a logical way. (**A**) Design is not uniquely Bayesian, but we include it here because we want to encourage the thoughtful development of mathematical models of ecological processes as a starting point for analysis (chapter 2). (**B**) The premise of this book is that mathematical models must be combined with data to allow us to learn about how ecological systems operate. Chapters 3, 4, 5, and 6 show how we specify models to include data. (**C**) A key idea is that a properly specified model provides all we need to know to implement the enormously powerful algorithm Markov chain Monte Carlo (MCMC). We provide a principled understanding of how and why MCMC works in chapter 7. (**D**) We then cover how to use output from MCMC as a basis for inference from single models (chapter 8) and from multiple ones (chapter 9). Finally, we return to the key process of model specification (**B**) in chapter 11 by providing a series of problems challenging you to formulate Bayesian models.

Crosscutting Themes

We err on the side of excessive explanation of notation and equations. We believe that a formidable impediment to understanding statistics by ecologists is that ecologists, for the most part, don't get up in the morning every day and write statistical models. The authors of most statistical textbooks *do*. Consequently, notation that is compact and efficient in the eyes of the practiced is murky for the rest of us. We promise to use a consistent notation in expressions that are fully explained, believing that clarity trumps elegance. We cannot avoid equations, but we can strive to make them understandable.

We use model diagrams[1] throughout the book. These sketches portray stochastic relationships among data, unobserved states, and parameters. We show how these diagrams, properly composed, provide a blueprint for writing out Bayesian models, and ultimately, for designing the samplers that allow us to obtain probability distributions of the quantities we seek to understand.

Unlike many existing books on Bayesian methods, ours will not emphasize computer code written in any specific language. We will teach algorithms but not coding. We choose this approach because there are an increasing number of software packages that implement algorithms for Bayesian analysis (e.g., Lunn et al., 2000; Plummer, 2003; INLA Development Team, 2014; Stan Development Team, 2014). The diversity of this software is likely to expand in the future. Including today's favorite flavor of code in our book assures that it will become obsolete as a new favorite emerges. Moreover, writing your own algorithms in the programming language of your choice is not all that difficult and offers the added benefit of forcing you to think through what you are doing. We make a single exception to this "no-code" rule in part II by including a script to illustrate the relationship between mathematical expressions and their implementation.

[1]Formally known as directed acyclic graphs or, more succinctly, Bayesian networks.

I Fundamentals

1 Preview

Art is the lie that tells us the truth.
—*Pablo Picasso*

All models are wrong but some are useful.
—*George E. P. Box*

Pablo Picasso was a contemporary of George Box's, a statistician who had an enormous impact on his field, writing influential papers well into his 90s. Both men sought to express truth about nature, but they used tools that were dramatically different—Picasso brushing strokes on canvas, and Box writing equations on paper. Given the different ways they worked, it is remarkable that Box and Picasso made such similar statements about the central importance of abstraction to insight. Abstraction plays a role in all creative human enterprise—in art, music, literature, engineering, and science. We create abstractions because they allow us to focus on the most important elements of a problem, those relevant to the objectives of our work, without being distracted by elements that are not relevant.

Scientific models are, above all else, abstractions. They are statements about the operation of nature that purposefully omit many details and thus achieve insight that would otherwise be discursively obscured. They provide unambiguous statements of what we believe is important. A key principle in modeling and statistics—in science for that matter—is the need to reduce the dimensions of a problem. A data set may contain a thousand observations. By reducing its dimensions to a model with a few parameters, we are able to gain understanding.

However, because models are abstractions and reduce the dimensions of a problem, we must deal with the elements we choose to omit. These elements create uncertainty in the predictions of models, so it follows that assessing uncertainty is fundamental to science. Scientists, journalists, logicians, and attorneys alike can rightly claim to make statements based on evidence, but only scientific statements include evidence tempered by uncertainty quantified. We know what is certain only to the extent that we can say, with confidence, what is uncertain. Sharpening our thinking about uncertainty and learning how to estimate it properly is a main theme of this book.

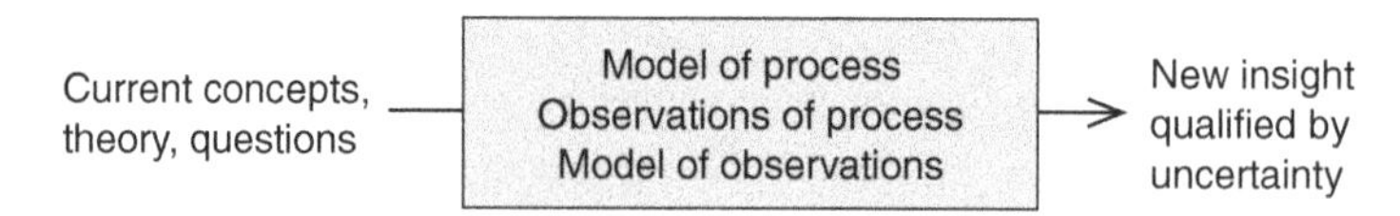

Figure 1.1.1. The fundamental challenge in ecological research is to establish a credible line of inference extending from concepts and theory to new insight tempered by uncertainty.

Your science will have impact to the extent that you are able to ask important questions and provide compelling answers to them. Doing so depends on establishing a line of inference that extends from current thinking, theory, and questions to new insight qualified by uncertainty (fig. 1.1.1). This book offers a highly general, flexible approach to establishing this line of inference. We cannot help you pose novel, interesting questions, but we can teach an approach to inference applicable to an enormous range of research problems, an approach that can be understood from first principles and that can be unambiguously communicated to other scientists, managers, and policy makers. We emphasize that understanding the principles of this framework allows you to customize your analyses to accommodate the inevitable idiosyncrasies of specific problems in research.

We sketch that framework in this chapter to give a general sense of where this book is headed, a preview we use to motivate the development of concepts and principles in the chapters that follow. There should be details of our approach that are unfamiliar, otherwise you probably don't need this book. Soon enough, we will explain those details fully. For now, we offer a somewhat abstract overview followed by a concrete example as an enticement to read on. It will be rewarding to return to this section after you have worked through the book. We hope you will be pleasantly surprised by your increased understanding of our small preview. The only part of this chapter that is essential for the remainder of the book is understanding our notation, which we describe in section 1.1.1.

1.1 A Line of Inference for Ecology

Virtually all research problems in ecology share a set of features. We want to understand how the state of an ecological system changes over time, across space, or among individuals. We seek to understand why those changes occur. Our understanding usually depends on a sample drawn from all possible instances of the state because we want to make statements about a system that is too large to observe fully. The observations in that sample are often related imperfectly to the true state. In the subsections that follow, we lay out an approach first described by Berliner (1996)[1] for modeling the imperfect observations that arise from a process we want to understand. It does not apply to all research problems, but it is sufficiently general and flexible that it applies to most.

1.1.1 Some Notation

Before we proceed, we must introduce some notation. Boldface lowercase letters will indicate vectors (e.g., $\boldsymbol{\theta}, \mathbf{a}$), and lightface lowercase letters, scalars (θ, a).[2] Bold capital letters will be used for matrices (e.g., $\mathbf{A}$). The symbol $\boldsymbol{\theta}$ will indicate a vector of parameters, and, of course, θ will indicate a single parameter.[3] The letter $\mathbf{y}$ will indicate a vector of data, $\mathbf{Y}$ a matrix, and y or y_i a single observation. Corresponding notation using $\mathbf{x}$, x, and x_i will be used for predictor variables, also called covariates. The notation $[a|b,c]$ will be used for the probability distribution of the random variable a conditional on the parameters b and c.[4] Deterministic models will be denoted by $g()$ with arguments necessary for the model within the parentheses. Notation will be added as needed, in context.

1.1.2 Process Models

Process models include a mathematical statement that depicts a process and a way to account for uncertainty about the process. To compose a process model we start by thinking about the true state (z) of an ecological system. That state could be the size of a population, the flux of nitrogen from the soils of a grassland, the number of invasive plants in a community, or the area of landscape annually disturbed by fire. We seek to understand influences on that state, the things that cause it to change. We write an equation,[5] a deterministic model that represents our ideas about the

[1]For elaboration, see Wikle (2003); Clark (2003b); Cressie et al. (2009), and Wikle et al. (2013).

[2]It might be useful to review scalars, vectors, and matrices. Most ecologists are familiar with matrices—rows and columns of numbers. Vectors are "one dimension" of a matrix, that is, a row or a column. Scalars are a single element. So, a matrix might be $\mathbf{A} = \binom{c\ d}{d\ f}$; a vector, $\mathbf{a} = (c, d)'$; and a scalar, c. We will use the notation $(\,)'$ to list the elements of a vector.

[3]We illustrate with the familiar example $y = \beta_0 + \beta_1 x$. The vector of parameters from the model is $\boldsymbol{\beta} = (\beta_0, \beta_1)'$. When discussing parameters generically, we will use $\boldsymbol{\theta}$.

[4]The symbol "|" reads conditional on.

[5]Or equations. To keep things simple, we focus on a single equation here, but we might build a system of equations making multiple predictions of states.

behavior of the state of interest and the quantities that influence it.[6] When we say the model is deterministic, we mean that for a given set of parameters and inputs, it will make precisely the same predictions. We use the notation $g(\boldsymbol{\theta}_p, \mathbf{x})$ to represent the deterministic part of a process model, where $g()$ is any mathematical function, $\boldsymbol{\theta}_p$ is a vector of parameters in the model, and $\mathbf{x}$ is one or more explanatory variables that we hypothesize influence the true state.

Our deterministic model is an abstraction, so it follows that we have omitted influences on the true state from the model, and we must deal with our omissions. If we model aboveground net primary production of a grassland as a function of growing season precipitation, we have brushed aside the influence of grazing intensity and precipitation that occurs during the dormant season. If we model reproductive success of individuals as a function of age and genotype, we have ignored variation contributed by their nutritional status. A model of harvest from a fishery based on observations of stock size and sea temperature omits the effect of variation in the food web. We recognize that these neglected influences shape the behavior of the true state by treating them stochastically, by estimating a parameter, σ_p^2, that subsumes all the unmodeled influences on the true state. Including this stochastic component allows us to estimate a statistical distribution (fig. 1.1.2A) for the true state,

$$\underbrace{\left[z|g\left(\boldsymbol{\theta}_p, \mathbf{x}\right), \sigma_p^2\right]}_{\text{process model}}, \tag{1.1.1}$$

where the bracket notation $[z|g(\boldsymbol{\theta}_p, \mathbf{x}), \sigma_p^2]$ means the distribution of z conditional on $g(\boldsymbol{\theta}_p, \mathbf{x})$ and σ_p^2.[7] If the notation is somewhat unfamiliar at this point, don't worry; equation 1.1.1 simply says that if we know the functional form $g()$ and the values of $\boldsymbol{\theta}_p$, $\mathbf{x}$, and σ_p^2, we can specify the probability distribution of the true state, z (fig. 1.1.2 A).

We want to determine the probability distribution of the true state as well as the probability distributions of the parameters in our model. Doing so

[6]As a tangible example, we might model the influence of phytoplankton biomass (x) on zooplankton biomass (z) with a linear model, $z = \theta_0 + \theta_1 x$. In this case $z = g(\boldsymbol{\theta}, x) = \theta_0 + \theta_1 x$.

[7]This notation was first introduced by Gelfand and Smith (1990) as a way to reduce clutter. It has become widely used by statisticians and ecologists. There is a small caveat needed here. We are using the arguments for probability distributions as if they were the mean ($g(\boldsymbol{\theta}_p, x)$) and variance ($\sigma_p^2$) of the distributions. The arguments for most of the statistical distributions are parameters that are functions of the mean and variance, a topic we treat in detail in section 3.4.4. For now, we take the liberty of treating the mean and variance as the needed parameters because they are familiar to ecologists.

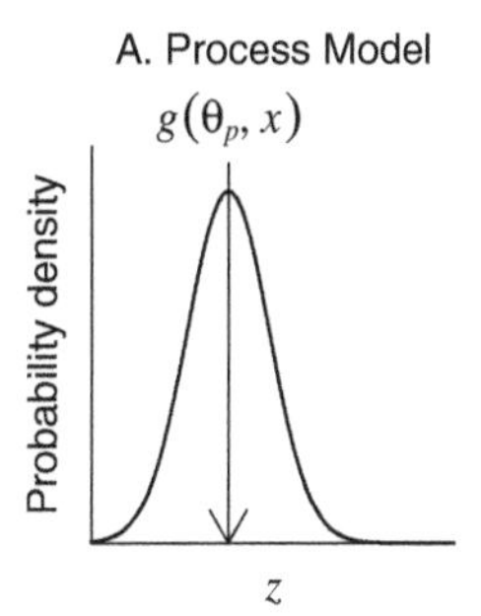

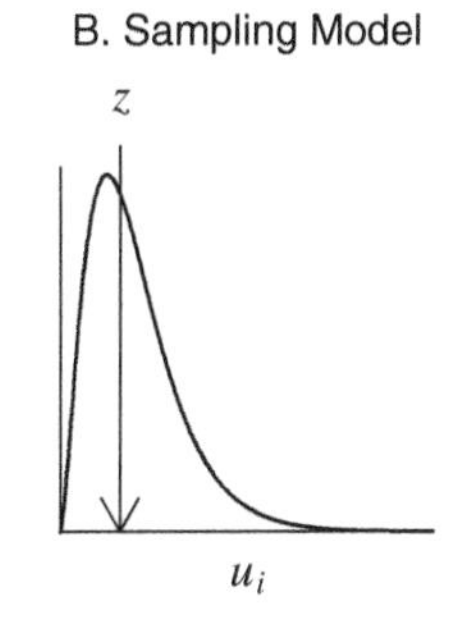

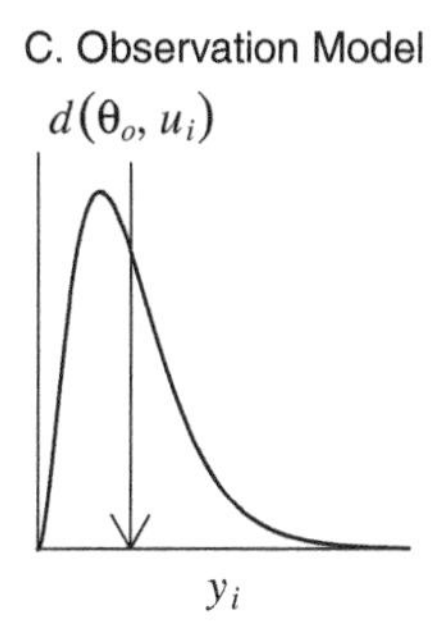

Figure 1.1.2. An observation on the operation of an ecological process (y_i) is linked to ideas about how the process works via three linked probability distributions. Hierarchical models consist of linked distributions. As we move from panel **A** to **C**, notice how the random variables (on the x-axis) becomes the the mean (or other central tendency) of the distribution in the next panel. **(A)** The deterministic model, $g(\boldsymbol{\theta}_p, \mathbf{x})$ predicts a true state, z, as a function of parameters $\boldsymbol{\theta}_p$ and predictor variables $\mathbf{x}$. There is uncertainty in the predictions, σ_p^2, that arises because there are influences on z that are not represented in the model. Thus, the distribution in **A** is $[z|g(\boldsymbol{\theta}_p, \mathbf{x}), \sigma_p^2]$. If the deterministic model predicts z well, then the distribution in **A** shrinks toward the $g(\boldsymbol{\theta}_p, \mathbf{x})$ arrow. **(B)** There are almost always more instances of z in nature than we can hope to observe. Observations of individual instances of z define a sampling distribution, $[u_i|z, \sigma_s^2]$, the breadth of which depends on σ_s^2. As variation among observations declines, the distribution in **B** shrinks toward the z arrow. Note that the mean of the distribution shown by the arrow is not at the peak of the distribution, because the distribution is skewed. **(C)** Observations (y_i) of instances of the true state u_i are often biased, such that $y_i \neq u_i$. A deterministic observation model, $d(\boldsymbol{\theta}_o, u_i)$, corrects for this bias. Uncertainty in this correction (σ_o^2) leads to the observation distribution $[y_i|d(\boldsymbol{\theta}_o, u_i), \sigma_o^2]$. As uncertainty in the observation model declines, the distribution in **C** shrinks toward the $d(\boldsymbol{\theta}_o, u_i)$ arrow.

requires evaluating the predictions of the process model against data. The data can be obtained in experiments or observational studies; they can be measurements we plan to collect or have already collected. This linkage between process models and observations is discussed next.

1.1.3 Sampling Models

We can rarely observe all instances of the true state in the system we study. Instead, we take a sample of $i = 1, \ldots, n$ observations of the true state and we notate the ith observation as u_i. This sample might be biomass from plots on a grassland landscape where we seek to understand the true state, aboveground productivity. It might be presence or absence of an exotic fungus on trees in a stand where we want to understand infestation of the stand. It might be classifications of zooplankton in aliquots from a stream where we want to estimate the stream's species richness. Uncertainty arises because our sample assuredly will not represent the true state perfectly. Again, we represent this uncertainty stochastically using a probability distribution relating the true state to an observation

$$\underbrace{\left[u_i|z,\sigma_s^2\right]}_{\text{sampling model}}, \tag{1.1.2}$$

where σ_s^2 represents sampling variation (fig. 1.1.2 B). Expression 1.1.2 implicitly assumes that we can observe instances of the true state without bias, which simply says that if we collect many observations, then the average (i.e., also called the *expected value* of the observations, $\text{E}(u)$) of the observations equals the mean of the distribution of the true state, $\text{E}(u) = z$. (Expectation will be treated in detail in chapter 3.) We realize samples of the true state is a nuanced concept—soldier on, things will become clear in the next section.

1.1.4 Observation Models

The assumption that we can observe the true state perfectly may not be reasonable. When we count animals, some are overlooked. When we use Lidar[8] to estimate the heights of 10,000 trees, we do not measure the height of each tree using a ladder and a meter tape (thankfully) but instead observe backscatter from a laser beam. When we measure nitrogen mineralization, we do not follow the fate of individual nitrogen atoms but measure the net change in the extractable soil ammonium pool over time. The mismatch between what we observe and the true state requires a model of the observations, which we notate as $d(\boldsymbol{\theta}_o, u_i)$, where $\boldsymbol{\theta}_o$ are parameters. It is important to understand that u_i is the quantity we would observe if we could *perfectly* observe the instance of the true state in a draw from all the instances, without any bias injected by our observation process. We use y_i to notate the actual measurements we have in hand, including error resulting from the way we observe the u_i. The observation model serves to eliminate the bias found in our observations, y_i relative to an instance (u_i) of the true state drawn from the distribution of z. The probability distribution of the observations (fig. 1.1.2 C) arising from the

[8]Lidar, light detection and ranging, a technique used in remote sensing.

observed instances of the true state is

$$\underbrace{\left[y_i|d(\boldsymbol{\theta}_o, u_i), \sigma_o^2\right]}_{\text{observation model}}, \tag{1.1.3}$$

where the σ_o^2 represents all the influences on the y_i that are not represented in $d(\boldsymbol{\theta}_o, u_i)$. As a simple example, the σ_o^2 could be the variance of the predictions of a regression model used to calibrate observations against true values. We emphasize that model $d(\boldsymbol{\theta}_o, u_i)$ is needed to offset bias in the y_i. If our observations are unbiased, then there is no need for an observation model, (i.e., $y_i = u_i$) and the uncertainty in our data arises solely from sampling variation (i.e., eq. 1.1.3). For simplicity, we have ignored the sampling variation and observation errors that might influence the **x**, but these could be handled in the same as we have done for the **y** (i.e., we would use probability distributions like eqs. 1.1.2 and 1.1.3).

1.1.5 Parameter Models

Because the approach we sketch is Bayesian, we also require models of the parameters expressing what we knew about the parameters when we began our investigation, that is, our *prior* knowledge. This knowledge is expressed in probability distributions, one for each parameter we seek to estimate[9]

$$\underbrace{[\boldsymbol{\theta}_p]\,[\boldsymbol{\theta}_o]\left[\sigma_p^2\right]\left[\sigma_s^2\right]\left[\sigma_o^2\right]}_{\text{parameter models}}. \tag{1.1.4}$$

These distributions must have numeric arguments that specify our current knowledge of the probability distribution of the parameters. The arguments can be chosen to make the distributions informative or vague, but as you will see, we will encourage you to make priors as informative as knowledge and scholarship allows. We might know a lot about a parameter or we might know very little.

1.1.6 The full Model

We are now equipped to write a mathematical expression representing our ideas about the operation of an ecological process linked to data in a way

[9]Equation 1.1.4 requires the assumption that the parameters $\boldsymbol{\theta}_p, \boldsymbol{\theta}_d, \sigma_p^2, \sigma_s^2$, and σ_o^2 are statistically independent. We will explain this idea in greater detail later.

that includes all sources of uncertainty—in the process, in our sample of the process, and in the way we observe it (fig. 1.1.2)

$$\left[\underbrace{z, \boldsymbol{\theta}_p, \boldsymbol{\theta}_o, \sigma_p^2, \sigma_s^2, \sigma_o^2, u_i}_{\text{unobserved}} \mid \underbrace{y_i}_{\text{observed}}\right] \propto$$
$$\underbrace{\left[y_i | d(\boldsymbol{\theta}_o, u_i), \sigma_o^2\right]}_{\text{observation model}} \underbrace{\left[u_i | z, \sigma_s^2\right]}_{\text{sampling model}} \underbrace{\left[z | g\left(\boldsymbol{\theta}_p, \mathbf{x}\right), \sigma_p^2\right]}_{\text{process model}} \underbrace{\left[\boldsymbol{\theta}_p\right]\left[\boldsymbol{\theta}_o\right]\left[\sigma_p^2\right]\left[\sigma_s^2\right]\left[\sigma_o^2\right]}_{\text{parameter models}}. \tag{1.1.5}$$

Equation 1.1.5 is Bayesian and hierarchical.[10] It is *Bayesian* because it treats the unobserved quantities as random variables. This treatment allows us to make statements about the probability distributions of all of the unobserved quantities based on the observed ones. It is *hierarchical* because the z and u_i are found on both sides of a conditioning symbol " | ", illustrating a powerful tool for simplifying problems that we will discuss more fully soon. The observation model includes our knowledge of the relationship between the true state and our observations of it and the uncertainty that occurs because that relationship is imperfect. The sampling model includes the uncertainty that comes from observing a subset of instances of the true state. The process model represents our hypotheses about the ecological process by specifying a probability distribution defining our knowledge and our uncertainty about the true state and the factors that control its behavior. The parameter models allow us to exploit previous estimates of parameters that we have made ourselves or that have been made by others. Together these models provide a line of inference extending from concepts to insight for a broad range of research problems (fig. 1.1.1). We can use equation 1.1.5 to obtain estimates of unobserved states, parameters, and quantities of interest derived from parameters and states. All of these estimates are properly tempered by uncertainty in a statistically coherent way.

In the remainder of this book, we develop the principles needed to understand equation 1.1.5 and to apply it to research problems in ecology. We tailor it to match the needs of the particular problem at hand. But first, we provide an example of its use.

[10]The symbol $\propto$ means "is proportional to." It may be confusing at this point that the observed $\mathbf{x}$ does not show up in the list of observed quantities on the left-hand side of the proportionality. Bear with us. We will make that clear as we proceed.

1.2 An Example Hierarchical Model

We are now going to apply the general framework we have previewed to a specific problem. Remember that we don't expect this section to be familiar to you. Although the application here focuses on learning about population dynamics of a large mammal from a time series of observations in East Africa, it could be about any topic, any research design, and any location. We urge you to draw analogies to your own work as the example develops.

The Serengeti wildebeest (*Connochaetes taurinus*) population migrates across the grasslands of Tanzania and Kenya in an annual cycle driven by availability of green plant biomass (Boone et al., 2006). During the late 1800s, the viral disease rinderpest created a panzootic, decimating the wildebeest and other wild and domestic ruminants. Numbers of these animals remained low until the 1950s when a campaign to vaccinate cattle in the pastoral lands surrounding the Serengeti eliminated the virus in wildlife (Plowright, 1982). Survivorship of wildebeest, particularly juveniles, doubled after rinderpest was eradicated, and the population grew rapidly until the mid 1970s when density-dependent mortality produced a quasi-equilibrium. This steady state appears to have been caused by intraspecific competition for green plant biomass during the dry season. This biomass varies in approximate proportion to annually variable rainfall (Mduma et al., 1999).

Developing informative models depends on a clearly stated question. An unambiguous question is vital because it guides the abstraction that we formulate; it informs our decisions about which variables and parameters we need to include in our model. In this particular example we ask the question: How does variation in weather modify feedbacks between population density and population growth rate in a population of large herbivores occupying a landscape where precipitation is variable in time? Answering this question requires a model that portrays density dependence, effects of precipitation, and their interaction.

1.2.1 Process Model

We pose the deterministic model portraying growth of the wildebeest population as

$$N_t = g\left(\boldsymbol{\beta}, N_{t-1}, x_t, \Delta t\right) = N_{t-1}e^{(\beta_0+\beta_1 N_{t-1}+\beta_2 x_t+\beta_3 N_{t-1}x_t)\Delta t}, \qquad (1.2.1)$$

where N_t is the unobserved, true abundance of wildebeest in year t (corresponding to z in eq. 1.1.5); x_t is a measure of the annual rainfall that influences population growth during $t - 1$ to t; and $\Delta t = 1$ year.[11] The meaning of the coefficients $\boldsymbol{\beta}$ requires some thought. A critical part of gaining insight from models and data is careful thinking about the biological meaning of the parameters in our models and their relationship to existing concepts and theory, the need for which is illustrated here.

We first consider the intercept, β_0, a logical place to start. If the units of precipitation, x_t, are centimeters per year, ranging from 0 to a large number, then algebra dictates that $e^{\beta_0 \Delta t}$ is the proportional change in the population during the period t to $t + 1$ that occurs when abundance is zero and rainfall is zero. The abundance equals zero part might seem a bit odd at first glance, because there is no population to grow when $N_t = 0$, but the parameter nonetheless serves to define the upper limit on population growth rate as numbers approach zero. There is no problem here. But the part about zero rainfall is troubling because it makes β_0 difficult to interpret biologically or at least makes it somewhat unuseful in its biological interpretation. Why would we want to estimate a parameter determining population growth when the population is low and rainfall is low? Alternatively, if we define x_t in terms of divergence from the long-term average by subtracting the mean rainfall from each year's observation, then the definition of β_0 becomes more sensible. In this case, $e^{\beta_0 \Delta t}$ is the proportional change in population size that occurs when the population is zero and rainfall is average. This definition allows us to relate our model to well-established theory; β_0 is analogous to the intrinsic rate of increase in the logistic equation.[12]

Once we have defined the intercept in a sensible way, the slope terms are easily interpreted. The parameter β_1 represents the strength of density dependence, that is, the change in the per capita population growth rate that occurs for each animal added to the population.[13] Again relating our model to classical theory, $\beta_1 = -(r_{max}/K)$, where K is the carrying capacity, that

[11]The identical form $\log(N_t) = \log(N_{t-1}) + (\beta_0 + \beta_1 N_{t-1} + \beta_2 x_t + \beta_3 N_{t-1} x_t)\Delta t$ might be more familiar to ecologists accustomed to a general linear modeling framework.

[12]This might require a bit of explanation. Equation 1.2.1 can be rearranged to $\log(N_t/N_{t-1}) = (\beta_0 + \beta_1 N_{t-1} + \beta_2 x_t + \beta_3 N_{t-1} x_t)\Delta t$. If we drop the $\beta_2 x_t$ and $\beta_3 x_t N_{t-1}$ terms and assume Δt is small, the resulting expression approximates $1/N \cdot dN/dt = r - rN/K$, where N is the population size, r is the intrinsic rate of increase, and K is the carrying capacity—the long-term average population size at which $1/N \cdot dN/dt = 0$. Thus, the portion of our model representing density dependence is a version of the logistic equation in discrete time, also known as the *Ricker equation*, where $\beta_0 = r$, and $\beta_1 = r/K$.

[13]You might be tempted to make the sign for β_1 negative, because density dependence reduces the per capita population growth rate as population size increases. There is a good reason to use addition. When the model is fit to data, we want to estimate the sign and the value of β_1 without needing to "back transform" the sign. Always use addition in additive models that you will fit to data.

is, the population size at which the long-term average population growth rate equals zero. The parameter β_2 expresses the strength of the effect of variation in rainfall, that is, the change in the per capita population growth rate per unit deviation from the long-term mean. Finally, the parameter β_3 determines the magnitude of the effect of rainfall on the effect of density.

Estimating the β's will inform the question we posed, but we don't pretend that they capture all the influences on wildebeest population dynamics. We recognize that there are other processes, for example, predation, poaching, and disease, that shape wildebeest numbers over time. We can acknowledge these other processes exist without portraying them explicitly. Instead, we lump these unmodeled effects into a single parameter, σ_p^2. The model of the process for a population at time t including deterministic and stochastic components is

$$\left[N_t | g\left(\boldsymbol{\beta}, N_{t-1}, x_t, \Delta t\right), \sigma_p^2\right]. \tag{1.2.2}$$

1.2.2 Sampling Model

The wildebeest population was estimated[14] on 20 occasions during 1961–2008 using spatially replicated counts of animals on georectified aerial photographs arrayed along transects, with each photograph covering a known area (Norton-Griffiths, 1973). For simplicity, we ignore the aspect of sampling contributed by the transects and treat the photographs as if they were a random sample from the area used by the population. Thus, we assume that each photograph provided a statistically independent estimate of population density calculated as the observed count divided by the area covered by the photograph. The stochastic sampling model representing the relationship between these observations and the true population size is

$$\left[y_{tj} \middle| \frac{N_t}{a}, \sigma_s^2\right], \tag{1.2.3}$$

where y_{tj} is the density of animals at year t on photograph j, and a is the total area from which the sample of photographs was drawn—the area used by the population we are modeling—which we assume to be a known constant. Note that equation 1.2.3 implies that there is no bias in the estimate of animal density on a given photograph, which is the assumption made

[14]Portions of this time series have been published in Hilborn and Mangel (1997) and Mduma et al. (1999). The most recent data were graciously provided by A.R.E. Sinclair, Ray Hilborn, and Grant Hopcraft. We use these data later to illustrate making inference from a single model.

by the researchers who collected the data. Although this assumption is reasonable for large animals counted in open habitat, we could obviate the need for this assumption by modeling the animals that were present and not observed.

As a purely hypothetical example, imagine that we fitted a sample of animals with high-resolution telemetry instruments that would allow us to know whether each animal was within a photograph. Imagine also that we marked them with highly visible neckbands to allow us to distinguish between animals that were fitted with instruments and those that were not. We could use these observations to estimate the probability (ψ) that an animal truly present was counted. In this case our combined observation and sampling model would be

$$\underbrace{\left[y_{tj}|\psi n_{tj}, \sigma_o^2\right]}_{\text{observation model}} \underbrace{\left[n_{tj} \middle| \frac{N_t}{a}, \sigma_s^2\right]}_{\text{sampling model}}, \tag{1.2.4}$$

where n_{tj} are animals that are truly present on photograph j during year t, and σ_o^2 is the uncertainty associated with the estimate of ψ. We could also deal explicitly with the likely cases in which it was not certain whether an animal was on a sampled photograph.

To simplify the example and to exploit published data, we make the heroic assumption that rainfall at time t, x_t, is measured without error; we are treating it as known, but we are not obliged to do so. We could develop sampling models and observation models for the rainfall data in the same way we did for the count data (i.e., eqs. 1.2.3 and 1.2.4).

1.2.3 Full Model

Combining the sampling model (eq. 1.2.3) and process model (eq. 1.2.2) with models for the parameters, we obtain the foundation for a full Bayesian analysis capable of estimating the parameters and the unobserved, true population size (fig. 1.2.1); however, work remains to be done. The full model written thus far (fig. 1.2.1) applies to a single year of observations and a single photograph, but, of course, we want to use all the years and all the observations within a year. We have not yet chosen specific probability distributions for the stochastic components, and we must do so in a sensible way. We may need to deal with correlation among photographs or among annual estimates in the time series of data, and with errors in the estimation of rainfall. We need to lay out a method for numerically estimating the parameters and unobserved states. However, although tasks remain, all

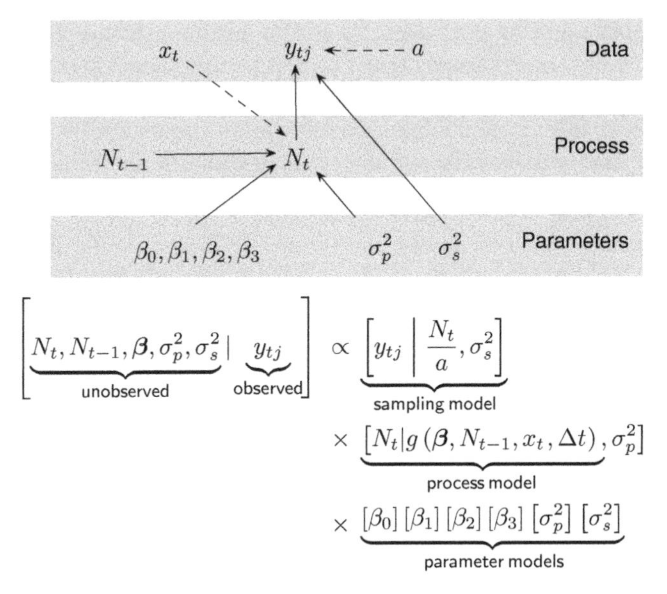

Figure 1.2.1. Hierarchical Bayesian model of the dynamics of the Serengeti wildebeest population. The true population size at time t is modeled using the deterministic model $g(\boldsymbol{\beta}, N_{t-1}, x_t, \Delta t) = N_{t-1}e^{(\beta_0+\beta_1 N_{t-1}+\beta_2 x_t+\beta_3 N_{t-1} x_t)\Delta t}$, which represents the effects of the true, unobserved population size (N_t); the dry season rainfall (x_t); and their interaction on population growth. The biological interpretation of the parameters $\boldsymbol{\beta}$ are given in the text. The parameter σ_p^2 represents all the effects on wildebeest abundance not represented by the $\boldsymbol{\beta}$. The y_{tj} is a single observation of population density obtained from one of $j = 1, \ldots, n_t$ photographs chosen from area $= a$ km^2. The parameter σ_s^2 represents sampling error. The arrow diagram is a Bayesian network, also called a *directed acyclic graph*, that can be used as a visual guide for properly writing the full model. The solid lines show stochastic relationships. The dashed lines show deterministic relationships, implying that the quantities at the tails of the arrows are known without error. Instructions for drawing diagrams like this one will be developed in subsequent chapters.

proceed in a logical way from the foundation we have built (fig. 1.2.1). All are manageable. Completing them allows us to reliably estimate unobserved states, parameters, and derived quantities of interest.

1.3 What Lies Ahead?

One of the aims of this book is to enable ecologists to write equations for models that allow data to speak informatively. Our goal is to provide an understanding of principles needed to accomplish this vital task. We will show how an enormous range of research problems in ecology can be decomposed into a set of sensible parts, as we have illustrated. The specific example we offered no doubt raised more questions than it answered, which was our intention. Notable among these questions might be the following:

1. Why does the Bayesian approach to analysis work? What are the probabilistic foundations of inference from models like this one?
2. How does a Bayesian analysis relate to analyses based on maximum likelihood?
3. How do we choose the appropriate statistical distributions for the stochastic components of the model (fig. 1.2.1)?
4. How can we incorporate multiple sources of data in estimates of parameters and states?
5. How do numerical methods allow us to implement the model and obtain results? How do these methods work?
6. How can we evaluate the model to assure it adequately represents the data?
7. What can we conclude about the operation of the ecological process based on estimates of the parameters and states? What do we do about derived quantities; for example, how do we make inferences about the population size where growth rate is maximum?
8. What if we want to evaluate the strength of evidence for this model (equation 1.2.1) relative to competing ones, for example, a model with nonlinear density dependence?
9. How do we predict future states with honest estimates of uncertainty?

Answering these questions in a clear and accessible way for a broad range of research problems in ecology is the goal of remainder of this book.

2 Deterministic Models

We write models to express our ideas about how ecological processes work (figure 0.0.1 A). We then use data to decide if those ideas are any good. Combining data with models provides insight about ecology by answering a fundamental question: What is the probability that I will observe the data if the model faithfully represents the processes that give rise to the data? Our starting point for answering this question will always be consideration of an ecological process. We formalize our thoughts by writing a deterministic equation or equations making predictions that can be compared with observations (fig. 0.0.1 A). There is no uncertainty in our predictions because the model is deterministic. In this chapter we talk about deterministic models as expressions of ecological hypotheses. In the next chapter we lay out the principles of probability that allow us to combine these models with data in a way that provides new knowledge qualified by uncertainty.

2.1 Modeling Styles in Ecology

We wrote this book to be broadly useful across all the subdisciplines of ecology, offering a way of working with models and data that is recognizably valuable, whether you study genes or ecosystems. To that end, we briefly discuss the different ways that deterministic models have been used in ecology. We want to ensure that ecologists who work on different kinds of problems recognize a familiar style of modeling, and have confidence that what lies ahead is relevant to their own research.

Modeling has formed a productive, diverse activity throughout the history of ecology (Kingsland, 1985; Golley, 1993). Subdisciplines of ecology have developed different modeling "cultures," as is plainly seen in textbooks about quantitative ecology, all of which teach how to use models of one type or another. Despite their ostensibly shared focus on modeling, it is striking how starkly *different* these books are; some lack a single topic in common in their tables of contents (e.g., compare Schwartzman and Kaluzny, 1987; Gotelli and Ellison, 2004; Otto and Day, 2007; Railsback and Grimm, 2012).

The marked divergence among these texts reflects fundamental differences among modeling traditions in ecology. We identify three main ones, which we call theoretical, empirical, and simulation (table 2.1). We do not argue that these modeling styles are distinct; of course, there is overlap among them. Nonetheless, categorizing them this way offers a useful heuristic that helps us organize our thinking about how models have been used in the past and how they might be used more fruitfully in the future if each style could exploit the strengths of the others.

2.1.1 Theoretical

Seminal ideas in ecology were motivated by mathematical analysis of systems of deterministic, nonlinear differential and difference equations (Levins, 1969; May, 1977; Anderson and May, 1979; MacArthur and Wilson, 2001). No data are required in this modeling style—insight comes directly from mathematical properties of the equations forming the model. Discovering equilibrium points and conditions for local stability—the workhorses of ecological theoreticians—are examples of this type of insight (Edelstein-Keshet, 1988; Otto and Day, 2007). The absence of data from these models and their simplicity are what allows them to be used to generalize about the entire range of possibilities for ecological states. When data are used to specify values for parameters, the rich range of behavior of these models is abruptly compressed to the specific behavior dictated by such data (Levins, 1966). The theoretical style depends on simple models, because nonlinear systems with many state variables[1] and parameters are mathematically intractable.

The great strength of the theoretical tradition in ecological modeling is that the parameters and state variables are defined in biological terms. They symbolize quantities in nature, and the model symbolically portrays how a process works. These models rigorously express ideas about *why*

[1] *State variables* are the quantities of interest in models that change over time or space. Parameters describe how those changes occur. For example, N is a state variable in the logistic equation, $dN/dt = rN(1 - N/K)$, and r and K are parameters.

TABLE 2.1
Comparison of Prevailing Modeling Styles in Ecology

	Theoretical	*Empirical*	*Simulation*
Number of variables	few	few	many
Requires data for insight?	no	yes	yes
Fields in which most often applied	theoretical ecology	all subdisciplines of ecology	ecosystem ecology, landscape ecology
Definition of parameters	as biological quantities	as relationships in data	as biological quantities
Basis for uncertainty	NA*	probability theory	ad hoc
Example texts	Edelstein-Keshet (1988); Otto and Day (2007); Diekman et al. (2012)	Sokal and Rohlf (1995); Gotelli and Ellison (2004)	Schwartzman and Kaluzny (1987); Railsback and Grimm (2012)

* Uncertainty refers to things that are unknown. Theoretical models may use stochastic, simulation, but they do not distinguish between quantities that are known and those that are unknown.

outcomes occur by abstracting the mechanism that produces them. Thus, the theoretical style is a particularly valuable way to develop hypotheses, which, in turn, are often tested using empirical models.

2.1.2 Empirical

Empirical models describe relationships in data. The preponderance of modeling in ecology has used this style in regression, analysis of variance, and related analyses. All these methods depend on a deterministic model at their core, but many ecological researchers rarely give these models much thought—linear models are a ubiquitous default. The parameters in these models customarily lack portable, biological interpretation; most are defined solely in terms of the data used to estimate them. Nonetheless, purely empirical models describing patterns in data— that is, models with no biological definition for their parameters—have motivated theoretical models that represent the mechanisms producing the patterns (Spalinger and Hobbs, 1992; Ritchie, 1998; Brown et al., 2002), forming a revealing interplay between modeling styles in ecology.

The great strength of the empirical style of modeling is its foundation in statistics, a foundation that allows rigorous statements about uncertainty. The close connection between empirical models and agronomic-style experimental designs means that these models are usually quite simple. Simplicity allows isolation of the influence of a few variables on an ecological state. However, the simplicity of empirical models means they cannot represent multiple interactions and the effects of composites of many variables. The need to represent the interaction of multiple states motivated the use of simulation models.

2.1.3 Simulation

The computing power of the entire U.S. Department of Defense in 1970 was less than that of a smartphone today. Before the arrival of fast, cheap computers, the theoretical and empirical styles of ecological modeling were the only modeling alternatives. A new, third style of modeling based on computer simulation emerged in the mid-1970s with the publication of the first model representing the dynamics of an entire ecosystem (Innis, 1978). The original simulation models did not differ qualitatively from the systems of differential equations that predominated in the theoretical tradition. As in the theoretical style, parameters and state variables included in the simulation approach enjoyed clear biological definitions. The important difference was in dimensionality—the number of state variables included—and as a result, in the analysis of the models. Gaining insight from nonlinear models with many state variables required assigning specific values to parameters and using computer simulation to visualize the change in state variables over time. The simulation approach has expanded dramatically since the 1970s to include various types of spatiotemporal models that exploit geographic information systems. A more recent development is the modeling of interactions among individuals (Judson, 1994; Railsback and Grimm, 2012).

It is often said that ecology is foremost the study of interactions. The strength of the simulation approach is its ability to represent multiple interactions extending across spatial scales and levels of organization, interactions that cannot be included in traditional statistical or mathematical analysis. Including these interactions is possible because simulation models aggregate multiple relationships among variables, relationships that are often developed using empirical and theoretical approaches. These models do not flinch at complexity, dispensing with the reductionism of the empirical and theoretical modeling styles. The primary weaknesses of the simulation approach are the absence of a statistically coherent way to deal with uncertainty and a somewhat cultural tendency to build models of unbridled complexity. Until recently, the problem of estimating uncertainty has been treated in a largely ad hoc fashion, without a principled, probabilistic foundation (but see Clark et al., 2011; Luo et al., 2011; Gudimov et al., 2012; Araujo et al., 2013; Braakhekke et al., 2013; Broquet et al., 2013; Fiechter et al., 2013; Pacella et al., 2013; Wikle et al., 2013).

2.1.4 What Is Needed?

The prevailing modeling styles in ecology have unique strengths: the theoretical approach has encouraged ecologists to write equations portraying how processes work; the simulation approach allows us to represent multiple interactions and dependencies; and the empirical approach reliably includes uncertainty. What is needed is a way to merge styles in a way that exploits the strength of each. The Bayesian hierarchical framework encourages that merger by admitting deterministic models of all types, by allowing use of multiple sources of data, and by providing a broadly useful framework for quantifying uncertainty in multidimensional and complex models. A key feature of this book is that it teaches how to incrementally add complexity to simple ecological models so that uncertainty is properly modeled at each increment in model development. This incremental approach to model building requires a set of building blocks—deterministic models to represent our ideas about the operation of processes, which we treat in this chapter, and probability models to get the uncertainties right, which we describe in the following chapter.

2.2 A Few Good Functions

For the remainder of this chapter we outline a few deterministic models widely used in ecology. We do not attempt a comprehensive treatment, because other sources cover this topic lucidly and in depth (Otto and Day, 2007; Bolker, 2008). However, we will develop familiarity with a set of core functions that are widely used to represent ecological process across all subdisciplines and that will reappear throughout the book, particularly in chapters 6 and 12.

In all areas of ecology we seek to understand processes. We cannot escape the use of models to achieve that understanding; even the empirical tradition depends on them. A deterministic model of an ecological process is a function, $g()$, that returns an output μ representing a true state z based on parameters ($\boldsymbol{\theta}_p$) and observations ($\mathbf{x}$):

$$\mu = g\left(\boldsymbol{\theta}_p, \mathbf{x}\right). \tag{2.2.1}$$

We distinguish the deterministic representation of the true state (μ) and its true value (z) because many influences that shape the behavior of the true state are not included in our deterministic model, as we will soon describe (chapter 3). The range of inputs to the function is called the *domain*, and the range of outputs is called the *range*. For simplicity, we assume the prediction of our model can be represented as a scalar, but it could be a vector, a matrix, or even a multidimensional array. We will use the function $g()$ to represent a deterministic model throughout the book, but the specific symbols chosen for arguments will vary. When more than one deterministic model is needed, we will call $h()$ into service.

2.2.1 Functions for Additive Effects

2.2.1.1 Simple Linear

The linear model

$$g(\boldsymbol{\beta}, \mathbf{x}) = \beta_0 + \beta_1 x_1 + \ldots + \beta_n x_n \tag{2.2.2}$$

and its generalizations are familiar tools in ecological modeling, particularly in the empirical tradition. The model is widely useful because it expresses effects that are additive, providing a simple way to represent the relationship between predictor variables[2] (the x-values) and a response. Interactions can easily be represented with product terms, and curvilinear effects can be portrayed by raising the x to powers, operations that should be familiar from training in basic statistics. We usually start with the additive, linear model if we lack a solid understanding of mechanisms.

We will soon learn that deterministic functions must have an appropriate range for their predictions to be linked with data in a statistically reliable way. For example, the range of equation 2.2.2 is all real numbers, so it would be fine for describing the change in biomass of a forest during an interval of time, because "change in biomass" is a continuous quantity that can be positive or negative. But it would not be strictly correct for describing the stand biomass at any point in time, because biomass cannot be negative. It would also not be appropriate for describing the proportion of a landscape that is burned, a quantity limited to values between 0 and 1. Equations must provide predictions matched to the ecological quantities we wish to understand.[3]

[2] Also called *covariates*.

[3] More formally, predictions must have the proper *support*, a concept we describe fully in the next chapter.

2.2.1.2 Exponential

We can model quantities that are nonnegative while preserving the additive structure of the simple linear model using

$$g(\boldsymbol{\beta}, \mathbf{x}) = \exp(\beta_0 + \beta_1 x_1 + \ldots + \beta_n x_n), \tag{2.2.3}$$

which maps the negative values of equation 2.2.2 to (0, 1) and maps the nonnegative values to $(1, \infty)$ (fig. 2.2.1 A,B). You may have seen this model written as

$$\log(\mu) = \beta_0 + \beta_1 x_1 + \ldots + \beta_n x_n, \tag{2.2.4}$$

using an algebraic rearrangement of equation 2.2.3. This version is widely used in the general linear modeling tradition, where $\log(\mu)$ is called a *link* function, in this case, a log link. When authors state they have log-transformed the response variable for a linear regression or are using a linear model with a log link, they are effectively using an exponential model (eq. 2.2.3).[4] We will use the untransformed version (eq. 2.2.3) in later examples because linear models are to be considered as one of many alternatives for functions used to represent ecological processes—those that are part of the general linear family, as well as those that cannot be transformed to a linear form.

2.2.1.3 Inverse Logit

Many quantities of interest in ecology are proportions—quantities that take on values ranging from 0 to 1. These can be modeled deterministically using the inverse logit function[5] (fig. 2.2.1 C,D),

$$g(\boldsymbol{\beta}, \mathbf{x}) = \frac{\exp(\beta_0 + \beta_1 x_1 + \ldots + \beta_n x_n)}{1 + \exp(\beta_0 + \beta_1 x_1 + \ldots + \beta_n x_n)}. \tag{2.2.5}$$

Equation 2.2.5 returns a real number less than or equal to 1 and greater than or equal to 0. As its name implies, the function is the inverse of the logit function, which, for a proportion p, is defined as

$$\text{logit}(p) = \log\left(\frac{p}{1-p}\right). \tag{2.2.6}$$

[4]Be sure you understand that this is the natural log. Statisticians almost never use the notation ln ( ), preferring log ( ) to mean the same thing.

[5]Also called the *logistic function*.

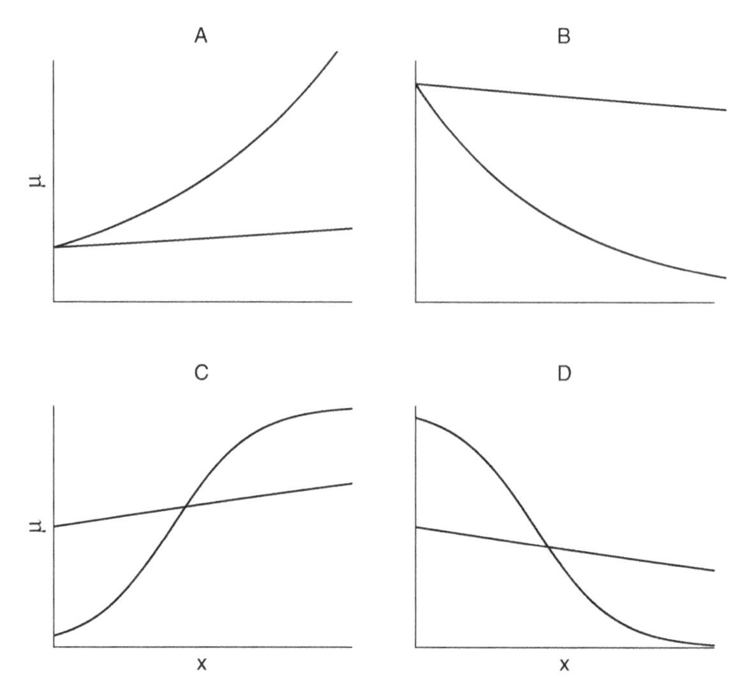

Figure 2.2.1. Examples of an exponential model for additive relationships, $\mu = \exp(\beta_0 + \beta_1 x)$, with a positive (**A**) and negative (**B**) slope. Examples of the inverse logistic model for additive relationships, $\mu = \frac{\exp(\beta_0+\beta_1 x)}{1+\exp(\beta_0+\beta_1 x)}$, with a positive (**C**) and a negative (**D**) slope. In each of the panels, the different curves show how varying the parameter β_1 controls the "straightness" and slope of the curve. Larger values of β_1 cause greater curvilinearity. Smaller values straighten the curve.

Solving for p in

$$\text{logit}(p) = \beta_0 + \beta_1 x_1 + \ldots + \beta_n x_n, \tag{2.2.7}$$

we obtain equation 2.2.5. The logit link function used in generalized linear modeling is equation 2.2.7. If we multiply equation 2.2.5 by a, it returns values between 0 and a maximum a,

$$g(\boldsymbol{\beta}, a, \mathbf{x}) = \frac{a \exp(\beta_0 + \beta_1 x_1 + \ldots + \beta_n x_n)}{1 + \exp(\beta_0 + \beta_1 x_1 + \ldots + \beta_n x_n)}. \tag{2.2.8}$$

Before we move to a second set of functions, it is worth reemphasizing a point we made in the preview (sec. 1.2), namely, it is best practice to compose all the preceding linear models using addition operators even if subtraction might seem to make greater sense biologically. This is because we want the *data* to determine the value of each of the coefficients in the models we fit, including the sign of the coefficient. Using subtraction can lead to confusion and errors, because then we must remember that a negative coefficient for a subtracted value means the sign of the coefficient is actually positive. If we use addition throughout, then the signs of the coefficients are directly interpretable without mental gymnastics.

2.2.1.4 Are These Really "Linear" Models?

When we introduce these models to students, they often respond, "These models are not really linear; both produce "curves" when I plot them. What do I do if I want to fit a straight line to strictly positive data or to data that range between 0 and 1?" The answer, of course, is that both functions can yield results that are "straight"[6] or curved depending on the values of the coefficients (fig. 2.2.1). Thus, we really have the best of both worlds – the exponential and inverse logit functions can have substantial curvature if we need nonlinearity, or can be straight if we don't.

2.2.2 Power Functions

The power function has seen broad application in ecology as a way to describe scaling, that is, how phenomena change as the dimensions of time, space, or mass change. There is an enormous amount of literature on this topic (reviewed by Peters, 1983; Pennycuick, 1992; Schneider, 1997; West et al., 2003; Marquet et al., 2005; Xiao et al., 2011). The power function is

$$g(\boldsymbol{\beta}, x) = \beta_0 x^{\beta_1} \tag{2.2.9}$$

and is often transformed to a linear form. Recall that the untransformed model predicts μ, so a linearized version is

$$\log(g(\boldsymbol{\beta}, x)) = \log(\mu) = \log(\beta_0) + \beta_1 \log(x). \tag{2.2.10}$$

[6]We put *straight* in quotation marks because the functions do not have constant slope. However, the changes in slope can be sufficiently minor that the model provides a fine approximation of a function that truly has a constant slope, as does $g(\boldsymbol{\beta}, x) = \beta_0 + \beta_1 x$.

It is possible to include multiple independent variables ($\mathbf{x}$) in power functions[7] using

$$g(\boldsymbol{\beta}, \mathbf{x}) = \beta_0 x_1^{\beta_1} x_2^{\beta_2} \ldots x_n^{\beta_n}, \tag{2.2.11}$$

or in linear form,

$$\log(\mu) = \log(\beta_0) + \beta_1 \log(x_1) + \beta_2 \log(x_2) + \ldots + \beta_n \log(x_n). \tag{2.2.12}$$

2.2.3 Asymptotic functions

Asymptotic processes are ubiquitous in ecology and for good reason—there are no quantities in biology that can increase or decline without limit. Asymptotes occur in models where the rate of change in the dependent variable slows with increases in the independent variable. These relationships are represented by a family of models, the best known of which, perhaps, is the Michaelis—Menten equation. It was originally formulated as a model of enzyme kinetics,

$$g(v_{max}, k, x) = \frac{v_{max} x}{k + x}, \tag{2.2.13}$$

where v_{max} is the maximum value of the independent variable (the true state, μ), approached asymptotically as x approaches infinity, and k is the half-saturation constant, the value of x where μ is half of its maximum. A commonly used rearrangement with parameters that are somewhat more interpretable biologically is

$$g(v_{max}, \alpha, x) = \frac{v_{max} x}{\frac{v_{max}}{\alpha} + x}, \tag{2.2.14}$$

where the parameter α is defined as the rate of change in μ at low values of x. Equation 2.2.13 has an intercept at $x = 0$, $\mu = 0$. This can easily be modified by adding a constant to x:

$$g(v_{max}, \alpha, c, x) = \frac{v_{max}(x + c)}{\frac{v_{max}}{\alpha} + (x + c)}. \tag{2.2.15}$$

[7] You will also see these written as $g(\boldsymbol{\beta}, \mathbf{x}) = \exp(\beta_0) x_1^{\beta_1} x_2^{\beta_2} \ldots x_n^{\beta_n}$ and $\log(\mu) = \beta_0 + \beta_1 \log(x_1) + \beta_2 \log(x_2) + \ldots + \beta_n \log(x_n)$.

If c is positive, the curve shifts left; if c is negative, the curve shifts right. Exponentiation of x and k in equation 2.2.13 creates the Hill function,[8]

$$g\left(v_{max}, k, q, x\right) = \frac{v_{max} x^q}{k^q + x^q}, \tag{2.2.16}$$

which represents sigmoidal processes with positive slopes when q is positive or negative slopes when q is negative. The magnitude of q controls the steepness of the slope.

The Michaelis–Menten equation is related to another commonly used asymptotic function, best known to ecologists as the model of Type II functional response first described by C. S. Holling (1959). If we substitute $v_{max} = 1/\gamma$ and $k = 1/\alpha\gamma$ in equation 2.2.13, we obtain

$$g\left(\alpha, \gamma, x\right) = \frac{\alpha x}{1 + \alpha\gamma x}. \tag{2.2.17}$$

Other asymptotic functions widely used in ecological models are the monomolecular function,

$$g\left(\alpha, \gamma, x\right) = \alpha\left(1 - e^{-\gamma x}\right), \tag{2.2.18}$$

and the Gompertz function,

$$g\left(\alpha, \gamma, x\right) = e^{-\alpha e^{-\gamma x}}. \tag{2.2.19}$$

2.2.4 Thresholds

Identifying thresholds[9] in ecological processes has emerged as an especially important challenge for ecology in a world experiencing rapid and manifold changes in physical and biological processes. The essential problem is to fit functions that take the form

$$\mu = \begin{cases} g_1\left(\boldsymbol{\theta}_1, x\right) & x < \tau \\ g_2\left(\boldsymbol{\theta}_2, x\right) & x \geq \tau \end{cases}, \tag{2.2.20}$$

where τ is a change point—a parameter we estimate that controls an abrupt change in the process. The change in the process is portrayed by

[8] A. V. Hill (1910) developed this model to represent oxygen binding to hemoglobin.

[9] Also known by the trendy term *tipping points*.

a shift between the two models, $g_1(\boldsymbol{\theta}_1, x)$ and $g_2(\boldsymbol{\theta}_2, x)$, which can be any deterministic function including all the ones we have discussed so far in this chapter. The challenge is to estimate $\boldsymbol{\theta}_1$ and $\boldsymbol{\theta}_2$. We point this out because Bayesian model fitting using the Markov chain Monte Carlo algorithm (chapter 7) allows us to accomplish this estimation in cases where maximum likelihood approaches have historically been defeated by the technical requirement for smooth functions with continuous derivatives. Abrupt changes like those depicted in equation 2.2.20 have been very difficult to fit using methods that require maximization or minimization using numerical methods, but they can easily be estimated using the approaches we describe later in the book.

3 Principles of Probability

3.1 Why Bother with First Principles?

Progress in science is made by comparing predictions of models with observations. All models make imperfect predictions of the operation of nature. A crosscutting theme in this book is that statistics help us solve the essential problem of science: gaining insight about phenomena of high dimension by the judicious reduction of their dimensions. Models reduce manifold influences on ecological processes to a few that can be understood. This means that models are inherently, deliberately approximate by virtue of the dimensions omitted.

To make a proper comparison between a model and observations, we need to understand the approximation inherent in models in terms of uncertainty. The output of the deterministic models we discussed in section 2.2 is a scalar or a vector; that is, for any given set of parameter values and input the model returns exactly the same result. In contrast, the output of a stochastic model is one or more probability distributions that reflect the uncertainty inherent in our model's predictions of a state and the way we observe it.

We use the term *stochastic* to refer to things that are uncertain.[1] Stochasticity arises in models in different ways, with different implications for sampling, experimental design, and forecasting. We mentioned these different sources of uncertainty earlier and will bring them up again and again.

[1] Stochastic models are used in the theoretical modeling tradition to represent random variation occurring over time and space not described by the deterministic core of the model. These models are often analyzed without reference to any data. We will not deal with these kinds of models, but we acknowledge that they are stochastic.

The main sources are the following:

1. **Process variance:** Process variance includes the uncertainty that results because our model fails to represent all the forces causing variation in the ecological quanties we seek to understand. Good models have low process variance because they capture most of the variation in the state they predict. Poor models have high process variance because they can't explain that variation. We often use the process variance to evaluate the fidelity of our model to the processes it represents, which means that it is critical to separate process variance from other sources of uncertainty. Failing to do so may lead to false conclusions about the model—we may have a great model and a poor system for observing the quantities it predicts. Lumping uncertainty about the process with uncertainty about the observations into the same "error term" makes it difficult to evaluate models and can lead to erroneous conclusions about the operation of ecological processes (Dennis et al., 2006; Hefley et al., 2013; Ahrestani et al., 2013).[2] The only way to reduce process variance is to improve our model. Expending more effort to observe the state of interest, improving our instruments, increasing our replications, and expanding the area we sample will do nothing to change process variance.
2. **Observation variance**: We rarely observe perfectly what we seek to understand. Observation variance quantifies these imperfections. There are usually two causes of observation variance. We seek to understand the true state of large areas or many individuals or many points in time, but we are forced by practicality to observe only a sample of them. Our sample is never a perfect reflection of the true state. In addition, we may need to correct our observing system for bias, and that correction is itself uncertain. In both cases, we can reduce uncertainty by taking more observations. Sampling variance asymptotically approaches zero as the number of observations increases. Models correcting for bias in our observations also become more certain with more observations.
3. **Variation among individuals:** Individual organisms differ because of their genetics and their individual histories—characteristics that may be hidden to us. These individual differences create uncertainty when we seek to understand responses of individuals to treatments or to

[2]Sometimes, particularly when designs are unreplicated, this separation is not possible, and lumping process and observation variance may be the best we can do. Section 6.3 covers this in detail.

environmental variation. The same idea can apply to spatial locations, which also have unique attributes.

4. **Model selection uncertainty:** The inferences we make based on a model depend on the model we choose from among many possible alternatives. The uncertainty that arises from our particular choice is called *model selection uncertainty.* We believe that the scientific objectives for the model trump formal model selection procedures, so our view is that sometimes we can ignore model selection uncertainty altogether, at other times we must quantify the uncertainty associated with using one model over another.

Dealing with uncertainty requires the proper tools, and primary among them are the rules of probability and an understanding of probability distributions. Equipped with these, ecologists can analyze the particular research problem at hand regardless of its idiosyncrasies. These analyses extend logically from first principles rather than from a particular statistical recipe. In the sections that follow, we describe these principles. Our approach is to start with the definition of probability and develop a logical progression of concepts extending from it to a fully specified and implemented Bayesian analysis appropriate for a broad range of research problems in ecology.

3.2 Rules of Probability

Ecological research requires learning about quantities that are unobserved from quantities that are observed. Any quantity that we fail to observe, including quantities that are observed imperfectly, involves uncertainty. The Bayesian approach treats all unobserved quantities as random variables to capture that uncertainty. A *random variable* is a quantity that can take on values due to chance—it does not have a single value but instead can take on a range of values. The chance of each value is governed by a probability distribution. We cannot underestimate the importance of this concept. Bayesian analysis is the only branch of statistics that treats all unobserved quantities as random variables. We will return to this idea in chapter 5.

All random variables have probability distributions even though these distributions may be unknown to us. The rules of probability determine how we gain insight about random variables from the distributions that govern their behavior. Understanding these rules lays a foundation for the remainder of the book. This material is not exactly gripping, but we urge you not to skip this section or rush through it unless you are already well grounded in formal principles of probability. Understanding these principles will serve you well.

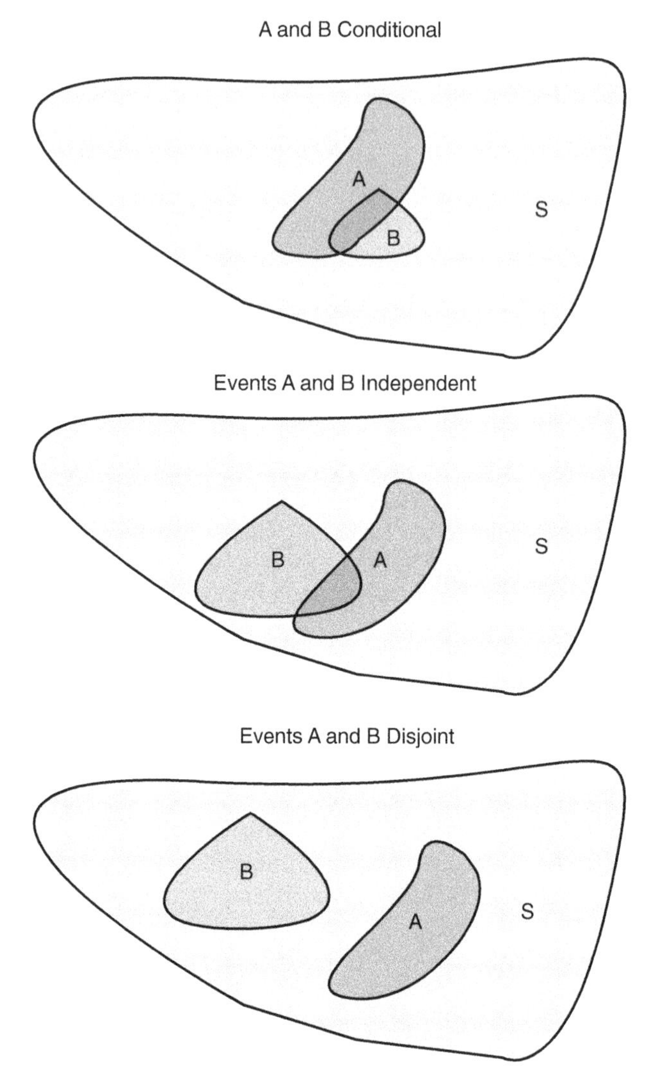

Figure 3.2.1. Illustration of conditional, independent, and disjoint probabilities. The area S defines a sample space including all the possible outcomes of a sample or an experiment. There are two sets of realized outcomes, A and B. The area of each event is proportional to the size of the set. The probability of A = area of A/area of S and the probability of B = area of B/area of S. Knowledge that event A has occurred influences our assessment of the probability of B when the intersection of the two events gives us new information about the probability of B. In this case, the probability of B is conditional on A, and vice versa (upper panel). In other cases the events intersect, but there is no new information. In this case the probability of B given A is

We start with the idea of a *sample space, S*, consisting of a set of all possible outcomes of an experiment or a sample, shown graphically as a polygon with a specific area (fig. 3.2.1). One of the possible outcomes of the experiment or sample is the random variable, "event A," a set of outcomes, which we also depict as a polygon (fig. 3.2.1). The area of A is less than the area of S because it does not include all possible outcomes. The area of A is proportional to the size of the set of outcomes it *does* include. It follows that the probability of A is simply the area of A divided by the area of S.

We now introduce a second event B to illustrate the concept of conditional, independent, and disjoint probabilities—concepts critical to understanding and applying Bayes' theorem to ecological models (chapters 5 and 6) Consider the case when we know that the polygon defining event B intersects the A polygon (fig. 3.2.1 upper panel) and, moreover, we know that event A has occurred. We ask, What is the probability of the new event B given our knowledge of the occurrence of A? The knowledge that A has occurred does two things: It shrinks the sample space from all of S to the area of A—if we know A has occurred, we know that everything outside of A has *not* occurred, so in essence we have a new, smaller space for defining the probability of A. Knowing that A has happened also affects what we know about B—we know that everything within B outside of A has not occurred (fig. 3.2.1). This means that

$$\Pr(B|A) = \frac{\text{area shared by } A \text{ and } B}{\text{area of } A}. \tag{3.2.1}$$

Dividing the numerator and denominator of the right-hand side by S we turn the areas into probabilities:[3]

$$\Pr(B|A) = \frac{\Pr(A \cap B)}{\Pr(A)} = \frac{\Pr(A, B)}{\Pr(A)}. \tag{3.2.2}$$

[3]The set operator $\cap$ means "intersection of."

Figure 3.2.1 (*Continued*).
the same as the probability of B, because area of B shared with A/area of A = area of B/area of S. In this case, we say that A and B are independent (middle panel, areas drawn approximately, but you really can't tell if one event is conditional on another by simply looking at the diagram unless you can see proportionality perfectly! If there is no intersection, then the events are disjoint. Knowing that A has occurred means that we know that B has not occurred (bottom panel). Thus, disjoint probabilities are a special case of conditional probability where $\Pr(A|B) = 0$.

Using the same logic, we obtain

$$\Pr(A|B) = \frac{\text{area shared by } A \text{ and } B}{\text{area of } B} = \frac{\Pr(A \cap B)}{\Pr(B)} = \frac{\Pr(A, B)}{\Pr(B)}. \quad (3.2.3)$$

The expression $\Pr(A|B)$ reads, "the probability of A conditional on knowing B has occurred." The bar symbol (i.e., |) reads "conditional on" or "given," expressing the dependence of event A on event B; if we know B, our knowledge changes what we know about A. It is important to note that $\Pr(A|B) \neq \Pr(B|A)$. The expression $\Pr(A, B)$ reads, "the *joint* probability of A and B" and is interpreted as the probability that both events occur. We will make important use of the algebraic rearrangement of equations 3.2.2 and 3.2.3 to expand their joint probability,

$$\begin{aligned} \Pr(A, B) &= \Pr(B|A)\Pr(A) \\ &= \Pr(A|B)\Pr(B). \end{aligned} \quad (3.2.4)$$

In some cases the area defining the two events overlaps, but no new information results from knowing that either event has occurred (fig. 3.2.1 middle panel). In this case the events are *independent*. Events A and B are independent if and only if

$$\Pr(A|B) = \frac{\text{area of } A \text{ shared by } A \text{ and } B}{\text{area of } B} = \frac{\text{area of } A}{\text{area of } S} = \Pr(A), \quad (3.2.5)$$

or equivalently,

$$\Pr(B|A) = \Pr(B). \quad (3.2.6)$$

Using equations 3.2.1 and 3.2.3, we can substitute for the conditional expressions in equations 3.2.5 and 3.2.6. A little rearrangement gives us the joint probability of independent events:

$$\Pr(A, B) = \Pr(A|B)\Pr(B) = \Pr(A)\Pr(B). \quad (3.2.7)$$

It is important to throughly understand the difference between the joint probability of events that are independent (eq. 3.2.7) and those that are not (eq. 3.2.4). When events are disjoint, there is no overlap between them (fig. 3.2.1 lower panel). In this case, knowing that one event has occurred means that we know the other event has *not* occurred. Thus, events that are disjoint are a special case of conditional probability: knowledge of one event gives us complete knowledge of the other event.

We may also be interested in the probability that one event or the other occurs (fig. 3.2.1), which is the total area of A and B not including the area they share[4], that is,

$$\Pr(A \cup B) = \Pr(A) + \Pr(B) - \Pr(A, B). \tag{3.2.8}$$

When A is independent of B,

$$\Pr(A \cup B) = \Pr(A) + \Pr(B) - \Pr(A)\Pr(B), \tag{3.2.9}$$

but if the events are conditional,

$$\Pr(A \cup B) = \Pr(A) + \Pr(B) - \Pr(A|B)\Pr(B), \tag{3.2.10}$$

or equivalently,

$$\Pr(A \cup B) = \Pr(A) + \Pr(B) - \Pr(B|A)\Pr(A). \tag{3.2.11}$$

If A and B are disjoint, then

$$\Pr(A \cup B) = \Pr(A) + \Pr(B), \tag{3.2.12}$$

which is simply a special case of equation 3.2.8 where $\Pr(A, B) = 0$.

The final probability rule we consider applies when we can partition the sample space into several nonoverlapping events (fig. 3.2.2). This rule is important because we will use it later to understand the components of Bayes' theorem (chapter 5). We define a set of events $\{B_n : n = 1, 2, 3, \ldots\}$, which taken together, cover the entire sample space, $\sum_n B_n = S$. We are interested in the event A that overlaps one or more of the B_n. The probability of A is

$$\Pr(A) = \sum_n \Pr(A \mid B_n)\Pr(B_n). \tag{3.2.13}$$

[4]The $\cup$ operator means "union of."

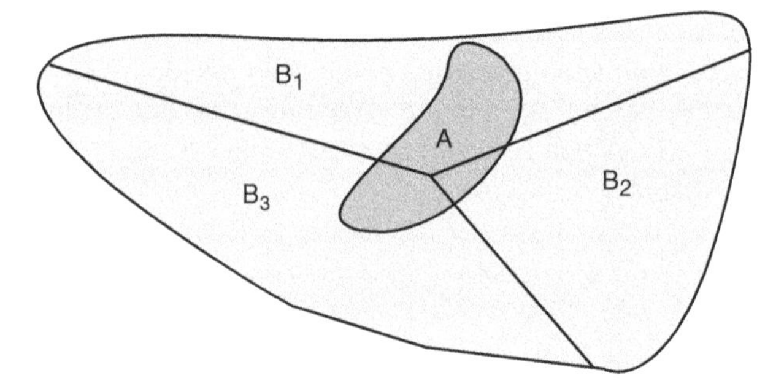

Figure 3.2.2. Illustration of the law of total probability.

Equation 3.2.13 is called the *law of total probability*. As the number of events approaches infinity and the areas of events become infinitesimally small, equation 3.2.13 becomes[5]

$$\Pr(A) = \int [A|B]\,[B]\,dB. \tag{3.2.14}$$

3.3 Factoring Joint Probabilities

It is hard to avoid a modicum of tedium in describing the rules of probability, but there is a very practical reason for understanding them: they allow us to deal with complexity. These rules permit us to take complicated joint distributions of random variables and break them down into manageable chunks that can be analyzed one at a time as if all the other random variables were known and constant. The importance of this idea and its implementation will be developed throughout the book, particularly in chapters 6 and 7. Here, we establish its graphical and mathematical foundation. What you learn here is critical to the model specification step in the general modeling process we described in the preface (fig. 0.0.1B). The material will become clear as we apply it to modeling problems.

[5]A bit about integral notation is needed here. Ecologists with some training in calculus likely recognize the definite integral $\int_L^U [A|B]\,[B]dB$ as the area under the curve $[A|B][B]$ from L to U. When no interval is specified (the $\int$ is "naked"), we denote integration over the domain of the function $[A|B][B]$. Thus, the notation $\int[A|B][B]dB$ means the definite integral of $[A|B]\,[B]$ over *all* possible values of B, the prevailing convention in statistics. It is important not to confuse this convention with that often used in mathematics where $\int[A|B][B]dB$ denotes the indefinite integral, that is, the antiderivative of $[A|B][B]$.

Consider the networks shown in figure 3.3.1. A Bayesian network (also called a *directed acyclic graph*) depicts dependencies among random variables. The random variables in the network are called *nodes*. The nodes at the head of the arrows are charmingly called *children*, and the tails, *parents*. Bayesian networks show how we factor the joint probability distribution of random variables into a series of conditional distributions and thereby represent an application of equation 3.2.4 to multiple variables (fig. 3.3.1). We use factoring to simplify problems that would otherwise be intractably complex.

Bayesian networks are great tools for thinking about relationships in ecology and for communicating them (e.g., fig. 1.2.1). They are useful because they allow us to visualize a complex set of relationships, thus encouraging careful consideration of how knowledge of one random variable informs us about the behavior of others. They lay plain our assumptions about dependence and independence. A properly constructed Bayesian network provides a detailed blueprint for writing a joint distribution as a series of conditional distributions: nodes at the heads of arrows are on the left-hand side of conditioning symbols, those at the tails of arrows are on on the right-hand side of conditioning symbols, and any node at the tail of an arrow without an arrow leading into it must be expressed unconditionally, for example, $\Pr(A)$. The network provides a graphical description of relationships that is perhaps easier to understand than the corresponding mathematical description, facilitating communication of ecological ideas underlying the network.[6]

The mathematics allowing factoring of joint distributions extends directly from the rules of probability we have already developed. Given the vector of jointly distributed random variables $\mathbf{z} = (z_1, \ldots, z_n)'$, the joint probability of the variables satisfies

$$\Pr(z_1, \ldots, z_n | p_1, \ldots, p_n) = \prod_{i=1}^{n} \Pr(z_i | \{p_i\}), \tag{3.3.1}$$

where $\{p_i\}$ is the set of parents of node z_i, and all the terms in the product are independent.[7] Independence of the terms in equation 3.3.1 is assured if the equation has been properly constructed from a Bayesian network, and the network shows relationships that are conditional and independent.

[6] At least Hobbs thinks so. Hooten prefers the equations.

[7] The operator $\prod_{i=1}^{n}$ says take the product of everything with the subscript i over $i = 1, \ldots, n$. It is the multiplicative equivalent of the summation operator, $\sum_{i=1}^{n}$, which may be more familiar to ecologists.

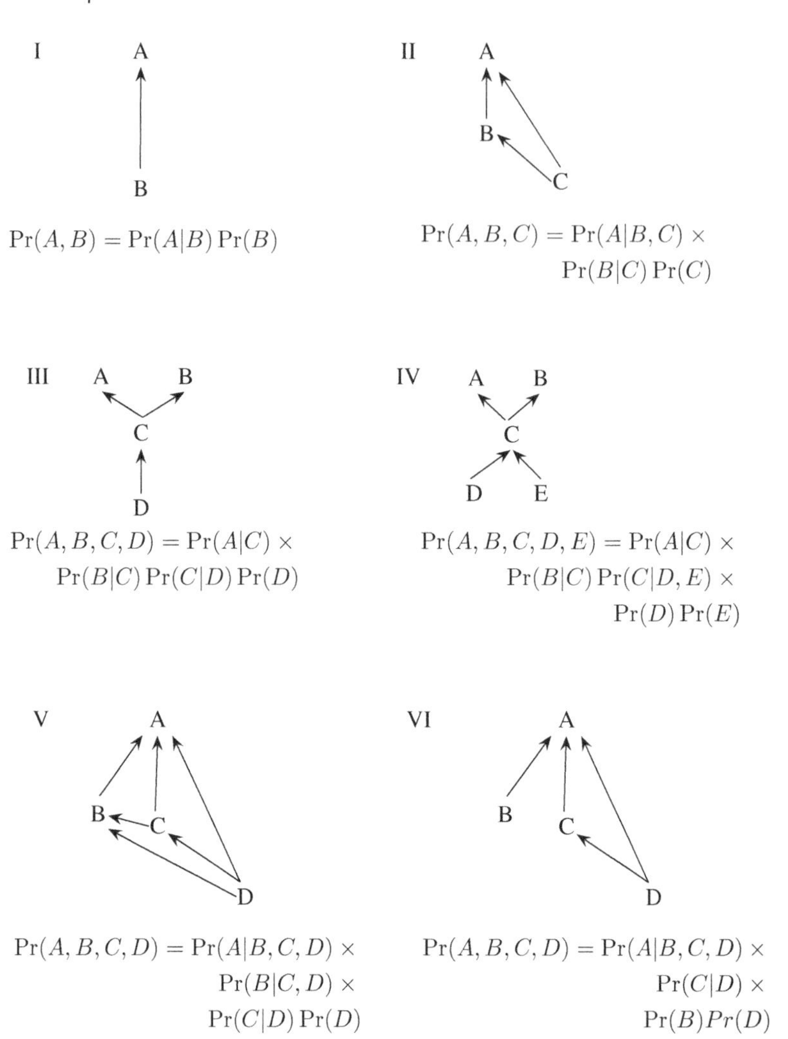

Figure 3.3.1. Bayesian networks specify how joint distributions are factored into conditional distributions using nodes to represent random variables and arrows to represent dependencies among them. Nodes at the heads of arrows must be on the left-hand side of conditioning symbols (|); nodes at the tails of arrows are on on the right-hand sides of conditioning symbols. Any node at the tail of an arrow without an arrow leading into it must be expressed unconditionally, for example, Pr(A). Some of the examples also indicate independence. The random variables A and B are independent after accounting for their mutual dependence on C in graphs III and IV; D and E are independent in IV, and B is independent of C and D in VI.

A somewhat more formal way to say the same thing is to generalize the conditioning rule of probability for two random variables (eq. 3.2.3) to factor the joint distribution of any number of random variables using

$$\Pr(z_1, z_2, \ldots, z_n) = \Pr(z_n|z_{n-1}, \ldots, z_1) \ldots \Pr(z_3|z_2, z_1)\Pr(z_2|z_1)\Pr(z_1), \tag{3.3.2}$$

where the components z_i may be scalars or subvectors of $\mathbf{z}$, and the sequence of the conditioning is arbitrary.[8] It is important to see the pattern of conditioning in equation 3.3.2. We can use the independence rule of probability (eq. 3.2.5) to simplify conditional expressions in equation 3.3.2 for random variables known to be independent. For example, if z_1 is independent of z_2, then $\Pr(z_1|z_2)$ simplifies to $\Pr(z_1)$. If z_1 and z_2 depend on z_3 but not on each other—which is to say they are conditionally independent—then

$$\Pr(z_1, z_2, z_3) = \Pr(z_1|z_2, z_3)\Pr(z_2|z_3)\Pr(z_3) \tag{3.3.3}$$

simplifies to

$$\Pr(z_1, z_2, z_3) = \Pr(z_1|z_3)\Pr(z_2|z_3)\Pr(z_3). \tag{3.3.4}$$

Another example of this kind of simplification is shown graphically and algebraically in figure 3.3.1 V and VI. Don't let the formalism in this paragraph put you off. It is simply a compact way to say what we have already shown graphically using Bayesian networks, which for many ecologists will be more transparent.

3.4 Probability Distributions

3.4.1 Mathematical Foundation

The Bayesian approach to learning from data using models makes a fundamental simplifying assumption: we can divide the world into things that are observed and things that are unobserved. Distinguishing between the observable and unobservable is the starting point for all analyses.

[8]We say the sequence is arbitrary to communicate the idea that the ordering of the specific z_i is not required for equation 3.3.2 to be true. In other words, z_n doesn't need to come first. However, the word *arbitrary* should not be taken to mean capricious. As we will learn, it is our understanding of the *biology* that determines what is conditional on what and ultimately governs the sequence of conditioning.

We treat all unobserved quantities as random variables governed by probability distributions, because the things we cannot observe have inherent uncertainty. It follows that understanding probability distributions forms a critical link between models and data in ecology. Becoming familiar with these distributions is the key to developing a flexible approach to the analysis of ecological models and data. Equipped with a toolbox of deterministic models (Otto and Day, 2007, chap. 2; Bolker, 2008) and the probability models described here, you will be able to thoughtfully develop a coherent approach to analyzing virtually any problem in research, regardless of its nuances. At the very least, you will be able to compose an approach that can be discussed productively with your statistical colleagues.

In this section we provide a compact description of distributions commonly used in developing models for ecological data. We first describe the features shared by all probability distributions and then outline features of specific distributions that we will use frequently in later chapters. We organize this section using the two types of random variables, discrete and continuous. Discrete random variables are those that take on discrete values, usually integers. It is possible, however, for discrete random variables to take on non-integer values. For example, a random variable might have support $\{0, \frac{1}{2}, 1\}$. In this case it is discrete but not integer-valued. All discrete random variables in this book will be integers, usually counts of things or membership in categories. In contrast, continuous random variables can take on an infinite number of values on any interval to represent length, mass, time, and energy, for example.

Probability mass functions and probability density functions are the fundamental link between models of ecological processes and observations of those processes. We first explain probability mass functions, then turn to probability density functions.

3.4.1.1 Probability Mass Functions

Assume we are interested in a discrete random variable z. That random variable might be the number of zooplankton in a 1L sample from a lake. The quantity $z = 304$ is a specific value that might be observed for that sample. The random variable could also be the number of individual plants in four categories: native annuals, native perennials, exotic annuals, and exotic perennials, for example, $\mathbf{z} = (4, 6, 18, 3)'$.[9] These examples illustrate an important point that we will return to more than once. Before we observe a quantity like the number of zooplankton in a sample, the quantity is a random variable whose value is governed by a probability distribution.

[9] We will use notation $\mathbf{u} = (a, b, c)'$ to indicate the vector $\mathbf{u}$ with elements a, b and c. The $(')$ following the () indicates that the variables within the parentheses are elements of a column vector.

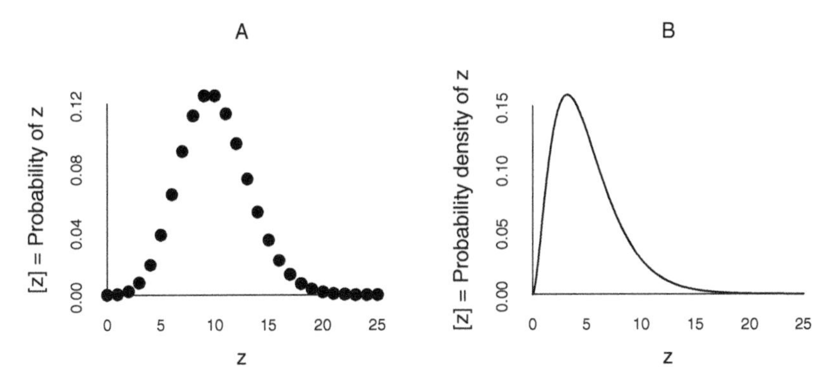

Figure 3.4.1. (**A**) A probability mass function computes the probability that a discrete random variable takes on a single value. Shown here is the distribution of a Poisson random variable with mean = 10. (**B**) A probability density function computes the probability density of a continuous random variable at a point. The integral of the function between two points gives the probability that the random variable falls in the interval between the points. Shown here is the distribution of a gamma random variable with mean = 5 and variance = 9.

After it is observed, the quantity becomes a known, specific value. A *probability mass function*,[10] $[z]$, for the random variable is

$$[z] = \Pr(z), \tag{3.4.1}$$

which simply says that given the argument z, the function $[z]$ returns the probability that the random variable will take on the particular value z (fig. 3.4.1, box 3.4).

All probability mass functions share two properties:

$$0 \leq [z] \leq 1, \tag{3.4.2}$$

$$\textstyle\sum_{z \in S} [z] = 1, \tag{3.4.3}$$

where S is the *support* of the random variable z, that is, the set of all values of z for which $[z] > 0$. Support is a vital concept because it defines the domain of the function $[z]$; values of z outside that domain have zero probability of occurrence and in some cases are not defined. Equation 3.4.2 says that the value of $[z]$ must be between 0 and 1, which of course makes sense if $[z]$ is a probability. Equation 3.4.3 says that the sum of $[z]$ over all its possible values must equal 1, which, is again, sensible for a probability mass function.

[10]Some statisticians use *probability function* synonymously with *probability mass function*.

Box 3.4 Notation for Probability Distributions

Statisticians read and write notation every day, allowing them to become accustomed to differences in notational styles that frequently occur in the literature. However, these differences can be confusing for ecologists who don't use notation as frequently as statisticians do. Here we explain the notation we will use to represent probability distributions and show how it relates to other widely used notations that mean the same thing.

Our Notation

First introduced in the seminal paper of Gelfand and Smith (1990), brackets have become a preferred notation for ecologists and statisticians using Bayesian methods, because they allow complex, multidimensional models to be written in compact form. We will use the notation $[z]$ to mean the probability of the random variable z if z is discrete and the probability density of z if z is continuous. Thus, $[z]$ denotes "z is distributed as." We will not include additional arguments within brackets when we refer to probability distributions broadly defined. If we are writing about specific distributions and want to refer to parameters, we will use $[z|\alpha, \beta]$ to denote the probability or probability density of z conditional on α and β.

We will often name the specific distribution; for example, we will write gamma $(z|\alpha, \beta)$ to denote that the random variable z follows a gamma distribution with parameters α and β. We will use $z \sim$ gamma (α, β) to mean the same thing. When we refer specifically to the probability of z (excluding probability density) we will use the notation $\Pr(z)$.

We will unapologetically use somewhat unconventional notation to achieve clarity when we want to specifically delineate the deterministic and stochastic components of models. So, for example, we might be interested in a deterministic model $\mu_i \equiv g(\boldsymbol{\theta}, x_i)$ that predicts the central tendency of the distribution of the random variable y_i, that is, $[y_i|\mu_i, \sigma^2]$, where σ^2 is a parameter that controls the dispersion of the distribution. (See chapter 1 for examples.) Equivalently, we will often use $\left[y_i|g\left(\boldsymbol{\theta}, x_i\right), \sigma^2\right]$. Of course, we realize that not all distributions have means and variances as parameters, so this notation runs some risk of being confused with the normal distribution when we intend it to refer to the full family of probability distributions. However, we have found in our teaching that it can be very helpful to call specific attention to the deterministic part and the stochastic

(continued)

(Box 3.4 *continued*)

part of models. We will reemphasize this point to prevent confusion with the normal distribution throughout the book. Moreover, we will teach how quantities representing central tendency and dispersion can be properly matched to specific parameters of distributions in section 3.4.4.

Notation Used by Others

Bracket notation for distributions is synonymous with the following. Often, mathematical statistics texts use the notation $P(Z = z) = f(z)$ or $P(Z = z) = f_Z(z)$ to mean the probability that random variable Z takes on a specific value z is given by the probability mass function $f(z)$. So, $P(Z = z) = f(z)$ is the same as $[z]$. Sometimes, authors reserve $p(Z = z)$ to refer to probability density, and $\Pr(Z = z)$ to refer to probability. Similarly, the notation $P(z|\alpha, \beta)$ means the same thing as $[z|\alpha, \beta]$, which is identical with $f(z|\alpha, \beta)$ and $f(\alpha, \beta)$ when f has been defined as a probability mass function or probability density function for z. We prefer the bracket notation because it is simpler.

Statisticians often write probability mass functions and probability density functions with only two arguments, the parameters, without specifying the random variable. For example they might write the distribution of the random variable z as gamma (α, β). To err on the side of clarity, we will include the random variable, that is, gamma$(z|\alpha, \beta)$.

3.4.1.2 Probability Density Functions

Probability density functions apply to random variables that are continuous, taking on values that are real numbers instead of integers. Given a continuous random variable z, a probability density function $[z]$ has the characteristics

$$[z] \geq 0, \tag{3.4.4}$$

$$\Pr(a \leq z \leq b) = \int_a^b [z]\, dz, \tag{3.4.5}$$

$$\int_{-\infty}^{\infty} [z]\, dz = 1\,, \tag{3.4.6}$$

which says that the probability density of z is nonnegative, that the probability of z falling between a and b is the integral of the density function from a to b, and that the integral of the density function over all real numbers equals 1.

It is important to understand that probability density functions do not return a probability, as probability mass functions do. Instead, they return a *probability density*. For continuous random variables, probability is defined only for a range of values, that is, $\Pr(a \leq z \leq b)$. The support for a continuous random variable includes all the values of z for which the probability density exceeds 0; that is, $[z] > 0$.

Some intuition for probability density can be gained by thinking about how we would approximate an integral of a probability density function $[z]$ over some range $\Delta z = b - a$ using a rectangular column with height $[(a+b)/2]$ and width Δz, remembering that the brackets indicate a function that returns the probability density of whatever is enclosed within them (fig. 3.4.2). Thus, we can think of $[(a + b)/2]$ as the average height of a bar over the interval from a to b. The probability of z over the interval a to b is $\Pr(a \leq z \leq b) \approx \Delta z[(a + b)/2]$. Thus, for very small Δz, the probability density of z is $[z] \approx \Pr(a \leq z \leq b)/\Delta z$.

We have found in our teaching that a potential point of confusion for many ecologists is that the y-axis for plots of probability mass functions always ranges between 0 and 1, while the y-axis for probability density functions can take on any value greater than 0. Students often ask, "How can the area under the probability density curve equal 1 if there are values on the y-axis that are greater than 1? Why is the axis so different for different random variables?" These questions arise from forgetting the definition of a definite integral, an easy thing to do for those of us who may use integrals infrequently. Recall that when we integrate, we are summing the area of bars (their heights $\times$ widths) under a curve where the number of bars approaches infinity, and the width of bars approaches zero. Thinking of integrals as the sum of the areas of many bars is the key to understanding the values on the y-axis of probability density functions (fig. 3.4.2, inset). The area depends on the height of the bars and the *widths*, which means that the scale of the y-axis of a probability density function depends on the scale of the x-axis. Any nonnegative value can appear on the y-axis for probability density, because those densities depend on the values of the random variable on the x-axis. This must be true to ensure that the integral over the entire support of x equals 1.

3.4.1.3 Moments

Important properties of probability distributions can be summarized succinctly using their *moments*. The first moment describes the central ten-

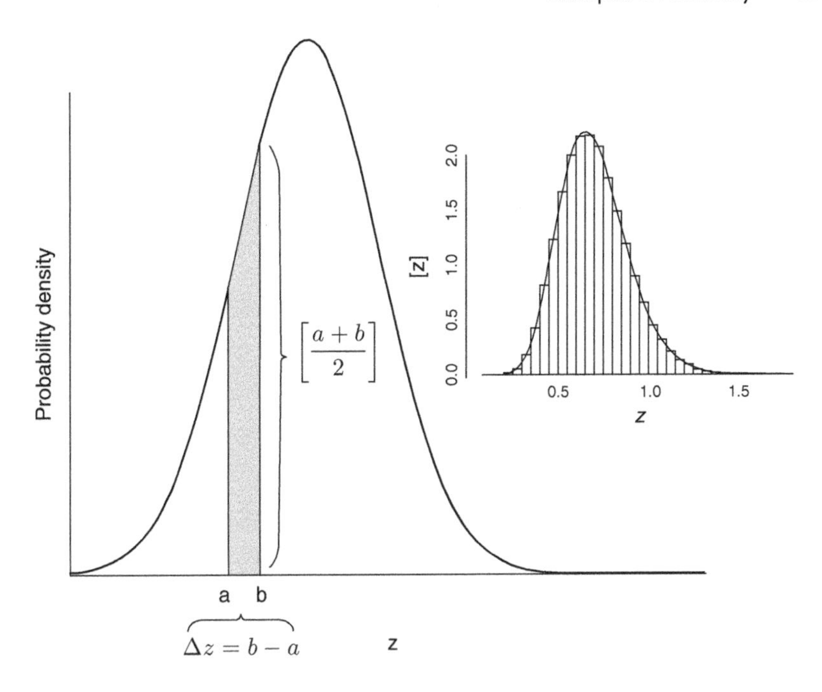

Figure 3.4.2. Illustration of probability and probability density for continuous random variables. When random variables are continuous, probability is defined only for intervals of values of the random variable. For example, the probability that the random variable z is within the interval a to b is the area of the shaded region, $\Pr(a \le z \le b) = \int_a^b [z]dz$. Probability density is the height of the shaded area at zero width, that is, $[z] = \Pr(a \le z \le b)/\Delta z$ with Δz infinitesimally small. Recall that as Δz approaches zero, Δz becomes dz in $\int_a^b [z]dz$. The inset demonstrates that the probability density, unlike probability can be greater than 1.

dency of the distribution, and the second central moment describes the dispersion or spread of the distribution. For the discrete random variable z, the first moment is the expected value of z, that is, the mean of its distribution:

$$\mu = \text{E}(z) = \sum_{z \in S} z[z]. \tag{3.4.7}$$

Equation 3.4.7 says the expected value of the random variable z is the sum of all possible values of z, each multiplied by its probability. Thus, the expected value is a weighted average of values of the random variable, where the weights are probabilities, $[z]$. The second central moment of the distribution of discrete random variables, the *variance*, is the expected value of the squared difference between the value of the random variable and the mean of the random variable

$$\sigma^2 = \mathrm{E}\left((z-\mu)^2\right) = \sum_{z \in S}(z-\mu)^2\,[z]\,. \tag{3.4.8}$$

For continuous random variables, we integrate rather than sum to obtain the moments:

$$\mu = \mathrm{E}(z) = \int_{-\infty}^{\infty} z[z]dz; \tag{3.4.9}$$

$$\sigma^2 = \mathrm{E}\left((z-\mu)^2\right) = \int_{-\infty}^{\infty} (z-\mu)^2[z]dz. \tag{3.4.10}$$

There are additional moments; skewness is the third, and kurtosis is the fourth, but we will not use them in the material that follows.

It is important to know how we approximate the first and second moments of random variables, continuous or discrete, using a technique called *Monte Carlo integration*. If we make many random draws from the distribution $[z]$, then its mean is approximately

$$\mu = \mathrm{E}(z) \approx \frac{1}{n}\sum_{i=1}^{n} z_i, \tag{3.4.11}$$

where n is the number of draws, and z_i is the ith value of the draw[11] of random variable z. In a similar way we can approximate the variance as[12]

$$\sigma^2 = \mathrm{E}\left((z-\mu)^2\right) \approx \frac{1}{n}\sum_{i=1}^{n}(z_i-\mu)^2, \tag{3.4.12}$$

where μ is estimated using equation 3.4.11. It is important to understand these approximations, because random draws from a distribution form the fundamental basis for learning about distributions of parameters and latent

[11]The idea of making draws from a distribution to approximate its mean or other moments will be very important in subsequent chapters, particularly 7 and 8. Be sure you understand this concept.

[12]Don't confuse this formula with the variance of a small sample in frequentist statistics, which you may remember as $\sigma^2 \approx \frac{1}{n-1}\sum_{i=1}^{n}(z_i-\bar{z})^2$, using $1/(n-1)$ instead of $1/n$. The version with n in the denominator provides the maximum likelihood estimate of σ^2, an estimate that is biased. An unbiased estimate for a small sample is obtained using $1/(n-1)$. However, in sampling from a distribution, we make so many draws (n) that using $1/(n-1)$ versus $1/n$ has no practical effect on the estimate of the variance.

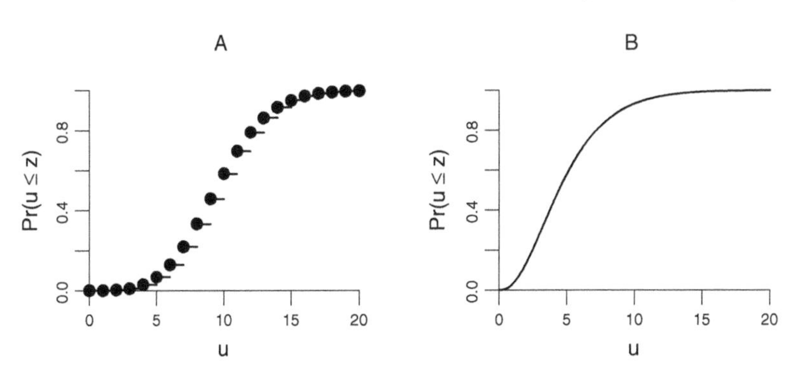

Figure 3.4.3. Cumulative distribution functions calculate the probability that a random variable takes on values less than or equal to a threshold, $F(z) = \Pr(u \leq z)$. Shown here are cumulative distribution functions for a Poisson random variable (**A**) with mean = 10 and a gamma random variable (**B**) with mean = 5 and variance = 9. Both functions asymptotically approach 1 as u approaches infinity. Corresponding probability mass and probability density functions are shown in figure 3.4.1.

quantities as well as quantities derived from them using Markov chain Monte Carlo, which we will treat in detail in chapters 7 and 8.

3.4.1.4 Distribution and Quantile Functions

We often wish to know the probability that a random variable takes on values above or below some threshold, for example, the probability that a population falls below 1000 individuals or that the proportion of plots containing a rare species exceeds 0.10. We do this by using the cumulative distribution function $F(z)$ for a discrete random variable[13] (fig. 3.4.3 A), which is defined as the probability that the value of the random variable is at most z:

$$F(z) = \sum_{u \leq z} [u]. \tag{3.4.13}$$

Given an argument z, the cumulative distribution function returns the probability that the random variable u is less than or equal to z, where $\Sigma_{u \leq z} [u]$ is the sum of all the probabilities of u for values of u less than or equal to z. For continuous random variables, the cumulative distribution

[13] Also called the *distribution function*.

function (fig. 3.4.3 B) is

$$F(z) = \int_{-\infty}^{z} [u]\, du. \tag{3.4.14}$$

The cumulative distribution function for a continuous random variable is the integral of the probability density function. It follows from the fundamental theorem of calculus that the density function is the derivative of the distribution function.

Finally, we introduce the *quantile function*, $F^{-1}(p)$, which we will often use for interval estimation of unobserved quantities. For a discrete random variable the quantile function returns the largest value of z for which $F(z) \leq p$, where p is the quantile of interest. For a continuous random variable, the quantile function is the inverse of the cumulative distribution function, $F^{-1}(p) = z$.

3.4.2 Marginal Distributions

Marginal distributions of random variables arise in many contexts in Bayesian modeling, so it is important to understand what they are. We start with the simplest case, two discrete random variables, x and y, that are jointly distributed:

$$[x, y] \equiv \Pr(x, y), \tag{3.4.15}$$

which simply means that $[x, y]$ is defined as ($\equiv$) the probability of the joint occurrence of x and y. The marginal distributions of x and y are most easily understood by example. Imagine that we are interested in a species for which births occur in pulses. We observe 100 females and record the age of each individual (as an integer) and also record the number of offspring she produced. We divide the frequency of each observed age and offspring combination by 100 to obtain the joint probabilities (see table 3.1). Cells in the table give the joint probability of age and number of offspring. The bottom row gives the marginal distribution of the number of offspring. The rightmost column gives the marginal distribution of age. Thus, y is "marginalized out" by summing to obtain the marginal distribution of y. The same idea applies to summing over x.

For a joint distribution of two random variables, we can focus on the probability of occurrence of one of them by summing over the probabilities of the other, effectively turning a bivariate distribution into a univariate one. If we are interested in the probability distribution of the number of offspring irrespective of the age of the mother, we ignore age by summing

TABLE 3.1
Example of Joint and Marginal Distributions for Age (x) and Number of Offspring (y)

	y = Number of Offspring			
x = Age	1	2	3	$\sum_y [x, y]$
1	0.1	0	0	0.1
2	0.13	0.12	0.02	0.27
3	0.23	0.36	0.04	0.63
$\sum_x [x, y]$	0.46	0.48	0.06	

down the columns in table 3.1 to obtain the *marginal distribution* of number of offspring. It is easy to see why this distribution is called marginal—it is based on the sums listed in the margins of the table. Thus, the marginal distribution of x is

$$[x] = \sum_y [x, y]$$
$$= \sum_y [x|y][y],$$

and the marginal distribution of y

$$[y] = \sum_x [x, y]$$
$$= \sum_x [y|x][x],$$

results that follow directly from the law of total probability (eq. 3.2.14). It is important to note that these are true probability distributions: $\sum_x [x] = 1$, and $\sum_y [y] = 1$.

Now, imagine that we add a third dimension to the data, sex of offspring, z, so that the joint distribution is now $[x, y, z]$, and the matrix of data in table 3.1 becomes a $3 \times 3 \times 2$ array. If we were interested in the probability distribution of male versus female offspring, we would sum over the probabilities of number of offspring and age, $[z] = \sum_x \sum_y [x, y, z]$. We could add any number of dimensions to the joint distribution and would follow the same procedure to focus on one of them. We sum over the probabilities of all the random variables except the one we are interested in. We obtain the distribution of the variables of interest by "marginalizing" over the distribution of the variables being discarded. The variables we leave out are said to have been "marginalized out."

We now consider random variables that are continuous rather than discrete (fig. 3.4.4). For example, we might be interested in the joint distribution of the mass of the mother, x, and the total mass of its offspring, y. When the joint random variables are continuous,

$$
\begin{aligned}
[x] &= \int [x, y] dy \\
&= \int [x|y][y] dy,
\end{aligned} \tag{3.4.16}
$$

where $\int [x|y][y]dy$ is the integral of $[x|y][y]$ over the support of y. Similarly,

$$
\begin{aligned}
[y] &= \int [x, y] dx \\
&= \int [y|x][x] dx.
\end{aligned} \tag{3.4.17}
$$

It may help you understand what this means by imagining subdividing the rows and columns in table 3.1 into increasingly smaller divisions and summing as before, except that now the numbers of rows and columns are infinite, requiring integration (fig. 3.4.4). But the concept illustrated above is exactly the same. Extending to multiple random variables, $[z_1, z_2, \ldots, z_n]$, we integrate over all the random variables except the one we seek to marginalize, an operation sometimes referred to as "integrating out." We accomplish the same thing as we did in the discrete case: we convert a joint distribution into a univariate distribution.

Integrals like those in equations 3.4.16 and 3.4.17 will appear frequently later in the book as a way to focus on the univariate distribution of unknown quantities that are parts of joint distributions that might contain many parameters and latent quantities. Thus, they are a vital tool for simplification. We urge you to be sure you understand what these integrals mean based on the simple example here before you proceed.

3.4.3 Useful Distributions and Their Properties

We now describe probability distributions that we have found to be most useful for modeling random variables in ecology. We describe key features of probability mass functions and probability density functions here, summarizing other aspects in appendix A. Most ecologists routinely use functions in software like R (R Core Team, 2013) to compute these functions. However, it is important to be familiar with the mathematical

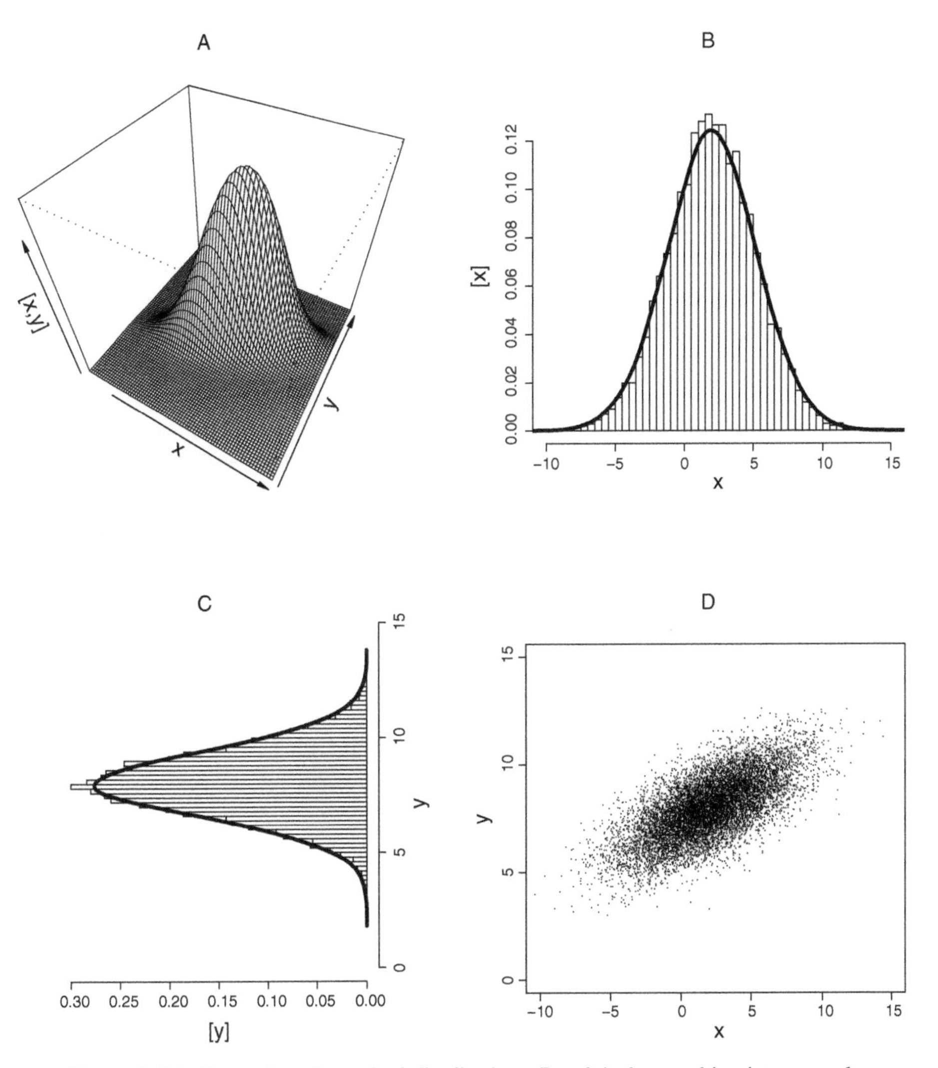

Figure 3.4.4. Examples of marginal distributions. Panel **A** shows a bivariate normal distribution for the correlated random variables x and y. The mean of x is 2, the mean of y is 8, and their covariance matrix is $\mathbf{\Sigma} = \begin{pmatrix} 10 & 3 \\ 3 & 2 \end{pmatrix}$. (If you are unfamiliar with covariance matrices, skip ahead momentarily to see box 3.4). Panel **D** shows 10,000 draws from the joint distribution of x and y. (**B**) Imagine that we "binned" all the observations of x to create a normalized histogram, ignoring the values of y. If the bins are infinitely small, we are "integrating out" y. (**C**) We similarly obtain the marginal density of y by integrating out x.

formulas that stand behind the computational versions of these functions because we will occasionally use these formulas later in this book, because mathematical expressions for distributions often appear in the literature, and because the expressions can be usefully manipulated, as will be illustrated with the exponential distribution later in this section.

One other concept is needed here. The material in section 3.4.1 treated probability distributions as if they had a single argument, z. This simplification made it easier to concentrate on the basic properties of probability mass functions and probability density functions undistracted by other arguments to the functions. However, specific distributions require *parameters* as arguments in addition to the random variable. *Parameters* are arguments to functions that give probability distributions a particular shape, that is, a central tendency, dispersion, and skew. Many ecologists are most familiar with the normal distribution for which the parameters and the first and second moments (i.e., the mean and variance) are the same. This is also true for the Poisson distribution, which has a single parameter, the mean, which is equal to the variance. For all other distributions, the moments and the parameters are not the same. Instead, the moments are *functions* of the parameters, and, hence, the parameters are functions of the moments. We will use these functional relationships between moments and parameters in a powerful way in section 3.4.4. For now, we highlight differences between random variables and parameters by using a consistent notation. We will denote random variables as z,[14] and we will use Greek letters to symbolize parameters.

3.4.3.1 Probability Mass Functions for Discrete Random Variables

Poisson. The Poisson distribution describes the probability of a number of events (z) occurring in a given unit of time or space assuming that the occurrence of one event has no influence on the probability of occurrence of the subsequent event (fig. 3.4.5). The distribution has a single parameter, λ, the average number of occurrences, also called the *intensity*. Note that λ is a positive real number, whereas z must be a nonnegative integer.[15]

[14]We typically use lowercase letters to denote univariate random variables throughout. Keep in mind that it is also common to see uppercase letters for random variables and lowercase letters for "realizations" of random variables (i.e., their numerical values). We use uppercase letters for matrices, so to avoid inconsistencies, we let the context dictate whether a lowercase letter denotes a random variable or a value. We find that our notation does the job most of the time and leads to substantially less confusion than the somewhat more conventional notation.

[15]Negative values for z are not defined because $z!$ is not defined for $z < 0$. However, functions in software, notably R (R Core Team, 2013), return 0 for negative arguments to the Poisson probability mass function.

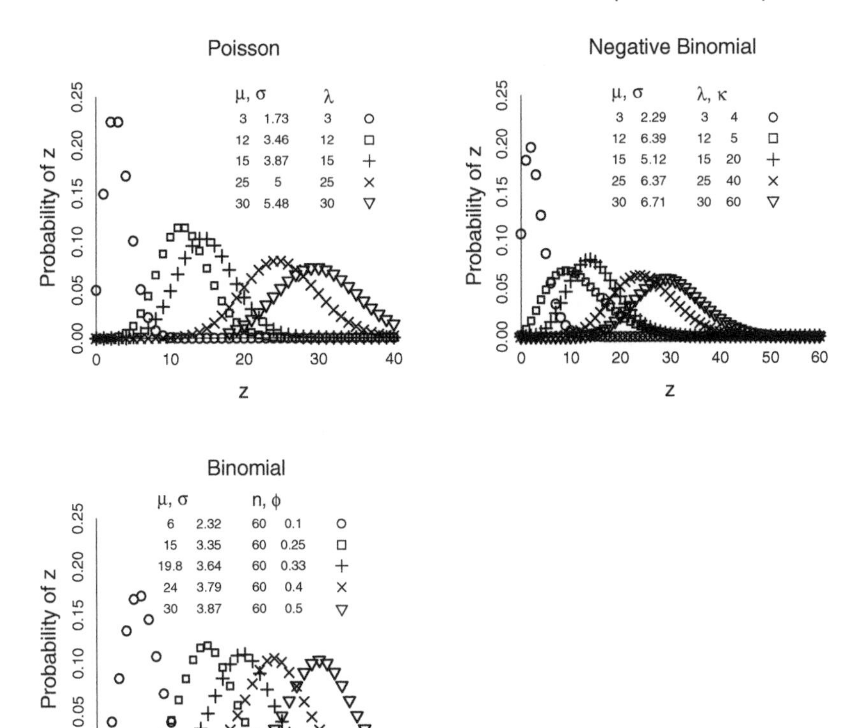

Figure 3.4.5. Example probability mass functions for discrete random variables with means (μ), standard deviations (σ), and parameters (λ, κ, n, ϕ).

The Poisson distribution applies to random variables for which the average number of events is the same as the variance in the number of events, that is, the first and second central moments are equal.

The Poisson probability mass function is

$$[z|\lambda] = \text{Poisson}(z|\lambda) = \frac{\lambda^z}{z!}e^{-\lambda}. \tag{3.4.18}$$

The function returns the probability of occurrence of z events conditional on the value of λ. You may also see the Poisson distribution written as

$$[z|\lambda] = \text{Poisson}(z|\lambda) = \frac{(\gamma\lambda)^z}{z!}e^{-\gamma\lambda}, \tag{3.4.19}$$

where γ specifies a specific interval of time, length, area, or volume; and λ is a rate, the number of occurrences per unit of time, length, area, or volume. Thus, the parameter γ sets the scale of the process. The quantity γ is entered into the equation as known (that is, as perfectly observed data)[16] and is called an *offset*. When the Poisson is expressed as equation 3.4.19, the parameter γ has units of time, length, area, or volume, and λ is a rate with units that are the reciprocal of the units of γ. When the Poisson is expressed as equation 3.4.18, then λ is unitless, because it refers to the average number of things. Units are implicit and are defined by the area, time interval, and so forth, over which counts are made.

Negative Binomial. Using the Poisson distribution requires the restrictive assumption that the mean of the distribution equals the variance. Sometimes we will wish to model the intensity of a number of events where the variance in the number of events exceeds the mean. In this case, the negative binomial distribution is a logical choice (fig. 3.4.5). It applies to the same type of count data as the Poisson but contains a second parameter, κ, controlling the dispersion of the distribution:

$$[z|\lambda, \kappa] = \text{negative binomial}(z|\lambda, \kappa) = \frac{\Gamma(z+\kappa)}{\Gamma(\kappa)\, z!}\left(\frac{\kappa}{\kappa+\lambda}\right)^{\kappa}\left(\frac{\lambda}{\kappa+\lambda}\right)^{z}. \tag{3.4.20}$$

The gamma function, $\Gamma(\;)$, may not be familiar to all readers. It is a function that interpolates a smooth curve connecting the points (x, y) given by $y = (x-1)!$ at the positive integer values for x. The mean of the negative binomial distribution is λ, and the variance is $\lambda + \lambda^2/\kappa$.

A second version of the negative binomial, less commonly used in ecology, gives the probability of a number (z) of failures that occur in a sequence of Bernoulli trials[17] before a target number of successes (k) is obtained:

$$[z|k, \phi] = \text{negative binomial}(z|k, \phi) = \frac{\Gamma(z+k)}{\Gamma(k) z!}\phi^{k}(1-\phi)^{z}. \tag{3.4.21}$$

The parameter ϕ is the probability of a success on a single trial. The parameter k is usually referred to as the *size* in this parameterization.

Software sometimes uses equation 3.4.21 in functions for the negative binomial, so you need to be careful using them if your intention is to use

[16]This is why it does not appear in the distribution $[z|\lambda]$. More about this later.

[17]A *Bernoulli trial* is an experiment with two possible outcomes. Coin flipping is the classical example of a Bernoulli trial.

3.4.20. A little algebra allows you to modify the arguments to functions implemented in software. To modify equation 3.4.20 so that the parameter k represents dispersion (as in eq. 3.4.20), substitute $\kappa/(\lambda+\kappa)$ for the function argument ϕ.

Binomial. The binomial distribution portrays counts that can be assigned to one of two possible categories, for example, alive or dead, present or absent, male or female, exotic or native, diseased or healthy (fig. 3.4.5). The distribution describes the probability of the number of "successes" out of n trials conditional on the parameter ϕ, the probability of a success on any single trial. "Successes" arbitrarily refers to one of the two categories such that successes + failures = number of trials. Trials most often represent observations of a specific number of individual organisms, locations, or sampling plots in the two categories. The probability mass function for a binomial random variable[18] is

$$[z|n,\phi] = \text{binomial}(z|n,\phi) = \binom{n}{z}\phi^z(1-\phi)^{n-z}. \qquad (3.4.22)$$

The random variable z and parameter n must be integers; the parameter ϕ is a real number, $0 \leq \phi \leq 1$. The function returns the probability of z successes conditional on n and ϕ. The mean of the binomial distribution is $n\phi$, and the variance is $n\phi(1-\phi)$.

Bernoulli. The Bernoulli is a special case of the binomial where the number of trials = 1. Its probability mass function is

$$[z|\phi] = \text{Bernoulli}(z|\phi) = \phi^z(1-\phi)^{1-z} \quad \text{for } z \in \{0,1\}. \qquad (3.4.23)$$

The parameter ϕ is the probability of success ($z = 1$) on a single trial. The function computes $z = 1$ with probability ϕ and $z = 0$ with probability $1-\phi$. The Bernoulli distribution has particularly important application in modeling presence or absence data and in occupancy modeling, where it is used to estimate the probability that a particular state is detected, allowing us to separate cases where a state of interest is "unoccupied" from the state "occupied but not observed." Examples of these models will be treated in sections 6.2.3 and 12.3.

[18]The term $\binom{n}{z} = n!/z!(n-z)!$.

Multinomial. The multinomial distribution is used to model random variables that fall into more than two categories on η trials:

$$[\mathbf{z}|\boldsymbol{\phi}, \eta] = \text{multinomial}(\mathbf{z}|\boldsymbol{\phi}, \eta) = \eta! \prod_{i=1}^{k} \frac{\phi_i^{z_i}}{z_i!}. \tag{3.4.24}$$

The symbol $\prod_{i=1}^{k}$ might be unfamiliar to some readers. It means, take a product over all the k quantities indexed by i. So, it is the multiplicative equivalent of the more familiar summation $\sum_{i=1}^{k}$.

The multinomial is the first multivariate distribution we have encountered. It is multivariate because it returns the probability of a *vector* of random variables ($\mathbf{z}$), conditional on a vector of parameters ($\boldsymbol{\phi}$) specifying the probability of occurrence in each category. The parameter η is the total number of occurrences, $\eta = \sum_{i=1}^{k} z_i$, where k is the number of categories. The mean of each category is $\eta\phi_i$, and the variance is $\eta\phi_i(1 - \phi_i)$.

The multinomial has many applications in ecological modeling because it is used to represent the number of individuals in a set of mutually exclusive states—a common output of ecological models. It is applied in capture-recapture analysis; in modeling movement of individuals among discrete locations; and in discrete-time matrix modeling of populations, communities, and ecosystems.

3.4.3.2 Probability Density Functions for Continuous Random Variables

Normal. The univariate normal distribution (also known as the *Gaussian distribution*) applies to continuous random variables that can take on values across the entire number line, $-\infty < z < \infty$ (fig. 3.4.6). It is widely used in statistics because it has properties allowing many results to be derived analytically, for example, least-squares estimates of parameters.[19] In addition, the normal distribution is widely used because of the central limit theorem, which states that the sum of a large number of samples from any distribution will be normally distributed. The variance of the normal distribution does not depend in any way on the mean of the distribution. The probability density function for a normally distributed

[19]It is easy to forget that the discipline of statistics developed well before the advent of fast computers. During much of that period of development, analytically tractable distributions were the sole route to useful results in research.

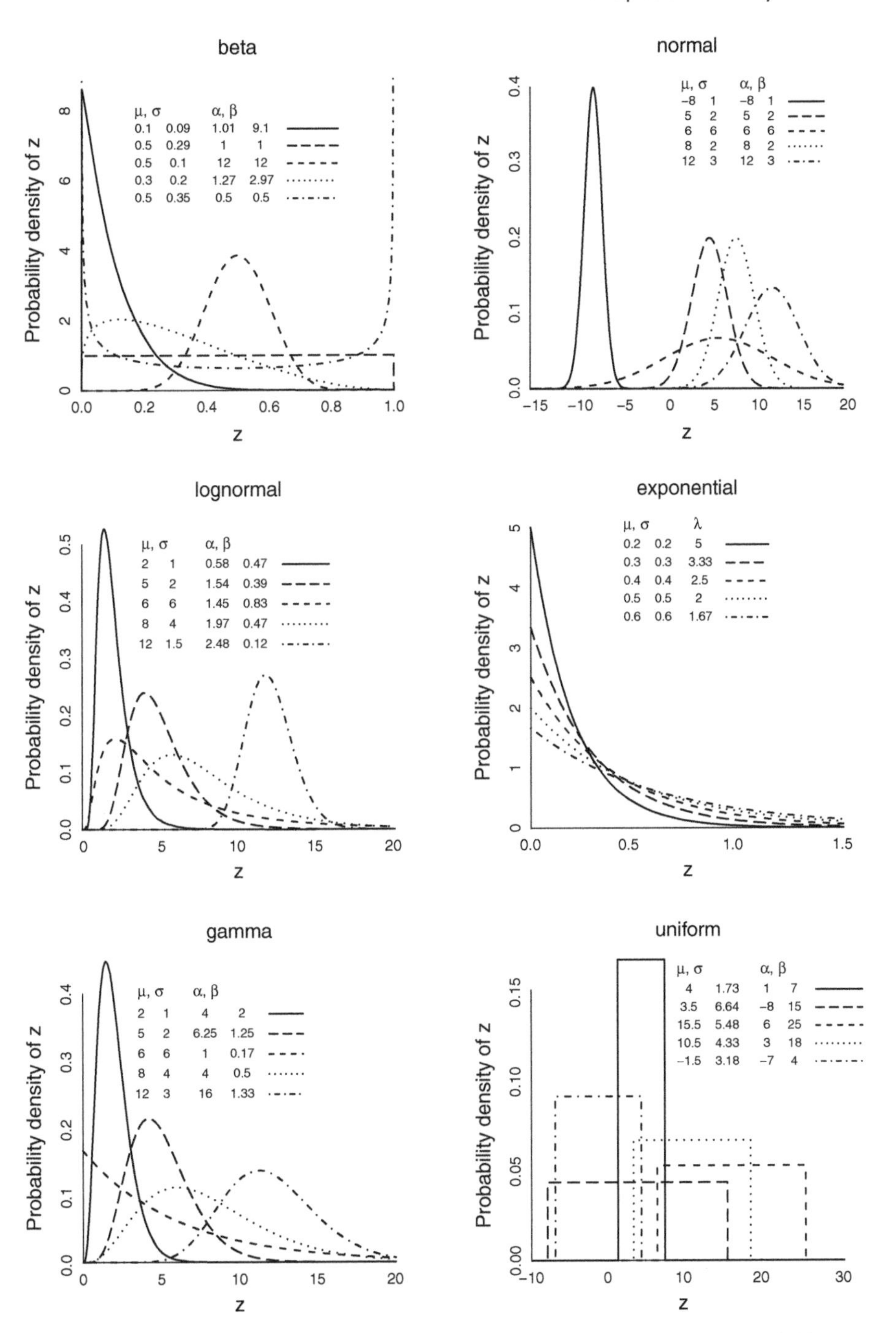

Figure 3.4.6. Example probability density functions for continuous random variables with means (μ), standardard deviations (σ), and parameters (α, β, λ).

random variable is

$$\left[z|\mu,\sigma^2\right] = \text{normal}(z|\mu,\sigma^2) = \frac{1}{\sigma\sqrt{2\pi}}e^{-\frac{(z-\mu)^2}{2\sigma^2}}. \tag{3.4.25}$$

It is important to note that despite its widespread use in traditional statistics, the normal distribution is often not an appropriate choice for modeling in ecology. The reason is that the preponderance of ecological quantities are nonnegative, which means that the support for these random variables should be zero for negative values (i.e., $\Pr(z \leq 0) = 0$), which of course is not true for the normal. Moreover, the normal is always symmetric and cannot represent the skewness that often arises in distributions of ecological quantities. Finally, the variance of a random variable often increases with the mean, but the normal has a constant variance. These problems are overcome by the lognormal and gamma distributions.

Lognormal. If a random variable z is normally distributed, then e^z is lognormally distributed; if a random variable z is lognormally distributed, then $\log(z)$ is normally distributed. The lognormal distribution (fig. 3.4.6) has properties analogous to those of the central limit theorem for the normal. If we take the product of large numbers of random variables, then the outcome is lognormally distributed regardless of the underlying distributions of the individual random variables. The lognormal distribution is widely used to represent growth of individuals or populations. If we define a growth process as $z_t = \alpha z_{t-1}$, then it follows that the random variable, z_t at time t represents the product of a constant α and the previous state of the random variable, z_{t-1}. The lognormal offers a good choice for modeling the process because it can be represented as a product of states and parameters. The probability density function for the lognormal distribution is

$$[z|\alpha,\beta] = \text{lognormal}(z|\alpha,\beta) = \frac{1}{z\sqrt{2\pi\beta^2}}e^{-\frac{(\log(z)-\alpha)^2}{2\beta^2}}, \tag{3.4.26}$$

where α is the mean of $\log(z)$, and β is the standard deviation of $\log(z)$. It would be tempting to think that the mean of the distribution is e^α, which instead gives the median. The mean depends on both parameters, such that

$$\text{E}(z) = \mu = e^{\alpha+\beta^2/2}. \tag{3.4.27}$$

Similarly, the variance of the lognormal distribution also depends on α and β:

$$\mathrm{E}\left((z-\mu)^2\right)=\sigma^2=(e^{\beta^2}-1)e^{2\alpha+\beta^2}. \tag{3.4.28}$$

The variance of the lognormal increases in proportion to the square of the mean for any given set of parameters α and β. Ecologists often observe continuous, nonnegative quantities for which the variance increases with the mean, which suggests that the lognormal or gamma can be used to model those quantities.

Gamma. The gamma distribution is broadly useful in ecology for modeling random variables that are nonnegative (fig. 3.4.6). Like the lognormal, the gamma distribution is well suited for representing random variables that are skewed. The gamma distribution was originally derived to model the time required for a specified number of events to occur in a Poisson process, that is, where events occur at average rate λ, and the occurrence of an event has no influence on the occurrence of a subsequent event. The distribution is also used to represent random variability in the mean, λ, of the Poisson distribution and thereby provides the basis for the derivation of the negative binomial.

The probability density function for the gamma distribution can take two forms. We will use

$$[z|\alpha,\beta]=\text{gamma}(z|\alpha,\beta)=\frac{\beta^\alpha}{\Gamma(\alpha)}z^{\alpha-1}e^{-\beta z}, \tag{3.4.29}$$

where α is called the *shape*, and β, the *rate*. Both parameters must be positive real numbers. The mean of the gamma distribution is

$$\mathrm{E}(z)=\mu=\tfrac{\alpha}{\beta}, \tag{3.4.30}$$

and the variance is

$$\mathrm{E}\left((z-\mu)^2\right)=\sigma^2=\frac{\alpha}{\beta^2}. \tag{3.4.31}$$

An alternative parameterization for the gamma distribution is

$$[z|k,\theta]=\text{gamma}(z|k,\theta)=\frac{1}{\Gamma(k)\theta^k}z^{k-1}e^{-\frac{z}{\theta}}, \tag{3.4.32}$$

where k is called the *shape*, and θ, the *scale*. This form is used to model the waiting time z for k events in a Poisson process where the average waiting time between events is $\theta = \beta^{-1}$. The distribution can also be parameterized with a shape, k, and a mean parameter, $k\beta^{-1}$. The same ideas about "waiting times" can be applied to space by using the gamma distribution to model the distance or area that must be covered before encountering k items that are Poisson distributed over space.

Exponential. The gamma distribution simplifies to the exponential distribution when $\alpha = 1$ or $k = 1$. The exponential distribution (fig. 3.4.6) gives the probability of times or distances between sequential events in a Poisson process,

$$[z|\lambda] = \text{exponential}(z|\lambda) = \lambda e^{-\lambda z}, \tag{3.4.33}$$

where λ is the average number of events per unit of time or space. For example, if prey are captured at an average rate λ, then the number of prey captured follows a Poisson distribution, and the time between captures follows an exponential distribution. The mean of the exponential distribution is λ^{-1}, and the variance is λ^{-2}.

The exponential distribution has direct application to ecological models composed of systems of differential equations in showing the relationship between rates (time^{-1}) and probabilities. Consider the differential equation

$$\frac{dq}{dt} = -kq, \tag{3.4.34}$$

which describes the instantaneous rate of change in a state variable, q. The quantity q can represent anything—number of individuals in a population, grams of nitrogen in the soil, area of landscape in forest. The usual metaphor for q is a "compartment," and equation 3.4.34 describes the rate of movement of particles (individuals, atoms, pixels) out of the compartment. The rate of change per particle is

$$\frac{dq}{dt}\frac{1}{q} = -k. \tag{3.4.35}$$

We can use the exponential distribution to translate the rate k (time^{-1}) into the probability that a particle leaves the compartment during a time interval Δt. We define the event "waiting time of a particle in the compartment" as the random variable z. The cumulative distribution function for the exponential distribution is

$$\int_{-\infty}^{z} \text{exponential}(u|\lambda)\,du = 1 - e^{-\lambda z}. \tag{3.4.36}$$

If we let $z = \Delta t$ and $\lambda = k$, where k is a continuous time rate constant (with dimensions of time^{-1}, equation 3.4.35), then the probability that a particle has waiting time $z < \Delta t$ is $1 - e^{-k\Delta t}$, which is the probability that it leaves the compartment during Δt. The probability that it remains in the compartment is the complement, $e^{-k\Delta t}$. It follows that the average number of particles that leave the compartment during t to $t + \Delta t$ is $q_t(1 - e^{-k\Delta t})$, and the average that remain is $q_t e^{-k\Delta t}$. For example, if the compartment represents "individuals that are alive," and k is the instantaneous mortality rate, then $e^{-k\Delta t}$ gives the probability that an animal survives from time t to $t + \Delta t$, $1 - e^{-k\Delta t}$ gives the probability that it dies, $q_t\left(1 - e^{-k\Delta t}\right)$ is the average number of deaths, and $q_t e^{-k\Delta t}$ is the average number of survivors. Of course, these population results can also be derived from the solution to the differential equation, $q_{t+\Delta t} = q_t e^{-k\Delta t}$.

Although the approach illustrated here uses constant relative rates (i.e., k), the same equations can be applied to nonlinear rates if Δt is sufficiently small. So, for example, if the rate of conversion of susceptible individuals, S, to infected ones in a population is controlled by the nonlinear rate $\frac{dS}{dt} = -\beta SI$, where I is the number of infected individuals, then per capita rate of infection is $\frac{dS}{dt}\frac{1}{S} = -\beta I$, and the probability that a susceptible becomes infected during a small interval of time Δt is $1 - e^{-\beta I \Delta t}$.

This type of logic forms a fundamental link between traditional, continuous-time state variable models in ecology, discrete-time state variable models, and models that are individual based. However, it is surprisingly absent from texts on individual-based modeling (e.g., Railsback and Grimm, 2012).

Inverse Gamma. The inverse gamma distribution (fig. 3.4.6) models the reciprocal of a gamma-distributed random variable. If $b \sim \text{gamma}(\alpha,\theta)$, and $z = b^{-1}$, then the probability density of z is given by

$$[z|\alpha, \beta] = \frac{\beta^{\alpha}}{\Gamma(\alpha)} z^{-\alpha-1} \exp\left(\frac{-\beta}{z}\right), \tag{3.4.37}$$

where the scale parameter $\beta = \theta^{-1}$. At the risk of getting ahead of ourselves, the inverse gamma is particularly useful in modeling the variance of the normal and lognormal distributions. We will study this application in sections 5.3 and 7.3.2.

The mean of the inverse gamma distribution is

$$\mathrm{E}(z) = \mu = \frac{\beta}{\alpha - 1} \quad \text{for } \alpha > 1, \tag{3.4.38}$$

and the variance is

$$\mathrm{E}(z - \mu) = \sigma^2 = \frac{\beta^2}{(\alpha - 1)^2(\alpha - 2)} \quad \text{for } \alpha > 2. \tag{3.4.39}$$

Beta. The beta distribution (fig. 3.4.6) models the distribution of random variables that can take on values between 0 and 1.[20] It is often used to model uncertainty in probabilities and proportions, making it an essential part of the toolbox of distributions needed by the ecological modeler. The probability density of a beta-distributed random variable, z, conditional on parameters α and β is

$$[z|\alpha, \beta] = \text{beta}(z|\alpha, \beta) = \frac{\Gamma(\alpha + \beta)}{\Gamma(\alpha)\Gamma(\beta)} z^{\alpha-1}(1 - z)^{\beta-1}. \tag{3.4.40}$$

The mean of the distribution is

$$\mathrm{E}(z) = \mu = \frac{\alpha}{\alpha + \beta}, \tag{3.4.41}$$

and the variance is

$$\mathrm{E}\left((z - \mu)^2\right) = \sigma^2 = \frac{\alpha\beta}{(\alpha + \beta)^2(\alpha + \beta + 1)}. \tag{3.4.42}$$

Beta distributions are widely used in analysis of survival, detection probability, and decomposition.

Uniform. The uniform distribution (fig. 3.4.6) returns a single probability density for all values of the random variable for which the probability density is greater than 0:

$$[z|\alpha, \beta] = \begin{cases} \frac{1}{\beta - \alpha} & \text{for } \alpha \leq z \leq \beta, \\ 0 & \text{for } z < \alpha \text{ or } z > \beta \end{cases}. \tag{3.4.43}$$

[20]Be careful not to confuse this with the Bernoulli, which models random variables that are 0 or 1.

The mean of the uniform distribution is $\mu = (\beta + \alpha)/2$, and the variance is $\sigma^2 = (\beta - \alpha)^2/12$. The uniform is especially useful for defining vague prior distributions in Bayesian analysis.

Multivariate Normal. The multivariate normal distribution is often used to represent a set of correlated real-valued random variables, each of which centers around a mean. It is particularly valuable for representing the probability distribution of data that are correlated over time or space and has important applications in regression. The multivariate normal is a generalization of the normal distribution for a single random variable to the distribution of a random vector of variables. A random vector is k-variate normally distributed if each linear combination of its elements has a univariate normal distribution. The probability density is

$$[\mathbf{z}|\boldsymbol{\mu},\boldsymbol{\Sigma}] = \text{multivariate normal}(\mathbf{z}|\boldsymbol{\mu},\boldsymbol{\Sigma}) = (2\pi)^{-\frac{k}{2}}|\boldsymbol{\Sigma}|^{-\frac{1}{2}}\, e^{-\frac{1}{2}(\mathbf{z}-\boldsymbol{\mu})'\boldsymbol{\Sigma}^{-1}(\mathbf{z}-\boldsymbol{\mu})}, \tag{3.4.44}$$

where $\mathbf{z}$ is a vector of random variables, $\boldsymbol{\mu}$ is a vector of means, and $\boldsymbol{\Sigma}$ is a variance-covariance matrix (box 3.4). The term $|\boldsymbol{\Sigma}|$ indicates the determinant[21] of $\boldsymbol{\Sigma}$.

Box 3.4 Covariance Matrices

We do not use covariance matrices often in this book, but it is important to understand them, because they are needed when modeling a vector of random variables rather than a single one. Each of these random variables has its own variance, but they also may *covary*. Covariance describes how two or more random variables deviate from their mean–if they covary, then they deviate in similar ways. The covariance of random variables x and y is formally defined as

$$\sigma_{xy} = \mathrm{E}\left((x - \mathrm{E}(x))(y - \mathrm{E}(y))\right). \tag{3.4.45}$$

(*continued*)

[21]Don't confuse $\boldsymbol{\Sigma}$ with the summation sign. Instead, it is uppercase boldface version of the scalar σ used to indicate a matrix. Determinants are quantities calculated on square matrices. The use of determinants is beyond the scope of this book, but we wanted to clarify this notation so that it is not confused with absolute value.

(Box 3.4 *continued*)

As you might expect, covariance is closely related to correlation (ρ_{xy}),

$$\rho_{xy} = \frac{\mathrm{E}\left((x - \mathrm{E}(x))(y - \mathrm{E}(y))\right)}{\sigma_x \sigma_y}, \tag{3.4.46}$$

where σ_x and σ_y are the standard deviations of the two random variables. When random variables covary, a scatter plot of their values tends to fall along a line, as in figure 3.4.4 D. When they do not covary, the values form a diffuse cloud.

A covariance matrix is a square matrix with the number of rows and columns equal to the number of elements in the vector being modeled. The diagonal elements are the variances of each random variable in the vector, and the off-diagonal elements i, j are the covariance of random variable i with random variable j. If we assume that all random variables in the vector have the same variance, σ^2, and do not covary, then $\boldsymbol{\Sigma} = \sigma^2\mathbf{I}$, where $\mathbf{I}$ is the identity matrix, consisting of 1s on the diagonal and 0s elsewhere. In this case, $\boldsymbol{\Sigma}$ has σ^2 on the diagonal, and 0s elsewhere.

Dirichlet. The beta distribution models a random variable that can take on a value between 0 and 1. The Dirichlet is the multivariate analog of the beta—it models random variables that are vectors of proportions summing to 1. Thus, it can be especially useful in modeling composition of populations, communities, and landscapes as well as the time or energy budgets of individuals. The probability density of a random vector $\mathbf{z}$ conditional on a vector of k parameters $\boldsymbol{\alpha}$ is

$$[\mathbf{z}|\boldsymbol{\alpha}] = \text{Dirichlet}(\mathbf{z}|\boldsymbol{\alpha}) = \frac{\Gamma\left(\sum_{i=1}^{k} \alpha_i\right)}{\prod_{i=1}^{k} \Gamma(\alpha_i)} \prod_{i=1}^{k} z_i^{\alpha_i - 1}, \tag{3.4.47}$$

where k is the number of elements in the vector. The mean of the ith element of the random vector $\mathbf{z}$ is $\mathrm{E}(z_i) = \mu_i = \frac{\alpha_i}{\alpha_0}$ with variance $\mathrm{E}((z_i - \mu_i)^2) = \sigma_i^2 = \frac{\alpha_i(\alpha_0 - \alpha_i)}{\alpha_0^2(\alpha_0 + 1)}$, where $\alpha_0 = \sum_{i=1}^{k} \alpha_i$.

3.4.4 Moment Matching

The concept of parameters that differ from moments is unfamiliar to ecologists trained in statistics classes emphasizing methods based on the normal distribution, which is to say, most ecologists. The two parameters of the normal distribution are its first and second central moments, the mean and the variance, motivating students and colleagues to ask us, "Why are parameters necessary? Why not simply use the moments as parameters for distributions?"

The answer is important, if not obvious. In the normal and multivariate normal, the variance does not change for different values of the mean. However, for other distributions we will use—the binomial, multinomial, negative binomial, beta, gamma, lognormal, exponential and Dirichlet—the variance is a function of the mean. Moreover, the parameters of these distributions are functions of both the mean *and* the variance, which allows the relationship between the mean and variance to change as the parameters change. The only time that the moments can be used as parameters is when the mean and the variance are the same, as in the Poisson, or are not related to each other, as is the case for the normal and multivariate normal. This creates a problem for the ecologist who seeks to use the toolbox of distributions that we have described so far, a problem that can easily be seen in the following example.

Assume you want to model the influence of growing season rainfall (x_i) on the mean aboveground standing crop biomass in a grassland at the end of the growing season (μ_i in kg/ha). You might be inclined to reach for the simple linear model $\mu_i = \gamma_0 + \gamma_1 x_i$ to represent this relationship. However, there are structural problems with a linear model, because it predicts values that can be negative, which makes no sense for biomass. Moreover, it predicts that growth increases infinitely with increasing rainfall, which clearly is not correct on biological grounds. So, using your knowledge of deterministic models (chapter 2), you choose

$$\mu_i = \frac{\kappa x_i}{\gamma + x_i}, \tag{3.4.48}$$

thereby deftly assuring that the model's estimate is nonnegative for nonnegative values of x_i and asymptotically approaches a maximum, κ.[22]

[22]If you wanted a model with "linear" components, you could also use $(\kappa \exp(\gamma_0 + \gamma_1 x_i))/(1 + \exp(\gamma_0 + \gamma_1 x_i))$, which also has a maximum at κ.

Equation 3.4.48 is purely deterministic. You would like to represent the uncertainty that arises because the model isn't perfect and because we fail to observe net primary production perfectly.[23] Your first thought about modeling the uncertainty might be to use a normal distribution, normal$(y_i|\mu_i, \sigma^2)$. So, your model predicts the mean (μ_i) of the distribution of observations of growth (y_i), and the uncertainty surrounding that prediction depends on σ^2. This is the traditional framework for regression. It is convenient because the prediction of the model is the first argument to the distribution. You are probably more familiar with the equivalent formulation, $y_i = \gamma_0 + \gamma_1 x_i + \epsilon_i$, $\epsilon_i \sim \text{normal}(0, \sigma^2)$. We avoid this additive arrangement for representing stochasticity because it cannot be applied to distributions that cannot be centered on zero.

However, informed by the section on continuous distributions, you decide that the normal is a poor choice for your model for two reasons. First, the support is wrong. Biomass cannot be negative, so you need a distribution for data that are continuous and strictly positive. Moreover, a plot of the data shows that the spread of the residuals increases with increasing production, casting doubt on the assumption that the variance is constant. As an alternative, you choose the gamma distribution[24] because it is strictly nonnegative and is parameterized such that the variance increases in proportion to μ^2.

This is entirely sensible, but now you have a problem. How do you get the prediction of your model, the mean prediction of biomass at a given level of rainfall (μ_i), into the gamma probability density function if the function doesn't contain an argument for the mean? How do you represent uncertainty by using the variance, σ^2?

The solution to this problem is *moment matching*. You need equations for the parameters in terms of the moments to allow you to use the gamma distribution to represent the uncertainty in your model. Equations for moments as functions of the parameters can be found in any mathematical statistics text. The converse is not true; it is uncommon to see the parameters expressed as functions of the moments. However, obtaining these functions is easy and useful. You simply solve two equations in two unknowns. On illustration of this solution using the gamma distribution with parameters

[23] In this case, these sources of uncertainty will be inseparable. Later, we will develop models that separate them.

[24] The lognormal would be another logical choice. The gamma or the lognormal would yield virtually identical estimates of parameters if you fit the model.

shape = α and rate = β is

$$\mu = \frac{\alpha}{\beta}, \tag{3.4.49}$$

$$\sigma^2 = \frac{\alpha}{\beta^2}, \tag{3.4.50}$$

so,

$$\alpha = \frac{\mu^2}{\sigma^2}, \tag{3.4.51}$$

$$\beta = \frac{\mu}{\sigma^2}. \tag{3.4.52}$$

You are now equipped to use the gamma distribution to represent the uncertainty in your model of net primary production:

$$\mu_i = \frac{\kappa x_i}{\gamma + x_i},$$

$$\alpha_i = \frac{\mu_i^2}{\sigma^2}, \tag{3.4.53}$$

$$\beta_i = \frac{\mu_i}{\sigma^2}, \tag{3.4.54}$$

$$\left[y_i|\mu_i, \sigma^2\right] = \text{gamma}(y_i|\alpha_i, \beta_i). \tag{3.4.55}$$

As a second example, imagine that you want to model the probability of survival of juvenile birds (μ_i) as a function of population density (x_i). Now, you need a model that makes predictions strictly between 0 and 1, so you might sensibly choose $\mu_i = (\exp(\gamma_0+\gamma_1 x_i))/(1+\exp(\gamma_0+\gamma_1 x_i)), 0 \leq \mu_i \leq 1$. But the gamma distribution is no longer appropriate for representing the uncertainty because it applies to random variables that can exceed 1. The normal is even worse because it includes negative values *and* values that exceed 1. A far better choice is the beta, which models continuous random variables with support on the continuous interval 0 to 1. Solving for the parameters in terms of the moments (using equations 3.4.41 and 3.4.42), you can now make a prediction of survival with your deterministic model and properly represent uncertainty using the beta distribution:

$$\mu_i = \frac{\exp(\gamma_0 + \gamma_1 x_i)}{1 + \exp(\gamma_0 + \gamma_1 x_i)}, \tag{3.4.56}$$

$$\alpha_i = \frac{\mu_i^2 - \mu_i^3 - \mu_i \sigma^2}{\sigma^2}, \tag{3.4.57}$$

$$\beta_i = \frac{\mu_i - 2\mu_i^2 + \mu_i^3 - \sigma^2 + \mu_i \sigma^2}{\sigma^2}, \tag{3.4.58}$$

$$\left[y_i | \mu_i, \sigma^2\right] = \text{beta}(y_i | \alpha_i, \beta_i). \tag{3.4.59}$$

Equations 3.4.51, 3.4.52 and 3.4.57, 3.4.58 are examples of moment matching. We use the functional relationship between the parameters and the moments to allow us to match the predictions of a model to the arguments of the distribution that is best suited to the model and the data. It is important to see how moment matching allows us to specify characteristics of distributions for which the variance is a function of the mean. These matching relationships are broadly useful for the ecological modeler because they allow use of all the distributions we have already described to represent the stochasticity regardless of the form of the arguments to those distributions. It is easy enough to derive the moment matching relationships yourself, but we saved you the trouble in appendix tables A.1 and A.2.

Up to now, we have matched both mean and variance to parameters. However, sometimes we need to match only the mean[25]. Using the beta distribution as an example, we have $\mu = \alpha/(\alpha + \beta)$, so $\alpha = \frac{\mu\beta}{1-\mu}$, allowing us to use $[y|\mu, \beta] = \text{beta}(y|\frac{\mu\beta}{1-\mu}, \beta)$.

3.4.5 Mixture Distributions

The distributions described in section 3.4.3 provide tremendous flexibility for representing uncertainty in ecological models. Sometimes, however, a single distribution fails to adequately portray the behavior of random variables in a way that is faithful to the process that gives rise to them. In this case, the ecological modeler may need to use mixtures of distributions.

[25]As you will learn soon (chapter 8), you can obtain the variance of the distribution of the mean from the output of a Markov chain Monte Carlo algorithm. If you don't need to know the variance of the distribution of the observations, then you don't need to moment match the observation variance.

The general form of a finite mixture distribution for discrete random variables is as follows. Given a finite set of probability distributions for the random variable z, $[z]_1, \ldots, [z]_n$, and weights $w_1, \ldots, w_n$, $w_i \geq 0$, $\sum_{i=1}^{n} w_i = 1$, the finite mixture distribution of z is

$$[z] = \sum_i w_i [z]_i. \tag{3.4.60}$$

Similarly, the general form of an infinite mixture distribution for the variable z depends on a probability density function $[z|\delta]$ with parameter δ. That is, for each value of z in some set δ, $[z|\delta]$ is a probability density function with respect to z. Given a probability density function $[\delta]$ (meaning that $[\delta]$ is nonnegative and integrates to 1), the function

$$[z] = \int [z|\delta]\,[\delta]\,d\delta \tag{3.4.61}$$

is an infinite mixture distribution for z.

We give two examples here. Suppose you study a species and you want to represent the distribution of the random variable, body mass of an individual, z. The sex of individuals is not easily determined in the field, but there are differences in body mass between sexes. How would you model the distribution of these observations?

Because body mass is strictly positive, a gamma distribution is a logical choice, but a single gamma probability density function is not up to the task of representing the two sources of variation in body mass arising from males and females. Instead, you might use

$$\begin{gathered}\phi \cdot \text{gamma}(z|\alpha_m, \beta_m) + (1 - \phi) \cdot \text{gamma}(z|\alpha_f, \beta_f),\\ \phi \sim \text{beta}(\eta, \rho)\end{gathered}$$

where ϕ is the probability that a draw from the population is male. This approach provides a weighted mixture of the male and female body mass distributions where the weighting is controlled by the probability that an individual is a male.

Ecologists often observe random variables that take on zero values more frequently than would be predicted by a single, unmixed distribution, for example a binomial, Poisson, negative binomial, or multinomial. An excessive number of zero-values for a discrete random variable is called *zero inflation*. A second, particularly useful example of a mixture distribution allows the ecological modeler to deal with this overdispersion.

To illustrate, imagine that we sampled many plots along a coastline, counting the number of species of mussels within each plot. In essence there are two sources of zeros. Some zeros arise because the plot was placed in areas that are not mussel habitat, while other zeros occur in plots placed in mussel habitat but that contain no mussels as a result of sampling variation. The Poisson distribution offers a logical choice for modeling the distribution of counts in mussel habitat,[26] but it cannot represent the variation among plots, because it can account only for sampling variation. It cannot portray the zeros that arise because plots were placed in areas where mussels never live.

We can represent these two sources of uncertainty by mixing a Poisson distribution with a Bernoulli distribution. Let z be a random variable representing the number of mussel species in a square meter plot, and w a random variable describing mussel habitat; $w = 1$ if a plot is located in mussel habitat, and $w = 0$ if it is located outside mussel habitat. The distribution of number of mussels in a plot is given by

$$z \sim \begin{cases} 0 & w = 0 \\ \text{Poisson}(\lambda) & w = 1 \end{cases}, \tag{3.4.62}$$

which you will also see written as

$$z \sim \text{Poisson}(z|\lambda w) \cdot \text{Bernoulli}(w|\phi), \tag{3.4.63}$$

$$\phi \sim \text{beta}(\eta, \rho), \tag{3.4.64}$$

where λ is the mean number of mussels per square meter in mussel habitat; ϕ is the probability that a plot contains mussel habitat, and η and ρ are parameters that control the distribution of ϕ. Mixture distributions like this one will be seen again in one of our examples of hierarchical models in sections 6.2.3 and 12.3.

[26]The negative binomial would also work and might be better than the Poisson if the sampling variation is overdispersed.

4 Likelihood

It is possible to learn Bayesian analysis with a bare-bones treatment of likelihood, but we include a full chapter on likelihood for two reasons. Likelihood forms the fundamental link between models and data in the Bayesian framework. Understanding this linkage is central to the aims of this book. In addition, *maximum likelihood* is a widely used alternative to Bayesian methods for estimating parameters in ecological models (Hilborn and Mangel, 1997; Bolker, 2008). It will be useful to understand the similarities and differences between Bayesian analysis and analysis based on maximum likelihood. They are more similar than you might think. Thus, although this is a book about Bayesian modeling, we believe it is important to provide an overview of likelihood and maximum likelihood before we turn to Bayesian inference.

4.1 Likelihood Functions

Until now, we have defined probability distributions in terms of a random variable, z, and parameters. This broad definition makes sense because the Bayesian approach views all things that are not observed[1] as random variables and thus requires a broad framework for treating many kinds of quantities. To explain likelihood, however, we begin by narrowing the arguments in probability distributions to include a data point y and parameters $\boldsymbol{\theta}$ in an ecological model. If we have a *known and fixed* value

[1]Data, before we observe them, are treated as random variables. After we observe them, they are fixed.

of $\boldsymbol{\theta}$, then we are able to calculate the probability (or probability density) of a variable observation y conditional on our model being a true representation of the process that gives rise to y. We use one of the probability distributions described in section 3.4 to make that calculation. A couple of simple examples illustrate how we do that.

EXAMPLE 1 You collect data on the number of tadpoles per volume of water in a pond. You observe 14 tadpoles in a 1 L sample. You *know* that the true average number of tadpoles per liter of water in the pond is 23. The probability of your observation is $[y_1|\lambda] = \text{Poisson}(y_1 = 14|\lambda = 23) = .0136$. A second sample contains 34 tadpoles. Conditional on the known, fixed average number of tadpoles in the pond ($\lambda = 23$), the probability of this additional observation is $[y_2|\lambda] = \text{Poisson}(y_2 = 34|\lambda = 23) = .0069$. The joint probability of the data conditional on λ is the product of the individual probabilities, $.0136 \times .0069 = 9.38 \times 10^{-5}$, assuming these observations are independent, which means that knowledge of one observation tells us nothing about the other. We could extend this calculation of joint probability by taking the product of the probabilities of any number of independent observations.

EXAMPLE 2 You are investigating decomposition of leaf litter over time using a simple model of exponential decay, $\mu_t = e^{-kt}$, where μ_t is the mean proportion of the initial mass of leaf litter remaining at time t (units: day), assuming that the proportion at $t = 0$ is 1, and k is the mass specific rate of decay (units: day^{-1}). Because the data are observed proportions, y_t, that can take on values continuously from 0 to 1, you choose a beta distribution to calculate the probability density of an observation given that the parameter k and the variance of the estimates of your model $\left(\sigma^2\right)$ are known and fixed. Using moment matching, you obtain

$$\alpha_t = \frac{\mu_t^2 - \mu_t^3 - \mu_t\sigma^2}{\sigma^2}, \tag{4.1.1}$$

$$\beta_t = \frac{\mu_t - 2\mu_t^2 + \mu_t^3 - \sigma^2 + \mu_t\sigma^2}{\sigma^2}, \tag{4.1.2}$$

$$\left[y_t|\mu_t, \sigma^2\right] = \text{beta}(y_t|\alpha_t, \beta_t). \tag{4.1.3}$$

Because $\mu_t = e^{-kt}$, you can also write $\left[y_t|e^{-kt}, \sigma^2\right] = \text{beta}(y_t|\alpha_t, \beta_t)$. Conditional on a *known*, fixed decay rate $\left(k = 0.01\ \text{day}^{-1}\right)$ and *known*, fixed $\sigma^2 = 6 \times 10^{-4}$, you calculate parameters for the beta distribution of the mass remaining on day 30: $\alpha_{30} = 236.33$ and $\beta_{30} = 82.68$. The probability density that an observation of $y_t = 0.7$ of the mass remains at time $t = 30$ is 4.040.

These examples were deliberately chosen to make things feel entirely backward. In both cases we know the value of the parameters—they are fixed—and we don't know the data, which are random variables. When the parameter is fixed and the data are random variables, we can calculate the probability of an observation (for discrete data) or the probability density of an observation (for continuous data) using the distributions described in section 3.4.

Alternatively, the usual case for ecological researchers is that the parameters are unknown and the data are fixed. That is, we have a set of observations in hand and we want to know what the observations tell us about parameters. We need a way to evaluate the evidence in the fixed data for variable parameter values, and we do this by using a *likelihood function* $\mathrm{L}(\theta|y)$, defined as

$$\mathrm{L}(\theta|y) = [y|\theta]. \tag{4.1.4}$$

Equation 4.1.4 simply says that the likelihood of the parameter given the data is equal to the probability (or probability density) of the data conditional on the parameter.[2] For n independent observations,

$$\mathrm{L}(\theta|\mathbf{y}) = \prod_{i=1}^{n} [y_i|\theta], \tag{4.1.5}$$

$$\log(\mathrm{L}(\theta|\mathbf{y})) = \sum_{i=1}^{n} \log[y_i|\theta]. \tag{4.1.6}$$

It is important to be clear on terminology. The left-hand side of equation 4.1.4 ($\mathrm{L}(\theta|y)$) is called a *likelihood function*. Bayesians refer to the distribution on the right-hand side ($[y|\theta]$) as a likelihood or a data model to differentiate it from other types of distributions used in Bayesian analysis, as will be discussed in the next chapter.

[2]Older sources on likelihood (Edwards, 1992; Azzalini, 1996; Royall, 1997) express equation 4.1.4 as a proportionality, $\mathrm{L}(\theta|y) = c[y|\theta]$. However, contemporary texts (Pawitan, 2001; Casella and Berger, 2002; Clark, 2007) drop the constant of proportionality c by assuming that $c = 1$. This is permissible because we use likelihood to make comparisons between models or parameter values, one relative to another. For these types of comparisons, the value of c is irrelevant and we simplify notation by letting $c = 1$.

4.2 Likelihood Profiles

The key difference between a probability distribution and a likelihood function is that the parameter is fixed and the data are random variables in a probability mass or density function, whereas in the likelihood function, the data are fixed and the parameters are variable. The relationship between likelihood functions and probability distributions can be most easily seen by plotting them (fig. 4.2.1). If we hold parameters constant and plot a probability density function $[y|\theta]$ as a function of different values of continuously valued y, the area under the curve equals 1 (fig. 4.2.1 A). However, if we hold y constant and plot the same probability density function as a function of θ (fig. 4.2.1 B), we obtain a *likelihood profile*[3] (Hilborn and Mangel, 1997; Bolker, 2008), where the area under the curve does *not* equal 1. A single point is shared between the two curves, showing that probability and likelihood are the same only when the parameter is treated as a fixed quantity.

We see the same type of relationship when the data are discrete. When we hold θ constant and plot a probability mass function $[y|\theta]$ for different values of y, then $\sum_y [y|\theta] = 1$. (fig. 4.2.1 C). However, when we plot the same probability mass function with y fixed as a function of a varying θ, we get a likelihood profile (fig. 4.2.1 D), and, again, the area under the curve does not equal 1.

The units of the y-axis of a likelihood profile are arbitrary and can be scaled to any quantity (because likelihood is strictly defined in terms of a multiplicative constant; see footnote 2). Often, the likelihood profile is scaled such that the peak of the curve equals 1, which is accomplished by dividing all the likelihoods by the likelihood at the peak of the profile, that is, the maximum likelihood. This provides a convenient scaling, but it does not change the relationship between likelihood and probability.

These plots (fig. 4.2.1) illustrate an important but somewhat subtle distinction between likelihood functions and probability distributions. Saying that the parameter θ is not fixed allows us to calculate $L(\theta|y)$ by allowing it to vary, but this does *not* mean that θ is treated as a random variable in the likelihood framework. It is not a random variable, because random variables are defined as quantities governed by probability distributions, and likelihood functions do not define the probability or probability density of θ. This distinction causes some authors to use notation aimed at preventing any confusion with conditional probability; that is, they use $L(\theta; y)$ instead of $L(\theta|y)$.

[3] Also called a "likelihood curve" (Edwards, 1992).

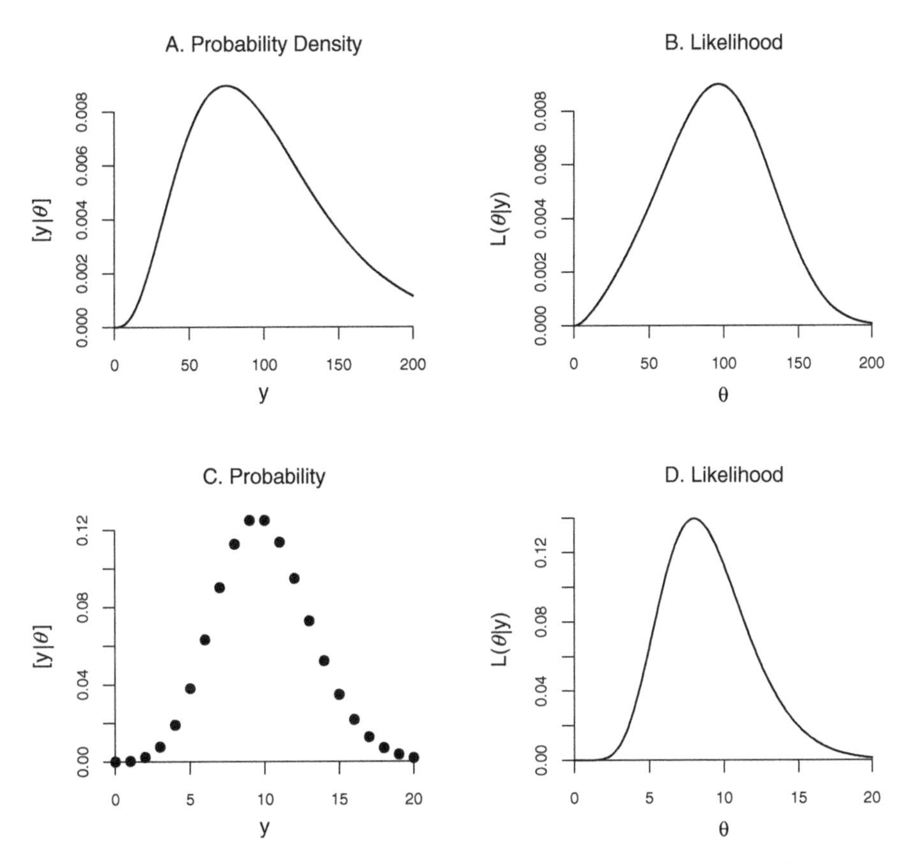

Figure 4.2.1. Illustration of relationships between probability distributions and likelihood profiles for continuous (**A, B**) and discrete (**C, D**) data. Assume we have continuous, strictly positive data with a fixed mean $\theta = 100$ and variance $\sigma^2 = 2500$. Panel **A** shows a gamma probability density function plotted over $y = 0, \ldots, 200$ conditional on the fixed parameters $\alpha = \theta^2/\sigma^2$, $\beta = \theta/\sigma^2$. The integral of this function taken from 0 to ∞ is 1. Assuming a fixed value of $y = 75$ and a fixed value of $\sigma^2 = 2500$, panel **B** shows a gamma probability density function (with moments matched to parameters) over $\theta = 0, \ldots, 200$. This curve is a likelihood profile, and it does not integrate to 1. A single point is shared between the two curves, $[y|\theta, \sigma^2]$, where $y = 75$, $\theta = 100$, and $\sigma^2 = 2500$, illustrating that probability and likelihood are the same only when the parameters are treated as fixed. In panel **C**, we assumed a fixed mean, $\theta = 10$, for a Poisson probability mass function of the data $y = 1, 2, 3, \ldots, 20$. The sum of the probabilities from 0 to ∞ equals 1. In panel **D**, a likelihood profile is plotted as the Poisson probability mass function, where its mean θ varies from 0 to 20 assuming a fixed $y = 8$. The integral of this curve over 0 to 1 does not equal 1.

4.3 Maximum Likelihood

A central tenet of likelihood is that all evidence is relative. The likelihood of a specific value of a parameter tells us nothing useful in the absence of a comparison with an alternative value (Edwards, 1992; Royall, 1997). The arbitrary scaling of the y-axis in a likelihood profile shows why this is true; it is impossible to use likelihood to say anything informative about a single parameter value or a single model, because the values of the likelihood profile can take on any value depending on the arbitrary value of c, as described in footnote 2. We learn about parameters only by comparing the likelihood of one parameter value with the likelihood of another parameter value. The constant of proportionality doesn't matter, because it is contained in both likelihoods.

We compare the evidence in data for alternative values of parameters using likelihood ratios. Imagine two specific values of parameters, θ_1 and θ_2. The evidence provided by the data for these values is contained in the likelihood profile (fig. 4.2.1 C,D) and is expressed as the likelihood ratio

$$\frac{\mathrm{L}(\theta_1|y)}{\mathrm{L}(\theta_2|y)} = \frac{[y|\theta_1]}{[y|\theta_2]}. \tag{4.3.1}$$

The likelihood principle dictates that all the information in a particular observation (or set of observations) relative to the merits of two, alternative values of parameters is contained in the likelihood ratio (eq. 4.3.1), and the likelihood ratio is interpreted as the degree to which the data support one value of a parameter over another (Edwards, 1992; Azzalini, 1996; Royall, 1997). The evidence in data for alternative values of parameters is often obtained using the natural logarithm of the likelihood ratio, which is formally defined as the *support* for one value of a parameter over another, conditional on the data.[4]

Thus, the ratio of the likelihoods or the difference between the log likelihoods provides the basis for evaluating the evidence in data for alternative values of parameters (fig. 4.3.1). In most practical problems in ecology, we are particularly interested in the value of the parameter θ that has the maximum support in data, which is found at the peak of the

[4]There is potential for confusion here. Edwards (1992) defined support as the log of the likelihood ratio. Statisticians also use support to mean the domain of a probability function or a probability density function, where the values of the probability or probability density of a random variable exceed 0. So, used this way, the support for the Poisson distribution is the set of all nonnegative integers; the support for the normal distribution is all real numbers. For the remainder of the book, we use the latter meaning.

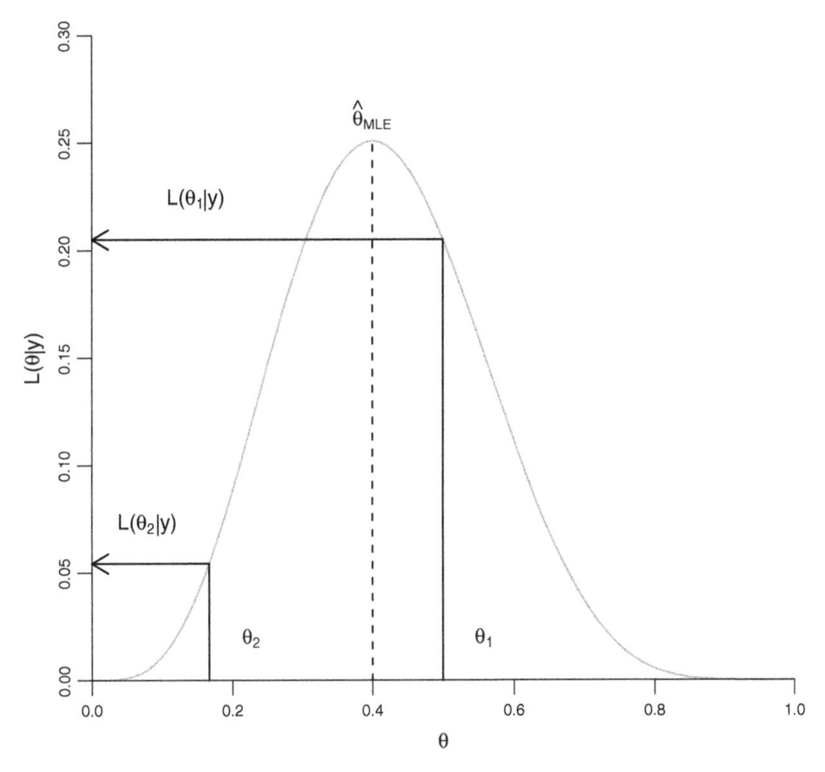

Figure 4.3.1. The evidence in data for alternative values of parameters is contained in the likelihood profile, illustrated by the curve. In the example here, there is greater evidence for $\theta = \theta_1$ relative to $\theta = \theta_2$, because $\mathrm{L}(\theta_1|y) > \mathrm{L}(\theta_2|y)$ (arrows). The maximum likelihood estimate for $\theta = \hat{\theta}_{MLE}$ is the value of θ for which $\mathrm{L}(\theta|y)$ is greater than for any other value (dashed line).

likelihood profile (fig. 4.3.1). This is the value of θ that maximizes the likelihood function (eq. 4.1.5) or the log likelihood function (eq. 4.1.6). We can find this maximum likelihood value of θ analytically for simple models or by numerical methods for more complex ones (Hilborn and Mangel, 1997; Pawitan, 2001; Bolker, 2008).

4.4 The Use of Prior Information in Maximum Likelihood

It is often heard that the difference between Bayesian analysis and likelihood is that Bayes uses prior information, although it is the treatment of unobserved quantities as random variables that truly distinguishes Bayes. We go into detail about the unique features of Bayes in the next chapter, but here show that using prior information does *not* set it apart from the other prevailing approach to parameter estimation, maximum likelihood.

The first step in estimating parameters using maximum likelihood is to choose a likelihood function, a choice that itself requires some prior knowledge of the parameters we seek to estimate. However, it is also possible to use information from previous studies in a more direct way by including this information in the likelihood function.

We showed earlier how multiple, independent observations from a data set can be used to calculate the likelihood (eq. 4.1.5) or the log likelihood (eq. 4.1.6) of a parameter. It is possible to include prior information on the parameter in the same way—the total likelihood is the product of the individual likelihoods of the data and the prior likelihood (Edwards, 1992; Pawitan, 2001), as follows. We define α and β as the parameters describing the probability distribution of θ based on prior research. We can include prior information on θ using

$$\begin{aligned} \mathrm{L}(\theta|\mathbf{y}) &= [\mathbf{y}|\theta]\,[\theta] \\ &= \underbrace{\prod_i [y_i|\theta]}_{a}\,\underbrace{[\theta|\alpha,\beta]}_{b}\,. \end{aligned} \tag{4.4.1}$$

We find the maximum likelihood value of θ, including current data and prior information, by finding the value of θ that maximizes the total likelihood, that is, the product of the probability or probability density of the data conditional θ (term a in eq. 4.4.1) with the probability density of θ conditional on parameters obtained in earlier studies (term b in equation 4.4.1). Edwards (1992, pg. 36) calls the log of term a the experimental support and the log of term b, the prior support. We show later (sec. 9.1.3.1) that equation 4.4.1 is a specific example of a more general statistical procedure called *regularization*.

5 Simple Bayesian Models

In this chapter we lay out the basic principles of Bayesian inference, building on the concepts of probability developed earlier (chapter 3). Our purpose is to use the rules of probability to show how Bayes' theorem works. We will make use of the conditional rule of probability and the law of total probability, so it might be useful to review these first principles (sec. 3.2) before proceeding with this chapter.

We begin with the central, underpinning tenet of the Bayesian view: the world can be divided into quantities that are observed and quantities that are unobserved. Unobserved quantities include parameters in models, latent states predicted by models, missing data, effect sizes, future states, and data before we observe them. We wish to learn about these quantities using observations. The Bayesian framework for achieving that understanding is applied in exactly the same way regardless of the specifics of the research problem at hand or the nature of the unobserved quanties.

The feature of Bayesian analysis that most clearly sets it apart from all other types of statistical analysis is that Bayesians treat all unobserved quantities as random variables.[1] Because the behavior of random variables is governed by probability distributions, it follows that unobserved

[1]There is some argument among statisticians about whether states of ecological systems and parameters governing their behavior are truly random. Ecologists with traditional statistical training may object to viewing states and parameters as random variables. These objections might proceed as follows. Consider the state "the average biomass of trees in a hectare of Amazon rainforest." It could be argued that there is nothing random about it, that at any instant in time there *is* an average biomass that is fixed and knowable at that instant—it is determined, not random. This is true, perhaps, but the practical fact is that if we were to attempt to know that biomass, which is changing by the minute, we would obtain different values depending on when and how we measured it. These values would follow a probability distribution. So, thinking of unknowns

quantities can be characterized by probability distributions like those we learned about in section 3.4. Bayesian analysis uses the rules of probability (sec. 3.2) to discover the characteristics of the probability distributions of unobserved quanties. Understanding those distributions enables the ecological researcher to make statements about processes tempered by honest specifications of uncertainty.

It is fundamental to Bayesian analysis to understand the distinctions among things that are known versus unknown, observed versus unobserved, and random variables versus fixed quantities. The first distinction is this: things that are *known* are not random variables but, rather, are treated as fixed. This might seem obvious, but it can be slippery. Numerical constants, for example π, are known. Things that are not observed, for example, parameters in a model, latent states, predictions, and missing data are unknown and are always modeled as random variables. But what about things we observe?

Observations of responses (i.e., the y) are always modeled as random variables. How can this be? How can something that we observe be random? The key idea here is that the y are random variables *before they are observed.* After we observe them, they are quantities in hand that represent one instance of a stochastic process. So, this one instance of observation is fixed, but if we repeated our observations of the response, we would not expect always to get identical values. The sources of stochasticity in responses will be treated in greater detail as we proceed.

What about observed predictor variables (i.e., covariates, the x)? Are they random or fixed? Rightly or wrongly (usually wrongly), ecologists often treat predictor variables as being observed perfectly—they are observations but they are treated as if they were known, fixed quantities. They are not random variables if we assume they are measured without error, but they *are* random variables if we assume they have measurement or sampling errors that we seek to include in our model.

---

as random variables is a scientifically useful abstraction with enormous practical benefits, benefits we demonstrate in later chapters. We leave arguments about whether states and parameters are "truly random" to metaphysics. As an aside, Ben Bolker (personal communication) points out that "The same traditionally trained ecologists who object to treating states as random variables don't mind using hypothesis tests that are grounded in the idea of a long-term frequency of observation in repeated observations, which don't sensibly exist in many cases . . . ."

5.1 Bayes' Theorem

The basic problem in ecological research is to understand processes that we cannot observe based on quantities that we can observe. We represent unobserved processes as models made up of parameters and latent states, which we notate here as θ. We make observations y to learn about θ. Before the data are observed, we treat them as random variables. The chance of observing the data conditional on θ is given by a probability distribution, $[y|\theta]$. Because θ is also a random variable, it is governed by the probability distribution $[\theta]$. We wish to discover the probability distribution of the unobserved θ conditional on the observed data, that is, $[\theta|y]$. Using the basic rules of conditional probability for two random variables, we have

$$[\theta|y] = \frac{[\theta, y]}{[y]}, \tag{5.1.1}$$

$$[y|\theta] = \frac{[\theta, y]}{[\theta]}. \tag{5.1.2}$$

Solving equation 5.1.2 for $[\theta, y]$ we have

$$[\theta, y] = [y|\theta]\,[\theta]. \tag{5.1.3}$$

Substituting the right-hand side of equation 5.1.3 for $[\theta, y]$ in equation 5.1.1 we obtain

$$[\theta|y] = \frac{[y|\theta]\,[\theta]}{[y]}. \tag{5.1.4}$$

Because y is conditional on θ, the law of total probability (eqs. 3.2.13 and 3.2.14) for discrete-valued parameters shows that

$$[y] = \sum_{\theta}[y|\theta][\theta], \tag{5.1.5}$$

where the summation is over all possible values of θ. For parameters that are continuous,

$$[y] = \int [y|\theta]\,[\theta]\,d\theta. \tag{5.1.6}$$

Substituting the right-hand side of equation 5.1.5 for $[y]$ in 5.1.4, we obtain Bayes' theorem for discrete-valued parameters,

$$[\theta|y] = \frac{[y|\theta]\,[\theta]}{\sum_\theta [y|\theta]\,[\theta]}, \tag{5.1.7}$$

and similarly substituting equation 5.1.6 for $[y]$ in 5.1.4, we find Bayes' theorem for parameters that are continuous,

$$[\theta|y] = \frac{[y|\theta]\,[\theta]}{\int [y|\theta]\,[\theta]\,d\theta}. \tag{5.1.8}$$

Bayes' theorem provides the basis for estimating the probability distribution of the unobserved quantities θ informed by the data y. A simple example illustrates these ideas graphically (box 5.1).

Box 5.1 Illustration of Bayes' Theorem

Imagine support for the parameter θ shown as the light-colored polygon labeled S. Assume that θ can take on three values, θ_1, θ_2, and θ_3. We assume for simplicity that these are the *only* possible values—they are mutually exclusive and exhaustive; that is, $\sum_i$ area of wedge$_i$ $= S$. The area of each θ_i wedge divided by the area of S reflects our prior knowledge of the parameter: area of wedge θ_i/area of S $=$ Pr(θ_i). If we have no reason to favor one value of θ_i over another, Pr(θ_1) = Pr(θ_2) = Pr(θ_3) = $\frac{1}{3}$

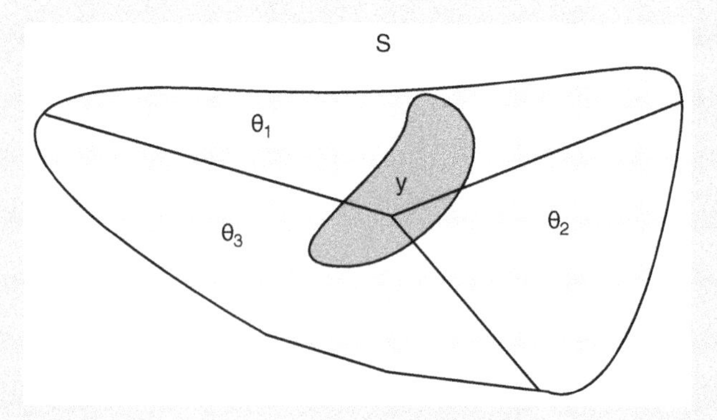

(*continued*)

(Box 5.1 *continued*)

We now collect some data, shown by the dark polygon y. The parameter θ controls how the data arise. So, for example, the data might be the number of survivors observed in a sample of n individuals during time Δt, where θ is the probability that an individual survives over the time interval. We want to use the data to update our knowledge of θ.

Given that we have data in hand, we can limit attention to the wedge of the θ_i contained within the data polygon. The probability of θ_i is $\Pr(\theta_i|y) = \frac{\text{area of } \theta_i \text{ within } y}{\text{area of } y} = \frac{\text{area of } \theta_i \text{ within } y/\text{area of } S}{\text{area of } y/\text{area of } S} = \frac{\Pr(\theta_i \cap y)}{\Pr(y)} = \frac{\Pr(\theta_i, y)}{\Pr(y)}$. Using the conditional rule of probability to substitute for $\Pr(\theta_i, y)$, we have $\Pr(\theta_i|y) = \Pr(y|\theta_i)\Pr(\theta_i)/\Pr(y)$. Using $\Pr(y) = \text{area of } y/\text{area of } S = \sum_j \Pr(y|\theta_j)\Pr(\theta_j)$, we find Bayes' theorem for discrete parameters:

$$\Pr(\theta_i|y) = \frac{\Pr(y|\theta_i)\Pr(\theta_i)}{\sum_j \Pr\left(y|\theta_j\right)\Pr\left(\theta_j\right)}. \tag{5.1.9}$$

The denominator is a normalizing constant assuring that $\sum_i \Pr(\theta_i|y) = 1$. As the number of wedges in S increases to infinity and their area decreases to 0, we have Bayes' theorem for continuous parameters:

$$\left[\theta|y\right] = \frac{[y|\theta][\theta]}{\int [y|\theta][\theta]d\theta}. \tag{5.1.10}$$

Understanding Bayesian inference and why it works requires that we understand each of its components, which we now explain for continuous parameters. The *likelihood* $\left[y|\theta\right]$ (fig. 5.1.1) plays a key role in Bayesian analysis by linking the unobserved θ to the observed y. It allows us to answer a central question of science: what is the probability that we will observe the data if our deterministic model ($g(\theta)$) accurately portrays the process that gives rise to the data? We have seen the likelihood before (eq. 4.1.4, fig. 4.1).

The *prior distribution* of the unobserved quantities, $[\theta]$, represents our knowledge about θ before we collect the data (fig. 5.1.1). The prior distribution can be informative, reflecting knowledge gained in previous research, or it can be vague, reflecting a lack of information about θ before we collected the data that are now in hand. We treat priors in greater detail in the next section; for now, we highlight prior distributions as one of the components of Bayes' theorem.

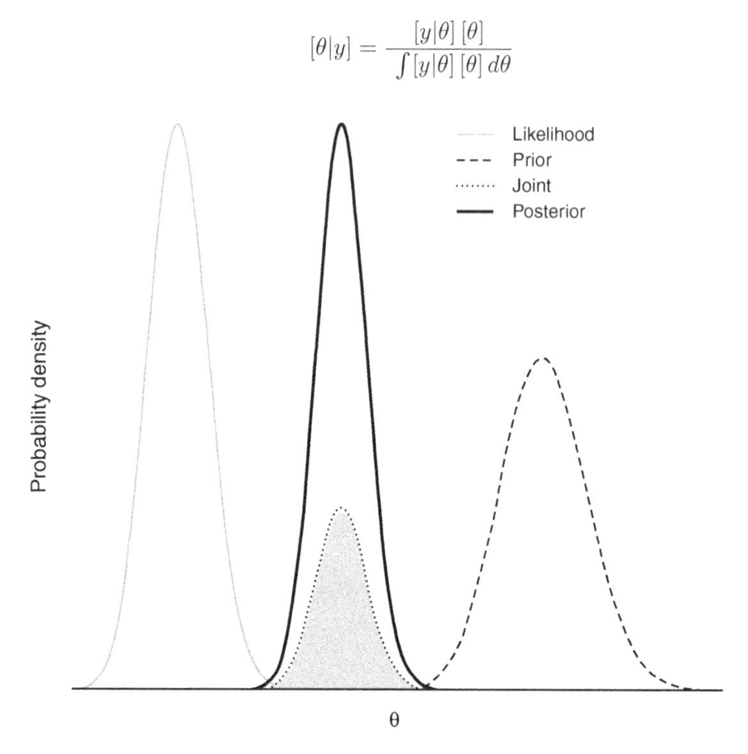

Figure 5.1.1. Illustration of Bayes' theorem for data (y) and unobserved quantities (θ). The likelihood ($[y|\theta]$, gray solid line) gives the probability that we will observe the data conditional on the value of the parameter. The prior ($[\theta]$, dashed line) specifies the probability of θ based on our knowledge of θ before the data are collected. The joint distribution (dotted line) is the product of the prior and the likelihood. The marginal distribution of the data $\left(\int [y|\theta][\theta]d\theta\right)$ is the integral of the joint distribution, shown here as the shaded area. (See sect. 3.4.2 for a review of the concept of marginal distributions.) The posterior is the distribution (black solid line) that results when we divide every point on the joint distribution curve by the area under the curve, effectively normalizing the joint distribution so that the area under the posterior distribution equals 1.

The product of the likelihood and the prior is the joint distribution[2] (fig. 5.1.1). We have seen this product $\left(\left[y|\theta\right][\theta]\right)$ before (eq. 4.4.1), and we learned that it does not define a probability distribution for θ because the area under the curve $\left[y|\theta\right][\theta]$ with respect to θ is not certain to equal 1.

[2]Recall that the joint distribution $[\theta, y] = [y|\theta][\theta]$.

The marginal distribution of the data,

$$[y] = \int [y|\theta] [\theta] d\theta, \tag{5.1.11}$$

is the area under the joint distribution curve (fig. 5.1.1). Dividing each point on the joint distribution $[y|\theta] [\theta]$ by $\int [y|\theta] [\theta] d\theta$ normalizes the curve with respect to θ, yielding the posterior distribution $[\theta|y]$. The posterior distribution is a true probability density function that meets all the requirements for these functions (sec. 3.4.1), including that $\int[\theta|y]d\theta = 1$. Dividing the joint distribution by $\int [y|\theta] [\theta] d\theta$ assures that the posterior distribution integrates to 1, which is why $[y]$ is often referred to as a *normalizing constant*.

Before the data are collected, y is a random variable, and the quantity $\int [y|\theta] [\theta] d\theta$ is a marginal distribution, a concept we will use frequently in later chapters (for review, see sect. 3.4.2). It is also called the *prior predictive distribution*—it tells us what we know about the *data* before they are collected. However, after the data are collected, $\int [y|\theta] [\theta] d\theta$ is a known, fixed quantity (a scalar). This means that

$$[\theta|y] \propto [\theta, y] \tag{5.1.12}$$

$$\propto [y|\theta] [\theta]. \tag{5.1.13}$$

We will make extensive use of this proportionality.[3] We can use equation 5.1.13 to learn about the posterior distribution from the joint distribution even when we cannot directly calculate $[y]$, as will often be the case. We call equation 5.1.13 a simple Bayesian model because it represents the joint distribution of the observed and unobserved quantities as the product of the likelihood and the prior distributions.

We could have developed the same ideas about discrete-valued parameters using sums rather than integrals.

5.2 The Relationship between Likelihood and Bayes'

The fundamental difference between inference based on maximum likelihood and inference based on Bayes' theorem is that Bayes' treats all

[3]The constant of proportionality is the reciprocal of the marginal distribution of the data, which is a constant after the data are observed.

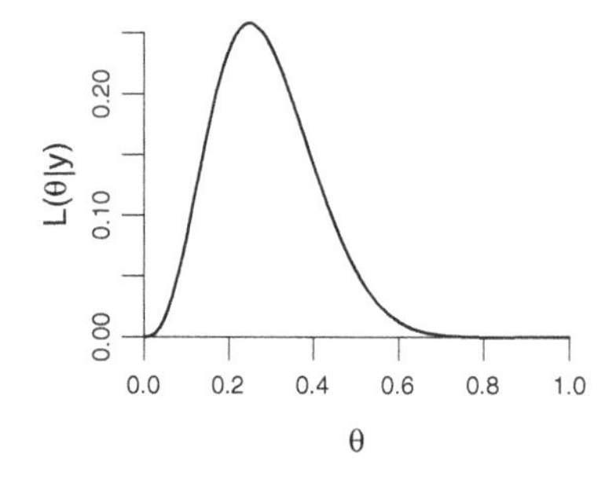

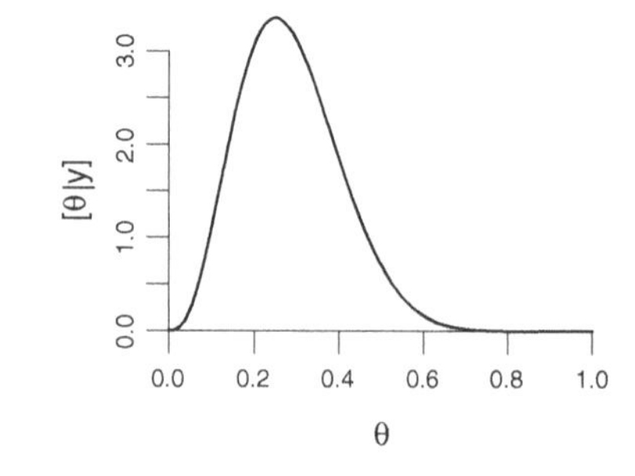

Figure 5.2.1. Likelihood profile (left panel) and posterior distribution (right panel) for the parameter probability of a success, θ, given the observation three successes on 12 trials with uninformative priors on θ. The shapes of the two curves are identical. The area under the likelihood profile does not equal 1. The area under the posterior distribution equals 1.

unobserved quantities as random variables governed by probability distributions. This treatment is possible because dividing the joint distribution by the marginal distribution of the data assures that the posterior distribution is a true probability distribution (fig. 5.2.1). This is a nontrivial result, because it allows Bayesian inference to make probabilistic statements about unobserved quantities of interest. In contrast, the likelihood profile is not a probability distribution—there is nothing that assures that the area under the curve equals 1 (fig. 5.2.1). Unknowns cannot be treated as random variables in the likelihood framework. Instead, likelihood depends on comparing the relative strength of evidence in data for one value of a parameter over another value. Prior information can be used in Bayesian and likelihood analysis with $[y|\theta][\theta]$. In likelihood, we find the values of θ that maximize $[y|\theta]\,[\theta]$. The normalization of this product by the marginal distribution of the data is what sets Bayesian inference apart from inference based on likelihood—it allows unobserved quantities to be treated as random variables.

5.3 Finding the Posterior Distribution in Closed Form

A simple Bayesian model contains a joint distribution expressed as a likelihood multiplied by a prior (or priors), $[y|\theta][\theta]$. There are special cases of this product where the posterior distribution $[\theta|y]$ has the same form as the prior, $[\theta]$. In these cases, the prior and the posterior are called *conjugate distributions* (or simply conjugates), and the prior is called a conjugate of the likelihood. Conjugate distributions are important for two reasons. For simple problems, they allow us to calculate the parameters of posterior distributions on the back of a cocktail napkin.[4] Moreover, the ease of calculation of parameters of the posterior for simple problems becomes important for complicated problems if we can break them down into parts that can be attacked one at time. We will learn about the role of conjugates in this process in the chapter on Markov chain Monte Carlo (chapter 7).

It is perfectly possible to make use of conjugate priors effectively without knowing how each one is derived. Seeing a single derivation (box 5.3) is adequate background for most ecologists who seek to use Bayesian methods. However, we offer a couple of examples here to provide intuition for conjugate relationships. More detailed treatment as well as tables showing the known conjugate distributions can be found in Bayesian textbooks (e.g., Gelman, 2006). The ones we use most frequently are shown in appendix table A.3.

Box 5.3 Derivation of the Posterior Distribution for a Beta Prior and Binomial Likelihood

We seek the posterior distribution of the parameter ϕ, the probability of a success conditional on n trials and y observed successes. The beta distribution is a conjugate prior for the binomial likelihood. We use Bayes' theorem to obtain

$$[\phi|y,n] \propto \underbrace{\binom{n}{y}\phi^{y}(1-\phi)^{n-y}}_{\text{binomial likelihood}}\,\underbrace{\frac{\Gamma(\alpha+\beta)}{\Gamma(\alpha)\Gamma(\beta)}\phi^{\alpha-1}(1-\phi)^{\beta-1}}_{\text{beta prior}}, \tag{5.3.1}$$

where α and β are the parameters of the beta prior distribution. By dropping the normalizing constants $\left(\binom{n}{y}, \frac{\Gamma(\alpha+\beta)}{\Gamma(\alpha)\Gamma(\beta)}\right)$ we obtain

$$[\phi|y,n] \propto \underbrace{\phi^{y}(1-\phi)^{n-y}}_{\text{binomial likelihood}}\,\underbrace{\phi^{\alpha-1}(1-\phi)^{\beta-1}}_{\text{beta prior}}. \tag{5.3.2}$$

Simplifying, we get

$$[\phi|y,n] \propto \phi^{y+\alpha-1}(1-\phi)^{\beta+n-y-1}. \tag{5.3.3}$$

(*continued*)

[4]It is embarrassing to do an elaborate numerical procedure to obtain results that can be obtained on a napkin.

(Box 5.3 *continued*)

Let $\alpha_{new} = y + \alpha$ and $\beta_{new} = \beta + n - y$. Multiplying equation 5.3.3 by the normalizing constant $\Gamma(\alpha_{new} + \beta_{new})/\Gamma(\alpha_{new})\Gamma(\beta_{new})$ we obtain the posterior distribution of ϕ, a beta distribution with parameters α_{new} and β_{new}:

$$[\phi|y, n] = \frac{\Gamma(\alpha_{new} + \beta_{new})}{\Gamma(\alpha_{new})\Gamma(\beta_{new})}\phi^{\alpha_{new}-1}(1-\phi)^{\beta_{new}-1}. \tag{5.3.4}$$

The following examples show how conjugate prior-likelihood relationships can be used to estimate posterior distributions easily and quickly. Imagine that you are studying infection of whitebark pine (*Pinus albicaulis*) with blister rust (*Cronartium ribicola*). You desire information on the proportion of individuals in a stand that are infected, that is, the prevalence of the disease, ϕ. You take a sample of 80 individuals and find 17 that are infected. What is the posterior distribution of ϕ? We will use the simple Bayesian model

$$[\phi|y] = \frac{[y|\phi]\,[\phi]}{[y]}. \tag{5.3.5}$$

We have no prior knowledge of disease prevalence in the stand, so a reasonable choice for a prior distribution of ϕ, a quantity that can take on continuous values between 0 and 1, is a beta distribution with parameters $\alpha_{prior} = 1, \beta_{prior} = 1$; that is, $\phi \sim \text{beta}(1, 1)$, which defines a uniform distribution over (0,1). A logical choice for the likelihood of ϕ is a binomial distribution with $y = 17$ successes given $n = 80$ trials, where we seek to know the probability of a "success"; that is, $y \sim \text{binomial}(80, \phi)$. Thus,

$$\text{beta}(\phi|\alpha_{posterior}, \beta_{posterior}) = \frac{\text{binomial}(y|\phi, n)\ \text{beta}(\phi|\alpha_{prior}, \beta_{prior})}{[y]}. \tag{5.3.6}$$

Using the beta-binomial conjugate prior relationship, we can calculate the parameters of the posterior beta distribution using $\alpha_{posterior} = \alpha_{prior} + y$ and $\beta_{posterior} = \beta_{prior} + n - y$. So, in this example, the posterior distribution of ϕ is beta$(1 + 17, 1 + 80 - 17)$, which has a mean of $\alpha_{posterior}/(\beta_{posterior} + \alpha_{posterior}) = 0.219$ and variance

$\alpha_{posterior}\beta_{posterior}/[(\alpha_{posterior} + \beta_{posterior})^2(\alpha_{posterior} + \beta_{posterior} + 1)] =$ 0.0021 (sec. 3.4.4 and app. table A.2). Using the quantile function for a beta distribution, we can calculate that the true value of ϕ lies between 0.137 and 0.315 with probability 0.95.

As a second example, suppose you are studying copepods in an Arctic lake during summer. You wish to estimate the posterior distribution of the mean abundance per unit volume using

$$[\lambda|\mathbf{y}] = \frac{[\mathbf{y}|\lambda][\lambda]}{[\mathbf{y}]}. \tag{5.3.7}$$

Prior research has shown that lakes like the one you are studying have a mean abundance of $\lambda_{prior} = 52$ individuals per liter with a standard deviation of 6.8. You take a sample of four scoops of 1L of water and count the individuals they contain, finding $\mathbf{y} = (64, 48, 59, 52)'$. What can we say about the abundance of copepods informed by the data and the prior estimate? A good choice for the likelihood in this example (i.e., $[\mathbf{y}|\lambda]$) is the Poisson, because the data are discrete and because the variance is approximately the same as the mean. A gamma prior distribution (i.e., $[\lambda]$) is conjugate to the Poisson likelihood, so the posterior distribution of the mean of the Poisson ($[\lambda|y]$ is also gamma). Thus,

$$\text{gamma}(\lambda|\alpha_{posterior}, \beta_{posterior}) = \frac{\prod_{i=1}^{4} \text{Poisson}(y_i|\lambda)\,\text{gamma}\left(\lambda|\alpha_{prior}, \beta_{prior}\right)}{[\mathbf{y}]}. \tag{5.3.8}$$

The parameters of a gamma posterior are $\alpha_{posterior} = \alpha_{prior} + \sum_{i=1}^{n} y_i$ and $\beta_{posterior} = \beta_{prior} + n$. To use the prior information we must first convert the prior mean and standard deviation to prior parameters using moment matching (sec. 3.4.4), $\alpha_{prior} = \mu^2_{prior}/\sigma^2_{prior} = 58.5$, and $\beta_{prior} = \mu/\sigma^2 =$ 1.12. It follows that the parameters of the gamma posterior distribution of the mean abundance are $\alpha_{posterior} = 58.5 + 64 + 48 + 59 + 52 =$ 281.5 and $\beta_{posterior} = 4 + 1.12 = 5.12$. The mean of the posterior is $\alpha_{posterior}/\beta_{posterior} = 55$ with variance $\alpha_{posterior}/\beta^2_{posterior} = 10.7$ and standard deviation 3.3. The upper .975 quantile for a gamma distribution with parameters $\alpha = 281.5$ and $\beta = 5.12$ is 61.5, and the lower .025 quantile is 48.7. Thus, the probability is .95 that the true mean number of individuals per liter is between 48.7 and 61.6.

5.4 More about Prior Distributions

We devote an entire section in this chapter to prior distributions because ecologists who have not received formal training in Bayesian methods will be especially unfamiliar with the use of priors, a concept that, in contrast with likelihood, has no parallel in traditional statistical training. We also include this section because ecologists often seek to minimize the influence of the prior on inference. This is a place where it is easy to make errors. Finally, we want to advocate the thoughtful use of informed priors in Bayesian modeling.

Some view the choice of a prior in Bayesian models as a contentious topic because it is a decision that can influence inference. However we will attempt to convince you of the following:

1. There is no such thing as a noninformative prior, but certain priors influence the posterior distribution more than do others.
2. Informative priors, when properly justified, can be tremendously useful in Bayesian modeling (and science, in general).

It is important to remember that one of the objectives of Bayesian analysis is to provide information that can inform subsequent analyses; the posterior distribution obtained in one investigation becomes the prior in subsequent investigation. Thus, we agree with the view of Gelman (2006) that "noninformative" priors are provisional. They are a starting point for analysis. As scientists, we should always prefer to use appropriate, well-constructed, informative priors on θ.

5.4.1 "Noninformative" Priors

We use quotation marks in this section title because there is no such thing as a noninformative prior. By that we mean that all priors will have some influence on the posterior distribution of some transformation of the parameter you may be interested in learning about. Let's begin, then, by studying potential priors for a very simple Bayesian model, a model for binary data. Consider the set of binary data (i.e., zeros and ones) denoted by y_i, for $i = 1, \ldots, n$. If we are interested in inference concerning the probability that a given observation will be one, $p = \Pr(y = 1)$, then we could formulate the parametric model

$$y_i \sim \text{Bernoulli}(p), \tag{5.4.1}$$

where $i = 1, \ldots, n$. In this case, a Bernoulli distribution is the "model" that we assume stochastically generated the data. The Bernoulli distribution contains the parameter p; thus, a complete Bayesian model requires a prior distribution for p. Let's examine a few priors for p as well as their influence on the posterior distribution for the following data set: $\mathbf{y} = (0, 0, 1, 0, 1, 0, 0, 0, 1, 0)'$.

Perhaps the most commonly chosen prior for p is the uniform distribution, such that $0 < p < 1$. The uniform is a specific case of the more flexible beta distribution; thus, it is common to select the prior

$$p \sim \text{beta}(\alpha, \beta), \tag{5.4.2}$$

where if $\alpha = \beta = 1$, this distribution becomes a uniform. The uniform distribution is commonly thought to be "noninformative" in this setting because all possible values of p are equiprobable. The uniform can be contrasted with a prior where larger values of p are more probable, such as when $\alpha = 4, \beta = 1$. We compare the posterior distributions arising from these two choices for a prior in figure 5.4.1. Notice how the prior in figure 5.4.1 B "pulls" the posterior toward the larger values, thus influencing it.

An alternative to the visual approach for assessing the influence of the prior on the posterior is to inspect the closed-form mathematical expression for the posterior (i.e., the result of conjugate relationships, sect. 5.3). For the Bernoulli-beta model[5] we are using in this example, the posterior distribution for p is

$$[p|\mathbf{y}] = \text{beta}\left(\sum_{i=1}^{n} y_i + \alpha, \sum_{i=1}^{n}(1 - y_i) + \beta\right). \tag{5.4.3}$$

In our simple example, for the data $\mathbf{y}$, the resulting beta posterior distribution has parameters $3 + \alpha$ and $7 + \beta$. Notice that larger values for the prior parameters α and β have more of an effect on these parameters in the posterior. Similarly, as both α and β get small, the posterior distribution appears to become less influenced by the prior (leaving only statistics related to the data in the posterior). Thus, a beta prior with $\alpha = 1, \beta = 1$ is less influential on the posterior than a beta prior with $\alpha = 4, \beta = 1$. This is a seemingly sensible result and one that is very commonly used to justify the specification of priors, especially for probabilities (i.e., p), regression

[5]We showed the derivation of the expression for the posterior distribution when the prior is beta and the likelihood is binomial in section 5.3. Recall that the Bernoulli is a special case of the binomial where n, the number of trials equals 1.

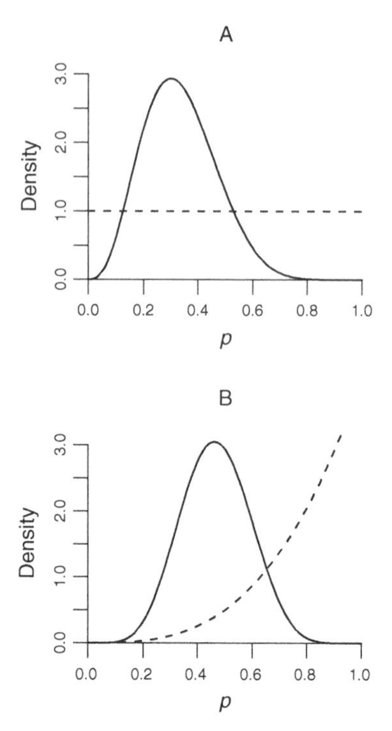

Figure 5.4.1. Prior (dashed line) and resulting posterior distributions (solid line) for a model with a Bernoulli likelihood and a beta prior with two prior specifications: (**A**) $\alpha = 1, \beta = 1$ and (**B**) $\alpha = 4, \beta = 1$.

coefficients (i.e., $\boldsymbol{\beta}$), and variance components (i.e., σ^2). Perfect flatness can be achieved only in bounded priors like the beta, but priors that *approach* flatness are often referred to as "flat" nonetheless. You will also see them called "diffuse," "weak," or "vague."

It is important to recognize that even the uniform prior for p technically has some influence on the posterior distribution, because prior parameters $\alpha = 1, \beta = 1$ yield the posterior parameters $3 + 1, 7 + 1$, which are not the same as $3, 7$, as would be the case if only statistics related to the data appeared in the posterior. Using this argument, one might be tempted to use $\alpha = 0, \beta = 0$ as prior parameters, but recall from the definition of the beta distribution that both parameters must be greater than zero to ensure a valid probability distribution. Furthermore, the sensibility of using very small values for α and β in the beta prior breaks down because, as we see in figure 5.4.2, a beta prior with $\alpha = 0.001, \beta = 0.001$ actually pulls the posterior distribution toward zero. Most of the mass of a U-shaped prior

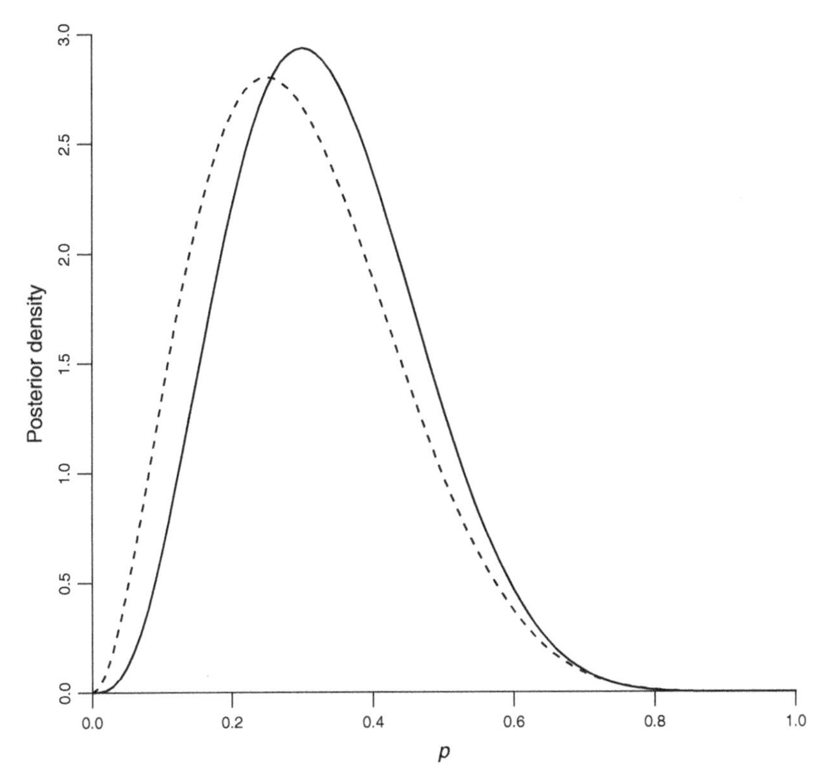

Figure 5.4.2. Resulting posterior distributions for the Bernoulli-beta model with prior specifications $\alpha = 1, \beta = 1$ (solid line) and $\alpha = 0.001, \beta = 0.001$ (dashed line).

distribution implied by the beta(0.001, 0.001) is near 0 and 1, suggesting that p is more likely to be large or small but not moderate (i.e., close to 0.5).

The take-home message is that all priors have an influence on the posterior distribution, and what might seem like a good trick to minimize the prior influence may not always do what you think it should. You can always overwhelm any amount of prior influence with enough data. In our example, if n gets large, then any prior values for α and β become inconsequential in the posterior; they will be very minimal compared with the large values for $\sum_{i=1}^{n} y_i$ and $\sum_{i=1}^{n}(1 - y_i)$. Thus, to some extent, the simplest way to minimize prior influence is to collect a larger data set!

Another caution in specifying priors that appear to minimize the influence on the posterior distribution pertains to "propriety." A proper probability distribution is a positive function that integrates to 1 over the support of its random variable (sec. 3.4). If the function does not integrate to 1, then it is termed "improper" and is not technically a valid probability distribution. That means we can't use it for statistical inference, because all statistical theory depends on the basic axioms about probability distributions. For example, continuing the previous discussion about how to make the beta distribution less influential, we would be tempted to use $\alpha = 0, \beta = 0$. However, because both parameters must be positive to guarantee a proper prior distribution, the beta(0, 0) is not a valid probability density function and thus its use is not advised. Interestingly, the resulting posterior, which we can still work out analytically, ends up being a beta(3, 7), which is proper in this specific case. Therefore, an improper prior can *sometimes* lead to a proper posterior, but that result has to be shown for the particular model being fit and almost always depends on the data. If you cannot mathematically show that your posterior is proper, then it's best to avoid improper priors.

Let's consider another situation. Suppose you have the same data and Bayesian model but are interested in obtaining inference related to the quantity p^2, rather than p. The seemingly benign uniform prior (i.e., beta(1, 1)) for p then becomes quite informative for p^2. To illustrate this point, we can find the implied prior distribution for p^2 using a Jacobian transformation technique.[6] In this case, if we use a uniform prior for p, the implied prior for p^2 (the quantity about which we desire inference) is proportional to $1/p$. Therefore, the values of p^2 under its implied prior are not equiprobable, as they are for p. Specifically, the uniform prior for p says that smaller values for p^2 are more probable than larger values. That result may not be what we had in mind when we chose the beta(1, 1) prior for p. A prior whose information about a parameter does not change when we transform the parameter is called "invariant to transformation." The *Jeffreys prior* was developed for this exact purpose, to help specify priors that are invariant to transformation.

The Jeffreys prior depends on the form of the likelihood (also called the *data model*). More specifically, the Jeffreys prior is proportional to Fisher information raised to the half power.[7] That is, if we can calculate the negative expectation of the second derivative of the log likelihood,

$$-E_y\left(\frac{d^2 \log[\mathbf{y}|p]}{dp^2}\right), \tag{5.4.4}$$

[6] The details of this technique are beyond the scope of this book but can be found in any graduate-level mathematical statistics book.

[7] This is the same Fisher information used to find asymptotic variance of an MLE.

then we have something proportional to the Jeffreys prior. The Jeffreys prior for our ongoing binary data example (eq. 5.4.1) is, perhaps surprisingly, a beta$(0.5, 0.5)$ distribution. This Jeffreys prior will contain the same information for p as it will for p^2, or any other transformation of p for that matter. Unfortunately, the Jeffreys prior is often called "noninformative," but for the same reasons cited earlier, it is not noninformative. We might use a Jeffreys prior when we don't know what else to use, in this case, because it happens to be invariant to transformation. For our example, the Jeffreys prior is U-shaped; not quite as extremely U-shaped as the beta$(0.001, 0.001)$ prior for p, but it will still give more prior preference to those values close to 0 and 1 than to $1/2$. The Jeffreys prior for this particular example turns out to be proper, but it is not guaranteed to be proper for all models.

You will commonly see a normal prior with large variance used as a prior distribution for a variety of parameters. A normal distribution with large variance (i.e., normal (0, 1000)) is often justified as an attempt to find a vague prior that is conjugate.[8] Given that the normal distribution is not bounded, it will be impossible to make it perfectly flat, so the large variance serves as a mechanism to at least spread it out.[9] A normal with infinite variance would be flat, but then it would also not be proper (i.e., would not integrate to 1). The use of a normal prior with large, but finite, variance seems to work well without complications for parameters that are means and where the data contain plenty of information. However, for other types of parameters, say transformations of probabilities such as logit(p), the normal prior with large variance can have a dubious influence on the posterior.

To illustrate our point, suppose we have the same Bernoulli model for the binary data we've been discussing in this section, and we use the prior for logit(p) such that

$$\text{logit}(p) \sim \text{normal}(0, \sigma_p^2) , \tag{5.4.5}$$

where, σ_p^2 is set to be a large number. The question is, what prior does this imply for p (rather than logit(p))? Simulating 10, 000 random draws from a normal distribution and taking the inverse logit transformation, we can see in figure 5.4.3 that a normal with $\sigma_p^2 = 100$ is much more informative than a normal with $\sigma_p^2 = 2$.

[8]Recall that conjugacy occurs when a prior and posterior have the same form. There can be many analytical and computational advantages to using conjugate priors, but they are not always the best choice.

[9]Keep in mind that the prior variance is always relative to the scale of the parameter. For example, if the data indicate that a parameter should be 1000, then a N(0, 100) prior for that parameter will probably be informative unless the sample size is huge, because a variance of 100 is small relative to 1000, as is the prior mean of 0.

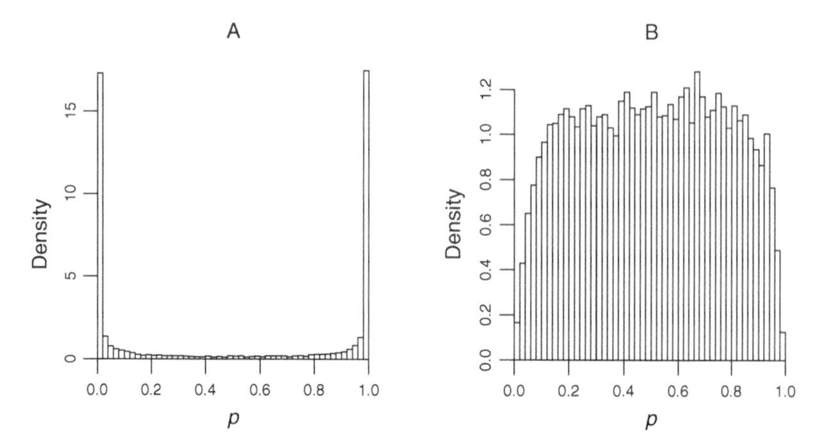

Figure 5.4.3. Histograms of p based on samples drawn from prior distributions for (**A**) logit(p) $\sim$ normal(0, 100) and (**B**) logit(p) $\sim$ normal(0, 2).

Priors with large variance might seem vague or less informative, but they are not always, thus it is a good idea to check the implied prior distribution in the transformation of the parameter for which you desire inference. You can do this by varying the values of the parameters for the prior and examining how that variation effects the posterior.

It's worth mentioning that the same methods are commonly used for choosing priors for variance components. In fact, we present models that contain such priors throughout this book. It is important to realize that such priors are not truly noninformative, for the same reasons we described earlier. For example, suppose we have data that can be sufficiently modeled with a normal distribution,

$$y_i \sim \text{normal}(\mu, \sigma^2), \tag{5.4.6}$$

for $i = 1, \ldots, n$, and where the mean μ is assumed to be known (for now). Our interest lies in obtaining inference about the variance, σ^2. A conjugate prior for the variance parameter is the inverse gamma distribution,

$$\sigma^2 \sim \text{inverse gamma}(\alpha, \beta), \tag{5.4.7}$$

which yields the posterior for σ^2:

$$\left[\sigma^2 | \mathbf{y}\right] = \text{inverse gamma}\left(\frac{n}{2} + \alpha, \frac{\sum_{i=1}^{n}(y_i - \mu)^2}{2} + \beta\right). \tag{5.4.8}$$

Notice that, much like the beta posterior discussed previously, here in the inverse gamma posterior for σ^2, if α and β get small, then the influence of the prior on the posterior is minimized. Thus, it is common to see priors for variance components specified as inverse gamma $(0.001, 0.001)$ in an attempt to minimize prior influence (but see Gelman, 2006). However, these priors are not "noninformative" and are not invariant to transformation. This sort of prior could be misleading for example, if, one was interested in obtaining inference about the standard deviation, σ, rather than the variance, σ^2.[10] As an alternative, the Jeffreys prior could be used for σ^2. For this model, the Jeffreys prior turns out to be proportional to $1/\sigma^2$, which has the form of an inverse gamma with $\alpha = 0$ and $\beta = 0$. This formulation for the inverse gamma does not yield a proper prior because both parameters (α and β) must be positive. However, the Jeffreys prior, as always, is invariant to transformation. As in the case with the Bernoulli model previously discussed, the Jeffreys prior for σ^2 yields a proper posterior as long as at least one observation is available (i.e., $n \geq 1$).

Finally, there is another approach to finding priors whose influence is minimal on the posterior; these priors are called *reference priors*. A reference prior is found by maximizing the Kullback-Leibler (K-L) divergence between the posterior and prior distributions.[11] The heuristic concept behind reference priors is that a prior which is as different as possible from the posterior may be desirable if you have no prior information or expertise and need a default prior to use just to obtain Bayesian inference.[12] Interestingly, for univariate parameters, the reference prior approach yields the Jeffreys prior! However, in multivariate situations, the reference prior must be found for each individual model where it is being used. However, calculating the correct reference prior can be quite challenging analytically and numerically. The field of objective Bayesian inference focuses on this task for various models.

5.4.2 Informative Priors

We have learned that all priors influence the posterior in some way but that we can often assess the amount of influence and sometimes even control it. But when formulating statistical models, we might ask

[10] This is more common than you might think, as it is easier to interpret the standard deviation, σ, than the variance, σ^2.

[11] The development of this concept is beyond the scope of this book, but in short, the K-L divergence provides a way to measure discrepancy between two distributions; it involves explicit integration and can be difficult to compute in practice, making this approach quite technical.

[12] Some argue that this very concept seems contrary to the Bayesian spirit by trying to avoid its biggest utility, the ability to properly account for previous research efforts in making scientific conclusions.

ourselves why we're trying to limit the influence of the prior on the posterior in the first place. The illusion of objectivity has been put on a pedestal in science, almost to the extent that we are to believe that only new data can be used to reach scientific conclusions. Extrapolating this concept to Bayesian statistics would then imply that we *should* be looking for priors that have no influence on the posterior (hence the previously mentioned subfield of objective Bayesian inference). However, a point not often recognized is that all parametric statistical modeling approaches are subjective, including maximum likelihood. The very fact that we have to choose a likelihood function implies that we have made a strong assumption about the data-generating mechanism. Nonparametric statistical approaches seek to minimize such assumptions, but they make their own set of strong assumptions based on their associated computational algorithms for providing inference. Any constraint we put on data or parameters to obtain inference imparts subjectivity. As we discuss in chapter 9, the various forms of regularization, including penalization methods and model selection, put extreme constraints on parameters, yet they are used throughout statistics and across all applied fields without much fanfare concerning their inherent subjectivity. More important, these approaches are recognized as being helpful in many ways!

Our view is that we would be remiss if we were to ignore decades of important scientific learning in the field of ecology and that there should be a way to rigorously incorporate this learning into our statistical approaches. Fortunately, the Bayesian framework provides such a mechanism. The posterior distribution itself is a formal, mathematically valid way to combine information from current as well as previous scientific studies. In that light, it is not difficult to see that the posterior distribution and Bayesian framework are a mathematical representation of the scientific method itself.

In the scientific method, we use existing data and expertise to formulate hypotheses about how the world works, then we make conclusions and update hypotheses using new data. In Bayesian statistics, we summarize our understanding of how the world works in a prior distribution and then "update" (i.e., compute the posterior distribution) our understanding using new data. Science would be completely haphazard if we threw out everything we knew about the world every time we began a new study. Haphazard is not even a strong enough word to describe science performed in a manner where we pretend to be completely ignorant about our study system; perhaps, lazy or irresponsible would be a better descriptor! In all seriousness, we challenge readers (and ourselves) to provide an example of a parameter in a statistical model they wish to fit knowing absolutely nothing about it—nothing at all. At a minimum, we should all know at least the support (i.e., the values the parameter can assume) for any parameter, but we often know quite a bit more than that. Ignoring prior information you have is like selectively throwing away data before an analysis.

Instead, we argue that science would be better off if we all took the time to carefully collect and represent our prior understanding of parameters in Bayesian models. Doing so can be hard work, as it sometimes requires a mathematical transformation of moments into natural parameters in the distribution we, as experts, value as best representing the data and parameters. It also could include being more responsible in our knowledge of preexisting scientific findings, for example, by more carefully reading the literature and translating those findings into quantitative information we can use in our prior. Formulating honest and responsible priors may also involve communicating with other experts on the topic under study, probing them for details that can be represented in probability distributions to serve as priors. Yes, this is beyond what we normally do in statistical analyses, but Bayesian methods provide the tools for incorporating such information, and we should be obligated to use them responsibly.

In addition to being helpful in accounting for the body of accumulated scientific knowledge when we make new inferences, strong so-called informative priors can be beneficial in the following ways:

- They allow us to benefit from several sources of information, including different data sources and expert knowledge. Given that priors are most influential when paired with small data sets, it can be incredibly helpful to obtain meaningful inference by having a formal mechanism for combining several smaller but independently collected data sets into a single modeling framework. An additional likelihood involving a separate data source can often be written as a prior in the original model containing the primary data source. We cover use of multiple likelihoods in the joint distribution in section 6.2.5.
- Informative priors stabilize computational algorithms. This benefit is not an inferential one but definitely a practical one. When statistical models accumulate parameters in such a way that the ratio of unknowns to knowns grows, the probability surfaces we need to explore during the fitting process can acquire pathological problems such as lack of identifiable[13] parameters, multicollinearity, and flat likelihood or posterior surfaces. Such issues can cause numerical approaches to become unstable (e.g., fail to converge). Stronger priors add definition

[13]Parameters are identifiable if they can be estimated given a large amount of data. They are unidentifiable if they cannot. See section 6.3 for a more complete definition of identifiability.

to the surfaces being explored by the statistical fitting algorithms and thus improve computational stability.

- Stronger priors offer a formal way to place constraints on the unknowns in statistical models. A seldom recognized fact is that such constraints are the basis for important inferential tools such as model selection. We cover this topic in great detail in chapter 9, but as a preview we note here that important concepts such as information criteria, penalized likelihood methods, ridge regression, Lasso, and cross-validation are used regularly in many fields and can all be considered as different ways to improve inference through the use of stronger priors. Most statisticians now recognize that imposing a constraint on an optimization problem (e.g., maximizing a likelihood) is the same concept as specifying a prior in a Bayesian model, and both can be helpful for the same reasons.

Excellent examples of the benefits of using informative priors can be found in Crome et al. (1996); McCarthy and Masters (2005); Elderd et al. (2006) and McCarthy et al. (2008).

Until now, we have considered informative priors as single distributions as if they were obtained from a single, previously conducted investigation. What do we do if we have multiple sources of prior knowledge informing a parameter θ? Recall the idea of a mixture distribution (sec. 3.4.5). We can compose a prior from multiple previous studies by mixing their estimates of θ. A prior on θ using information from L different studies can be written as

$$[\theta] = \sum_{l=1}^{L} w_l \, [\theta]_l \,, \tag{5.4.9}$$

$$\sum_{l=1}^{L} w_l = 1, \tag{5.4.10}$$

where the w_l are weights, and $w_l \geq 0$. If we believe that all studies were conducted equally well, then we would choose the w_l to be equal. As an example, assume we had three studies of the intercept β_0 in a regression with an associated variance that we wished to combine in a prior. We might reasonably use

$$[\beta_0] = \tfrac{1}{3}\text{normal}\left(\beta_{0,1}, \sigma_1^2\right) + \tfrac{1}{3}\text{normal}\left(\beta_{0,2}, \sigma_2^2\right) + \tfrac{1}{3}\text{normal}\left(\beta_{0,3}, \sigma_3^2\right). \tag{5.4.11}$$

Now that we can see the potential value of priors informed by single or multiple studies, we need to know how to represent existing scientific knowledge in the form of a probability distribution. There are different approaches for manifesting expert knowledge about a parameter into a prior distribution, but rather than cover each one generically, we present the following example.[14]

5.4.3 Example: Priors for Moth Predation

A particular species of nocturnal moth rests during the day on tree trunks, and its coloration acts as camouflage to protect it against predatory birds. A study was conducted to evaluate predation of a common moth species. Suppose that n sites (for $i = 1, \ldots, n$) were selected, and a varying number of dead moths, N_i, were glued to tree trunks at each site. After 24 hours, the number of moths that had been removed, y_i, presumably by predators, was recorded. A reasonable data model for the moth counts would be a binomial with N_i "trials" per site such that

$$y_i \sim \text{binomial}(N_i, p), \tag{5.4.12}$$

where the parameter p corresponds to the probability of predation and is the unknown about which we desire inference. Consider the following three scenarios in formulating an appropriate prior distribution for this model:

1. We desire a relatively vague prior that contributes information equivalent to two additional "placed" moths and an expected prior probability of predation of .5.
2. We desire an informative prior based on a previous observational study that reported an average of 10% (standard deviation of 2.5%) of the moths in a population were eaten by predators in a 24-hour period.
3. We desire an informative prior based on a pilot study that suggests the proportion removed in any given 24-hour period is unlikely to exceed 0.5 or be less than 0.1.

In scenario 1, the fact that we do not feel we have much prior information pertaining to p means that we want to spread out the probability mass in the prior between 0 and 1 such that our prior has a mean of 0.5 but no strong preference for any range of values. A beta distribution could work well here such that $p \sim \text{beta}(\alpha, \beta)$. But how do we assess the information content of

[14]This example is gratefully modified from the excellent text of Ramsey and Schafer (2012) using ideas from Kiona Ogle.

the prior in terms of an effective increase in sample size? The answer comes from looking at the form of the posterior distribution for this model:

$$[p|\mathbf{y}] = \text{beta}\left(\sum_{i=1}^{n} y_i + \alpha, \sum_{i=1}^{n}(N_i - y_i) + \beta\right). \tag{5.4.13}$$

In equation 5.4.13 we can see a form for the posterior very similar to the one in the Bernoulli model discussed previously, where each of the updated posterior parameters contains a sum of two components, one from the data and one from the prior. The first parameter, $\sum_{i=1}^{n} y_i + \alpha$, is the sum of y_i over all sites—that is, the total number of moths placed that were preyed on—plus the prior parameter α. The second posterior parameter, $\sum_{i=1}^{n}(N_i - y_i) + \beta$, is the total number of moths not removed by predation, plus prior parameter β. Thus, if we set $\alpha = 1$ and $\beta = 1$, it is equivalent to adding two moths to the sample size in such a way that the information does not impose any preference for predation. In this case, the implied prior is a beta(1, 1) or a uniform distribution. Of course, we could have started with a uniform, but it is instructive to see that the prior parameters α and β can be thought of as augmenting the sample size if that helps specify prior information. Given the preceding, what prior would be induced if we had the equivalent of 10 extra moths, worth of prior information such that 60% was in favor of predation and the other 40% was against predation? The answer could easily be visualized by plotting a beta probability density function with parameters $\alpha = 4$ and $\beta = 6$.

In scenario 2, we have information from a former study on moth predation. That study provides inference pertaining to the mean and standard deviation of the proportion of moths preyed on. In translating this information into our beta prior, we can consider the mean and variance equations associated with a beta random variable:

$$\text{E}(p) = \frac{\alpha}{\alpha + \beta}, \tag{5.4.14}$$

$$\text{Var}(p) = \frac{\alpha\beta}{(\alpha + \beta)^2(\alpha + \beta + 1)}. \tag{5.4.15}$$

Setting $\text{E}(p) = 0.1$ and $\sqrt{\text{Var}(p)} = 0.025$, we can backsolve for α and β to find the appropriate prior (as discussed in sect. 3.4.4). We then arrive at $\alpha = 14.3$ and $\beta = 128.7$ as parameters in our prior.

Scenario 3 is slightly more involved, but entirely realistic, in that it is common for prior information to arise as bounds on likely values for a parameter. In this scenario, if we assume that the term “unlikely” implies that p should fall between a lower bound and an upper bound with high probability (e.g., 95%), then we need to take an approach similar to the moment matching technique, but instead of relating moments to the results of a pilot study, we relate quantiles of the distribution to the results of a pilot study. That is, we must solve the system of equations

$$\int_0^{0.1} \text{beta}(p|\alpha, \beta)dp = .025 \tag{5.4.16}$$

$$\int_{0.5}^{1} \text{beta}(p|\alpha, \beta)dp = .025 \tag{5.4.17}$$

for α and β, where equations 5.4.16 and 5.4.17 integrate the beta probablity density function. This calculation would be quite difficult to perform analytically (i.e., with pencil and paper) but could be approximated numerically using an optimization algorithm in a mathematical or statistical software package. We used the function `optim()` in R (R Core Team, 2013) to find the appropriate prior parameter values $\alpha = 4.8$ and $\beta = 12.7$ by minimizing the difference between the output of the beta cumulative distribution function (i.e., `pbeta()` in R) and .025.

Thus, there are a variety of ways to convert preexisting scientific information and expertise into probability distributions for use as priors in Bayesian models. These informative priors can be very useful in many ways, but only when care is taken to appropriately specify them. It is a common concern that if Bayesian models fell into the wrong hands, they could be misused by those seeking to mislead science or policy. However, even under such dubious circumstances, the priors would have to be clearly spelled out in any scientific communication and would be scrutinized just as any other scientific finding is scrutinized during peer review. Furthermore, those with villainous intentions have much easier ways to mislead science or the general public, for example, by outright fabrication of scientific studies. We feel that carelessness by well-intentioned scientists (in the field, in the lab, or in specifying inappropriate likelihoods or priors) is probably a much more common cause of erroneous inference than is mischief.

5.4.4 Guidance

We admit that the cautionary statements in this section could make the choice of priors seem complicated and difficult; however, that is not our aim. We feel that priors can be an important component of science and can be helpful in obtaining useful models for inference. Our goal in this discussion of priors is to instill a sense of awareness about the decisions being made in the model-building process. If you are more thoughtful about specifying priors and the associated consequences after reading this section, then we have done our job.

The fact is, few of these details are made clear in other texts on applied Bayesian statistics, and we wrote this section, at least in part, as a reminder to ourselves to think deeply about how we can incorporate prior scientific knowledge in the form of a probability distribution for use as a prior. You'll notice that we commonly use default priors in examples throughout this book. It would seem that by doing so, we encourage this practice, but in reality we don't claim to be experts in all the applied subjects in the diverse examples we offer. Thus, it is with a touch of "do as we say, not as we do" that we suggest that our model specifications throughout are only placeholders for a model that might actually be used by an expert in the relevant field. This section also serves as a prelude to chapter 8, where we give a concrete example of the value of prior information and to chapter 9, where we describe ways that priors are an example of regularization, an approach widely used in statistics to improve model fit.

Although we have provided several approaches for specifying priors for specific models in this section, a list of all possible options for all possible models would be too lengthy. Thus, we echo the guidance provided by Seaman et al. (2012) and leave you with a few further general diagnostics and remedies to consider when specifying priors in Bayesian models:

- Bear in mind that one of the objectives of Bayesian analysis is to provide knowledge that can inform subsequent analyses; the posterior distribution obtained in one investigation becomes the prior in subsequent investigation. Thus, we agree with the view of Gelman (2006) that vague priors are provisional—they are a starting point for analysis. As scientists, we should always prefer to use appropriate, well-constructed informative priors.
- Visualize the prior you choose in terms of the parameters for which you desire inference. We did this earlier for the logit(p) (i.e., fig. 5.4.3). Sometimes you can do this analytically (i.e., with pencil and paper, using calculus), but it's often easier just to simulate values from your prior, then transform them to represent the desired quantity and plot a histogram.
- Perform a prior sensitivity analysis. Try several different priors, maybe by simply choosing different prior variances, and see how much the posterior distribution moves around as a result. Often, you'll see little posterior sensitivity to priors when there is a high ratio of data to parameters. However, if the posterior is sensitive to the prior and you

truly desire a prior that is only weakly informative, you will need to rethink your prior by changing its form or parameters. Alternatively, you must carefully justify your choice of prior in relation to the inference you seek.

- An influential prior might be indicated if the posterior inference differs greatly from maximum likelihood inference. Of course, this can be confirmed only in models where both approaches can be implemented easily, so it may not be practical for more complicated Bayesian models. Still, in some Bayesian models, inference will approach what would be obtained with inference based on maximum likelihood if certain priors are used.
- Dependent parameters should probably have priors that acknowledge the dependence. We are often lured into thinking that we can simply specify independent priors for parameters, but a prior that is a joint multivariate distribution or conditional distribution for one parameter given another is often more appropriate. An example in regression is $y_i \sim \text{normal}(\beta_0 + \beta_1 x_{1,i} + \beta_2 x_{2,i}, \sigma^2)$ for $i = 1, \ldots, n$. In this case, it is common to use the independent normal priors $\beta_1 \sim \text{normal}(0, \sigma_\beta^2)$ and $\beta_2 \sim \text{normal}(0, \sigma_\beta^2)$, but it can be helpful to use a multivariate normal prior for both regression coefficients simultaneously, $\boldsymbol{\beta} \sim$ multivariate normal($\mathbf{0}$, $\boldsymbol{\Sigma}$), where $\boldsymbol{\beta}$ is the vector containing β_1 and β_2, $\mathbf{0}$ is a vector of zeros, and $\boldsymbol{\Sigma}$ is a covariance matrix.
- Keep in mind that even with large sample sizes there may be not be enough information in the data to tease apart different parameters, regardless of their priors. This is more of an identifiability problem rather than a problem with the prior, and the form of the model itself should be reconsidered. For example, with binomial data $y_i \sim$ binomial(N, p) for $i = 1, \ldots, n$, where N and P are unobserved, there are not enough data in the world to learn about both N and p individually, but a strong prior on one of the two parameters (if warranted) can help focus the inference on the other. However, without sufficient prior information this is not a useful model in an inverse (i.e., statistical) setting.

6 Hierarchical Bayesian Models

It is worthwhile to review the key points covered thus far. We started with the first principles rules of probability (sec. 3.2). We used those rules to develop Bayes' theorem (sec. 5.1) and to show how we can factor joint distributions of observed and unobserved quantities into parts based on our knowledge of conditioning and independence (sec. 3.3). We learned about priors and their influence on the posterior (sec. 5.4).

We now apply what we have learned to ecological examples of hierarchical Bayesian models. These models offer unusually revealing and broadly useful routes to insight because they allow us to decompose complex, high-dimensional problems into parts that can be thought about and analyzed individually. We can use the same approach for virtually any problem, regardless of its particular features.

This chapter has two objectives: (1) to explain hierarchical models and how they differ from simple Bayesian models and (2) To illustrate building hierarchical models using mathematically correct expressions. We illustrate the first two sets of steps in the general modeling process that we introduced in the preface (fig. 0.0.1 A,B).

We begin with the definition of hierarchical models. Next, we introduce four general classes of hierarchical models that have broad application in ecology. These classes can be used individually or in combination to attack virtually any research problem. We use examples to show how to draw Bayesian networks that portray stochastic relationships between observed and unobserved quantities. We show how to use the network drawings as a guide for writing posterior and joint distributions.

6.1 What Is a Hierarchical Model?

A statistical model is Bayesian if the unobserved quantities we seek to understand are random variables whose probability distributions are estimated from observations and prior knowledge.[1] Recall from chapter 5 that a Bayesian model is simple if it represents the joint distribution of those random variables as the product of the likelihood multiplied by the prior distributions. For example,

$$\Big[\underbrace{\theta_1, \theta_2, z}_{\text{unobserved}} \mid \underbrace{\boldsymbol{y}}_{\text{observed}}\Big] \propto \underbrace{[\theta_1, \theta_2, z, \boldsymbol{y}]}_{\text{joint}} \tag{6.1.1}$$

$$\propto \underbrace{[\boldsymbol{y}|\theta_1, \theta_2, z]}_{\text{likelihood}} \underbrace{[\theta_1][\theta_2][z]}_{\text{priors}} \tag{6.1.2}$$

is a simple Bayesian model[2] involving the unobserved quantities θ_1, θ_2, and z, and the observations $\boldsymbol{y}$. It is important to remember that we factor the joint distribution using the rules of probability (sec. 3.3) to obtain the product of the likelihood and priors. The model is not hierarchical because there is no conditioning beyond the dependence of the data on the unobserved quantities. This means that every quantity that appears on the right-hand side of the conditioning symbol in the likelihood is found in a prior. The posterior distribution is proportional to the joint distribution because we have omitted the denominator of Bayes' theorem, the marginal distribution of the data $\left(\int\int\int [\boldsymbol{y}|\theta_1, \theta_2, z][\theta_1][\theta_2][z]\, d\theta_1 d\theta_2 dz\right)$, which is a scalar with a fixed value after we have observed the data. At the risk of getting ahead of ourselves, we are expressing the posterior as proportional to the joint distribution, because this proportionality is all we need to properly develop an algorithm for estimating the parameters and latent state, which we will cover in chapter 7 on the Markov chain Monte Carlo algorithm.

A Bayesian model is *hierarchical* whenever we use probability rules for factoring (sec. 3.3) to express the joint distribution as a product of conditional distributions. For example,

$$\begin{aligned}[\theta_1, \theta_2, z|y] &\propto [\theta_1, \theta_2, z, y] \\ &\propto [y|\theta_1, z][z|\theta_2][\theta_1][\theta_2]\end{aligned} \tag{6.1.3}$$

[1] Including the "knowledge" that little is known.

[2] Strictly speaking this assumes that θ_1, θ_2, and z are independent a priori. This is a common assumption in Bayesian models. Inference is rarely sensitive to this assumption.

is hierarchical because we factored $[\theta_1, \theta_2, z, y]$ to produce $[y|\theta_1, z]\,[z|\theta_2]\,[\theta_1]\,[\theta_2]$, assuming that θ_1 and θ_2 are independent a priori. We can quickly see that the model is hierarchical because the unobserved quantity z appears on the right-hand side of the "|" in the likelihood $[y|\theta_1, z]$ and on the left-hand side of the "|" in the distribution $[z|\theta_2]$. Note that there is no prior distribution[3] for z because it is conditional on a quantity for which there *is* a prior distribution, θ_2. The factoring of joint distributions into products of conditional distributions is not arbitrary but, rather, is based on our knowledge of an ecological process, how we observe it, and the assumptions we can use to simplify it, as we illustrate next.

6.2 Example Hierarchical Models

Hierarchical models are most often applied in ecological research to deal with four commonly encountered challenges:

1. Representing variation among individuals arising, for example, from genetics, location, or experience.
2. Studying phenomena operating at more than one spatial scale or level of ecological organization.
3. Estimating uncertainty that arises from modeling a process as well as uncertainty that results from imperfect observations of the process.
4. Understanding changes in states of ecological systems that cannot be observed directly. These states arise from "hidden" processes.

These broad challenges are not mutually exclusive; more than one often appears within the same investigation. Hierarchical models can be used to create a robust and flexible framework for analysis that is capable of meeting these challenges as they arise.

In the following examples we illustrate different types of hierarchical models. At the same time we show how to graphically represent relationships between observed and unobserved quantities in Bayesian networks, also called *directed acyclic graphs* (DAGs), a concept introduced in section 3.3. Bayesian networks form a template for writing properly factored expressions for joint distributions. Our purpose in this chapter is to emphasize writing mathematical expressions as the proper first step in modeling. For now, we postpone considering how we might implement or evaluate the model. The examples we offer here will be supplemented by

[3]Some would call $[z|\theta_2]$ a hierarchical prior and $[\theta_2]$ a hyper prior, but this perspective is somewhat unconventional.

worked problems in model building in part III, problems that will challenge you to diagram and write models.

As you read the following sections it will be especially useful to notice three themes that recur in the examples. The first theme is the one-to-one relationship between diagrams of stochastic relationships and the mathematical expressions for the posterior and joint distributions. This is a critical insight. Next, it will be helpful to see how we compose stochastic models by combining deterministic functions with probability distributions. Hierarchical models are often developed by substituting a model for a parameter, so it is especially instructive to see how we add detail to models and exploit additional explanatory data by "modeling parameters." This process illustrates how models of high dimension can be composed, even though the examples here are relatively simple. The final crosscutting theme to note in the examples is how we partition uncertainty into multiple sources. In particular, we often use a particular factoring of the joint distribution first proposed by Berliner (1996) and later elaborated by Wikle (2003); Clark (2005); Cressie et al. (2009), and Wikle et al. (2013):

$$\left[\boldsymbol{\theta}_p, \boldsymbol{\theta}_o, \mathbf{z}|\mathbf{y}\right] \propto \underbrace{\left[\mathbf{y}|\mathbf{z}, \boldsymbol{\theta}_o\right]}_{\text{data}} \underbrace{\left[\mathbf{z}|\boldsymbol{\theta}_p\right]}_{\text{process}} \underbrace{\left[\boldsymbol{\theta}_o\right]\left[\boldsymbol{\theta}_p\right]}_{\text{parameters}}. \tag{6.2.1}$$

We decompose the joint distribution this way because it represents such a broad range of problems in ecological research. There is a "true" ecological state of interest $\mathbf{z}$, a state that is not observable. We relate that state to the observable data ($\mathbf{y}$), using a model with a vector of parameters $\boldsymbol{\theta}_o$, including parameters representing uncertainty in our observing system. The behavior of the true state is predicted with a model parameterized by $\boldsymbol{\theta}_p$, including parameters representing stochasticity in the process.[4] This model represents our hypothesis about how an ecological process works.

6.2.1 Understanding Individual Variation: Fecundity of Spotted Owls

Understanding variation in processes caused by variation among individual organisms forms a central challenge in population and community ecology. Our first example is fashioned after Clark (2003a), who studied the effects of individual differences in fecundity on population growth rate of northern

[4]You may wonder, Where's the x? What happened to observations of predictor variables? Suspend disbelief for a moment. We will deal with this question in the next section.

spotted owls, *Strix occidentalis caurina.* In this example, we are interested in estimating the average number of offspring annually produced by each breeding female, that is, their average fecundities, as well as the average fecundity for the population.

A simple Bayesian model requires the assumption that all owls have the same average fecundity. This means that variation among individuals occurs from year to year because fecundity is a random variable, but a sample of many years would have the same average reproductive output for all individuals. We can represent these ideas in a Bayesian network (fig. 6.2.1 A).

Recall that Bayesian networks (fig. 3.3.1) are drawings that depict probability distributions graphically. These drawings are particularly useful for showing the dependencies in hierarchical models. The nodes in the diagrams represent random variables; solid arrows represent stochastic relationships among the random variables, and the tails of the arrows specify the parameters defining the distribution of the random variable at the heads of the arrows.

Here is an illustration. Assume we have a single observation (y_i) representing the number of offspring of owl i. We could model average fecundity (λ) using

$$\underbrace{[\lambda|y_i]}_{\text{posterior}} \propto \underbrace{\underbrace{\text{Poisson}(y_i|\lambda)}_{\text{likelihood}}\,\underbrace{\text{gamma}(\lambda|.001, .001)}_{\text{prior}}}_{\text{joint}}. \tag{6.2.2}$$

We will use a Poisson distribution for the likelihood—a logical place to start when modeling count data when we can assume that the variance is approximately equal to the mean.[5] We use a gamma distribution for the prior on λ because it is conjugate to the Poisson. We give numeric arguments to the gamma distribution to make it minimally informative.

Writing the posterior distribution is easy. We simply write a distribution with the unobserved quantities on the left-hand side of the "|" and the observed quantities on the right-hand side. Composing expressions for the joint distribution is guided by the diagram in figure 6.2.1 A. The nodes at the heads appear on the left-hand side of a "|", and the nodes at the tails of the arrows appear on the right-hand side. Any node at the tail of an arrow that does not have an arrow leading into it is expressed as a prior

[5]If this assumption doesn't hold, then the negative binomial distribution would be a better choice. Later (sec. 8.1), we will learn methods to evaluate the assumptions we make in choosing distributions.

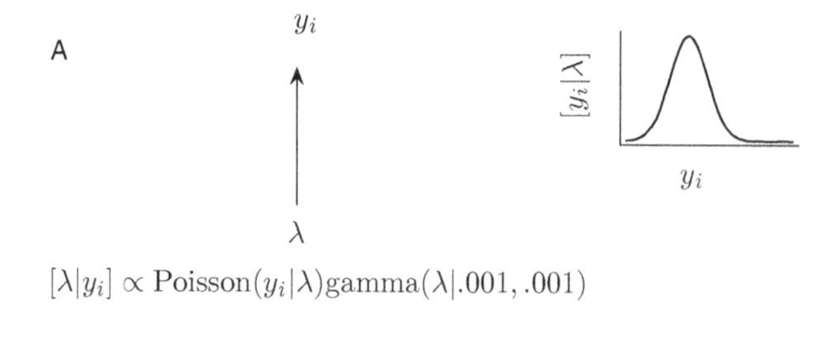

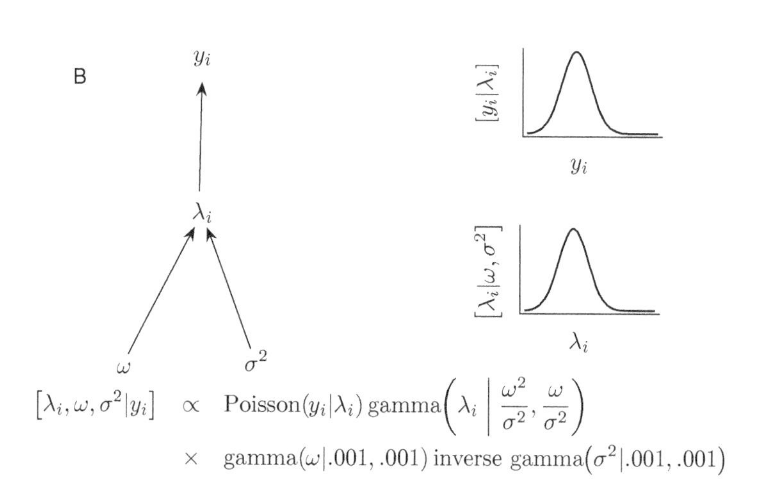

Figure 6.2.1. Bayesian networks for simple (**A**) and hierarchical (**B**) Bayesian models of fecundity of spotted owls assuming a single observation, y_i. There are only two levels in the simple model (**A**) because the joint distribution is a product of the likelihood and the priors. In this case, we assume the data arise from a Poisson distribution with a mean fecundity (λ) that is the same for all owls. There are three levels in the hierarchical model (**B**) because the joint distribution is a product of two conditional distributions and the priors. In this case, we assume that each owl has its own average fecundity (λ_i) that is drawn from a gamma distribution with mean ω and variance σ^2. Note the correspondence between the heads of arrows and random variables on the left-hand side of conditioning symbols in the joint distribution and the tails of arrows and random variables on the right-hand side of conditioning symbols. Any random variable at the tail of an arrow without an arrow leading into it requires a prior distribution. The equations and the diagrams represent distributions (right column), where the heads of the arrows are the random variables shown on the x-axis, and the tails of the arrows are the moments (or the parameters) that define the distributions.

distribution. The prior distributions must have numeric arguments for their parameters. Because the parameters of priors are constant (i.e., they are not random variables) they do not appear as nodes in the diagram. Remember, nodes represent random variables.

It may strike you that diagrams are superfluous when you are writing simple Bayesian models, and your impression is correct. However, these diagrams become more useful in helping visualize and write hierarchical relationships. They are especially helpful (at least for ecologists, if not for statisticians) when there are complex, multilevel relationships among observed and unobserved quantities, as we will soon see.

We now model the case where *each* owl has its *own* mean fecundity. Variation in average fecundity among individuals might occur because of differences in genetics or age or variation in the quality of habitats where they establish territories. In this example, we are not trying to determine the causes of individual variation but simply acknowledge that it exists and include it in our model. This is a key idea.

Consider a network with an additional level in the hierarchy (Figure 6.2.1 B) We now treat the average fecundity of each individual (λ_i) as a random variable drawn from a gamma distribution with mean ω and variance σ^2. We use the diagram in fig. 6.2.1 B as a template to write the posterior and joint distributions:

$$\left[\lambda_i, \omega, \sigma^2 | y_i\right] \propto \text{Poisson}(y_i|\lambda_i)\,\text{gamma}\left(\lambda_i \,\middle|\, \frac{\omega^2}{\sigma^2}, \frac{\omega}{\sigma^2}\right) \tag{6.2.3}$$

$$\times\ \text{gamma}(\omega|.001, .001)\text{inverse gamma}(\sigma^2|.001, .001). \tag{6.2.4}$$

The likelihood is the same as in our previous model except for the subscript on λ_i indicating that each individual has a fecundity—the observations for owl i will vary from year to year, but over the long term the observations on owl i will average λ_i. The important difference between the simple Bayesian model (eq. 6.2.2) and the hierarchical one (eq. 6.2.3) is the addition of a model for the λ_i; that is, $\lambda_i \sim \text{gamma}(\lambda_i|\frac{\omega^2}{\sigma^2}, \frac{\omega}{\sigma^2})$. Assuming that individual owls have fecundities that are drawn from a distribution treats fecundity as a *random effect*, whereas assuming all individuals have the same average fecundity treats fecundity as a *fixed effect* (box 6.2.1). We choose a gamma prior distribution for the population mean fecundity ω because it is continuous and nonnegative. We choose an inverse gamma prior for σ^2 because it is a variance.[6] Both priors have values of parameters chosen to make them minimally informative. No prior is required for λ_i because it occurs on the left-hand side of a conditional—its distribution is determined by the parameters ω and σ^2, which do have priors.

Box 6.2.1 Random Effects

The terms *random effect* and *fixed effect* are used in the scientific literature in ways that can be confusing. Gelman and Hill (2009, pg. 245) offer several examples of inconsistent use of the terms. They recommend dispensing with the use of the term "random effects" altogether and replacing it with *group-level effects*. This is a sensible suggestion, because all "effects" are considered to be random variables in the Bayesian framework. However, "random effects" is widely used, sometimes pertaining to individuals rather than to groups. We will use the term later in the book and explain it here.

In Bayesian hierarchical modeling, random effects are used to describe variation that occurs beyond variation that arises from sampling alone. For example, imagine that you wish to estimate the average aboveground biomass in a grassland. You take a sample of biomass in several 0.25 m^2 plots. If the biomass is randomly distributed across the area you sample, then a reasonable way to model the variation in the biomass in the ith plot (y_i) would be

$$y_i = \mu + \epsilon_i,$$
$$\epsilon_i \sim \text{normal}(0, \sigma^2),$$

which is the same as

$$y_i \sim \text{normal}(\mu, \sigma^2),$$

where μ is the mean biomass per plot, and σ^2 is the variance among plots. We generally prefer the latter notation, because not all variation is additive. If a random variable like μ is strictly positive, then adding a normal random variable (ϵ_i) to it to represent uncertainty makes no sense, because μ cannot be

(*continued*)

[6]Several other distributions (e.g., uniform, inverse gamma, and Cauchy) can be used as priors on variances. However, as you will see later in the book, the inverse gamma is often the distribution of choice because of conjugate relationships (sec. 5.3) for normally distributed random variables. Conjugacy can facilitate model implementation (as we describe in sec. 7.3.2.3). Gamma distributions are used for the inverse of the variance, $1/\sigma^2$, the precision, for the same reason. See Gelman (2006) for a thoughtful discussion of priors for variances in hierarchical models.

(Box 6.2.1 *continued*)

negative. Alternatively, the notation $[y_i|\gamma, \beta]$ works for any random variable, regardless of its support. We are using a normal distribution for clarity here, but because biomass is strictly positive, a better choice might be lognormal or gamma. However, this would somewhat complicate the example, so to keep things simple and familiar, we chose the normal.

Now, imagine that you sampled at five different locations, indexed by j. If we treat location as a *fixed effect*, our model doesn't change, because we assume that the variation is due entirely to sampling, that is, $y_{ij} \sim \text{normal}(\mu_j, \sigma^2)$. When we do this we are treating the μ_j as *fixed* across the locations. (A pooled model would have a single mean regardless of site, i.e., $y_{ij} \sim \text{normal}(\mu, \sigma^2)$.) Alternatively, we might more reasonably assume that there are differences in productivity among sites arising from any number of different sources—soil type, depth to the water table, topography, level of herbivory, and so on. In this case, we allow each location to have its own mean biomass drawn from a distribution of means with *hyperparameters*: mean of means equal to α, and variance of means equal to ς^2. Our model then becomes

$$y_{ij} \sim \text{normal}(\mu_j, \sigma_j^2), \qquad (6.2.5)$$

$$\mu_j \sim \text{normal}(\alpha, \varsigma^2). \qquad (6.2.6)$$

In this case, we are treating the effect of location as random, an effect that varies randomly according to sources of variation that we acknowledge exist but that we are not attempting to explain. You will also see this written as

$$y_{ij} = \mu_j + \epsilon_{ij}, \qquad (6.2.7)$$

$$\mu_j = \alpha + \eta_j, \qquad (6.2.8)$$

$$\epsilon_{ij} \sim \text{normal}(0, \sigma_j^2), \qquad (6.2.9)$$

$$\eta_j \sim \text{normal}(0, \varsigma^2). \qquad (6.2.10)$$

We have used the problem of estimating a mean to illustrate random effects, but the same idea applies to any parameter in any model. For example, a common use of random effects is to allow the intercepts of regressions to vary by location or some other grouping variable, such as

$$y_{ij} \sim \text{normal}(\beta_j + \beta_1 x_{ij}, \sigma^2), \qquad (6.2.11)$$

$$\beta_j \sim \text{normal}(\mu, \varsigma^2). \qquad (6.2.12)$$

Notation that might be puzzling is seen in the parameters for the gamma distribution, $\frac{\omega^2}{\sigma^2}$ and $\frac{\omega}{\sigma^2}$. Where did these come from? The parameters for a gamma distribution are α, the shape, and β, the rate. Recall from section 3.4.4 on moment matching that the mean of the gamma distribution is α/β with variance α/β^2, allowing us to solve for α and β terms of the mean and variance; that is, $\alpha = \omega^2/\sigma^2$, $\beta = \omega/\sigma^2$. The average fecundity for the population is $\omega = \alpha/\beta$.

These clarifications make an important point about drawing Bayesian networks and converting them into mathematical expressions. Recall that the heads of arrows in Bayesian networks are random variables governed by a distribution defined by the parameters at the tails of the arrows (i.e., fig. 3.3.1). Thus, it is possible to define these distributions in terms of means and variances or in terms of parameters. It follows that it would have been perfectly correct[7] to write the model as

$$[\boldsymbol{\lambda}, \alpha, \beta|\mathbf{y}] \propto \prod_{i=1}^{n} \text{Poisson}(y_i|\lambda_i)\ \text{gamma}(\lambda_i|\alpha, \beta) \times \text{gamma}(\alpha|.001, .001)\,\text{gamma}(\beta|.001, .001). \tag{6.2.13}$$

The point is that Bayesian networks are thinking tools—graphical aids for properly writing models. In some cases it will be most helpful to think about stochastic relationships in terms of the moments of distributions; in other cases it will be more useful to think in terms of parameters. Moment matching allows these approaches to be interchangeable. We can be flexible in our use of tools.

We now embellish the example from Clark (2003) to illustrate how we might add parameters and explanatory observations (i.e., covariates) to our model to explain variation among individuals in fecundity. Reproductive success for many species of vertebrates rises to a peak during midlife before declining (Part and Forslund, 1996; Hamel et al., 2012) as individuals grow old. Thus, it might be reasonable to model λ_i as a quadratic function of "reproductive age," defined as time after the animal is capable of reproduction ($x_i = \widetilde{x}_i - x_{0,i}$) where $\widetilde{x}_i$ is the chronological age of the ith individual, $x_{0,i}$ is the age of first reproduction. Thus, an animal is $x_i = 0$ when it first reproduces. Defining age this way makes it convenient to interpret the intercept.

[7] Actually, most statisticians would prefer this form, because it focuses attention on the parameters.

We now model the process "change in fecundity with age" $g(\alpha, \boldsymbol{\beta}, x_i)$ as

$$g\left(\alpha, \boldsymbol{\beta}, x_i\right) = \alpha + \beta_1 x_i + \beta_2 x_i^2, \tag{6.2.14}$$

$$\begin{aligned}\left[\boldsymbol{\lambda}, \boldsymbol{\beta}, \alpha, \sigma_p^2 | \mathbf{y}\right] \\ \propto \prod_{i=1}^{n} \text{Poisson}(y_i|\lambda_i)\text{gamma}\left(\lambda_i \,\middle|\, \frac{g\left(\alpha, \boldsymbol{\beta}, x_i\right)^2}{\sigma_p^2}, \frac{g\left(\alpha, \boldsymbol{\beta}, x_i\right)}{\sigma_p^2}\right) \quad (6.2.15) \\ \times \prod_{j=1}^{2} \text{normal}(\beta_j|0, 100)\ \text{normal}(\alpha|0, 100) \\ \times \text{inverse gamma}(\sigma_p^2|.001, .001),\end{aligned}$$

where α is the average reproduction of an owl at reproductive age 0, β_1 and β_2 are parameters that control the change in fecundity with age, and σ_p^2 is process variance. It is important to understand that process variance includes all the influences that create variation in fecundity beyond the effect of the bird's age. It is also important to see that we have replaced the parameter ω with a model $g\left(\alpha, \boldsymbol{\beta}, x_i\right)$ that exploits observations on an owl's age and our understanding of the relationship between age and fecundity.

Again, we choose a gamma distribution for λ_i because it is continuous and nonnegative. We could also use other continuous distributions with nonnegative support, for example, the lognormal. The distribution for λ_i could be viewed as a "prior" informed by our process model. We use an inverse gamma distribution for the flat prior on σ_p^2 because it is a variance (but see Gelman 2006). We choose a normal distribution for the β's because they are continuous random variables that can take on any real value. To minimize the information contained in the priors for the β's we center them on 0 and assign a variance that is very large relative to their values.

You might reasonably ask, why doesn't the data set $\mathbf{x}$ appear in the posterior distribution in the same way that $\mathbf{y}$ does? After all, both are observed quantities. The short answer is this. The $\mathbf{x}$ are not treated as random variables in this formulation. You are right that both the $\mathbf{x}$ and the $\mathbf{y}$ are observed, but in this case we are assuming that the $\mathbf{x}$ data are observed *perfectly*,[8] while the data $\mathbf{y}$ are random variables. This means the $\mathbf{x}$ are known, fixed quantities, treated no differently than the constant π in the normal distribution. They are not random variables, and hence, they

[8]You may recall the assumption of conventional linear regression, customarily but often wrongly ignored by ecologists, that the predictor variables are measured without error.

should not appear in the expression for the posterior distribution which, by definition, is composed of random variables. The predictor variables correctly appear as arguments to the deterministic function $g(\alpha, \boldsymbol{\beta}, x_i)$. Sometimes, the predictor variables *do* appear in the posterior distribution, and we describe these cases in a subsequent example and in box 6.2.2.

What else might influence fecundity? We might reasonably hypothesize that the fecundity of each owl at reproductive age 0 should increase with decreasing territory size (e.g., Elbroch and Wittmer, 2012), which is to say that territory size shifts the curve $g(\alpha, \boldsymbol{\beta}, x_i)$ up or down. A reasonable deterministic model of this process is $h(\gamma, \nu, u_i) = \gamma e^{-\nu u_i}$, where u_i is the observed area of the territory of the ith individual, γ is the maximum potential fecundity during the first reproduction, and ν controls the decline in mean fecundity that occurs as territory area increases. We can include this process in our model by allowing each individual to have a different intercept in the "change in fecundity with age" model $g(\alpha_i, \boldsymbol{\beta}, x_i)$, where

$$\alpha_i \sim \text{gamma}\left(\frac{h(\gamma, \nu, u_i)^2}{\varsigma_p^2}, \frac{h(\gamma, \nu, u_i)}{\varsigma_p^2}\right). \tag{6.2.16}$$

The parameter ς_p^2 is the process variance associated with the territory model, including all the influences on an individual's fecundity at first reproduction that are not determined by territory size.[9]

We can now see the relationship between this model and the general template we outlined in equation 6.2.1:

$$[\boldsymbol{\theta}_p, \boldsymbol{\theta}_o, \mathbf{z}|\mathbf{y}] \propto \underbrace{[\mathbf{y}|\mathbf{z}, \boldsymbol{\theta}_o]}_{\text{data}} \underbrace{[\mathbf{z}|\boldsymbol{\theta}_p]}_{\text{process}} \underbrace{[\boldsymbol{\theta}_o][\boldsymbol{\theta}_p]}_{\text{parameters}} \tag{6.2.17}$$

$$\underbrace{[y_i|z_i, \boldsymbol{\theta}_o]}_{\text{data}} = \text{Poisson}(y_i|\lambda_i)$$

$$\underbrace{[z_i|\boldsymbol{\theta}_p]}_{\text{process}} = \text{gamma}\left(\lambda_i \,\middle|\, \frac{g(\alpha_i, \boldsymbol{\beta}, x_i)^2}{\sigma_p^2}, \frac{g(\alpha_i, \boldsymbol{\beta}, x_i)}{\sigma_p^2}\right)$$

$$\times \text{gamma}\left(\alpha_i \,\middle|\, \frac{h(\gamma, \nu, u_i)^2}{\varsigma_p^2}, \frac{h(\gamma, \nu, u_i)}{\varsigma_p^2}\right)$$

[9]We point out that this is the first time you have seen the symbol ς, which is called *variant sigma*. We will use it often in the remainder of the book when we use more than one variance term in a model.

$$\underbrace{[\boldsymbol{\theta}_o]\,[\boldsymbol{\theta}_p]\,[\gamma][\nu]}_{\text{parameters}} = \prod_{j=1}^{2} \text{normal}(\beta_j|0, 100)\ \text{gamma}\,(\gamma|.001, .001)$$
$$\times \text{gamma}(\nu|.001, .001)\ \text{inverse gamma}(\sigma_p^2|.001, .001)$$
$$\times \text{inverse gamma}(\varsigma_p^2|.001, .001).$$

Again, notice that the predictor variables **x** and **u** do not appear in the posterior distribution because we assumed they are known.

Birds were marked and followed throughout their life, so it is reasonable to assume that age was measured perfectly. But this is not a reasonable assumption for territory size. Assume that we have $j = 1, \ldots, 3$ observations of territory size for each bird. We can now think of an observation of territory size as a random variable arising from $[u_{ij}|\tau_i, \sigma_{o,i}^2]$, where τ_i is the true, unobserved territory size of the ith bird, and $\sigma_{o,i}^2$ is the observation variance. Modeling the predictor variables this way means that the u_{ij} *is* a random variable and must be included in the expression for the posterior distribution. The full model predicting owl fecundity is shown as a Bayesian network and an expression for the posterior and joint distributions in figure 6.2.2. We provide general guidance on when to include predictor variables in posterior distributions in box 6.2.2.

It is useful to think about the relationships between the equations we used to construct the model and to consider where uncertainty arises. We have observations of a process that includes sampling error in our estimates of the fecundity of individual owls.[10] In our first hierarchical model (eq. 6.2.13), we have a single term for uncertainty that arises in the process of reproduction because different owls have different mean fecundities resulting from differences in age, location, genetics, and all other sources of variation. In our second model (eq. 6.2.15), we seek to reduce that uncertainty about the process by including additional knowledge—the age of each owl—and by using a model that explains variation in fecundity in a biologically sensible way. In our third model (eq. 6.2.17), we seek to reduce uncertainty further by modeling the average reproduction at reproductive age 0, the intercept in the "effects of age" model, as a function of territory size. We include all the variation in the true, average fecundity that is not explained by our model in the stochastic terms σ_p^2 and ς_p^2. It is important to understand that our deterministic models $g\,()$ and $h\,()$ could have taken any functional form—linear or nonlinear. In the fourth model (fig. 6.2.2),

[10]Remember, the observation variance in this case equals the mean. We could use a different distribution, for example, a negative binomial, if we wanted to estimate the observation variance separately, but doing so would probably require repeated observations of the fecundity of each individual.

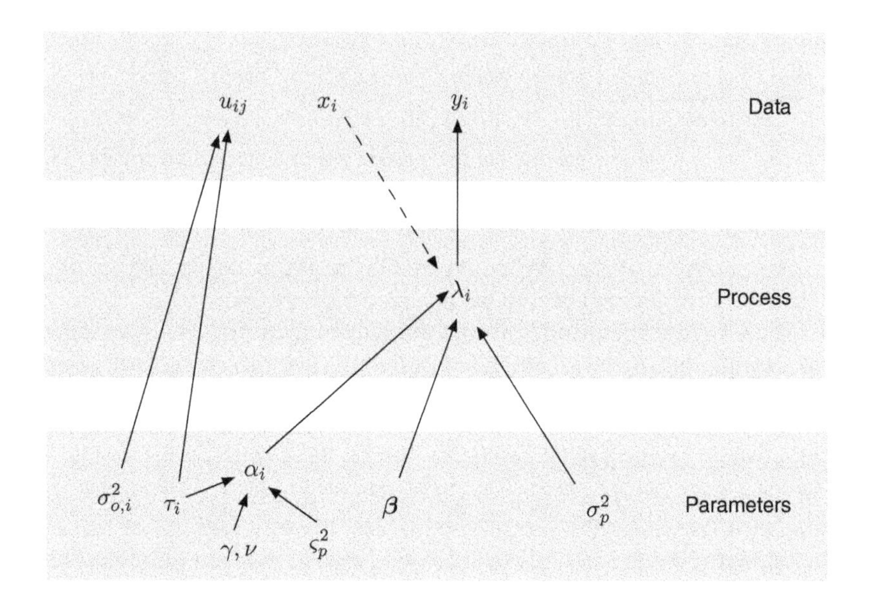

$$\begin{aligned}
g(\alpha_i, \boldsymbol{\beta}, x_i) &= \alpha_i + \beta_1 x_i + \beta_2 x_i^2 \\
h(\gamma, \nu, \tau_i) &= \gamma e^{-\nu \tau_i} \\
[\boldsymbol{\lambda}, \boldsymbol{\alpha}, \boldsymbol{\beta}, \gamma, \nu, \boldsymbol{\tau}, \boldsymbol{\sigma}_o^2, \sigma_p^2, \varsigma_p^2 | \mathbf{y}, \mathbf{u}] &\propto \prod_{i=1}^{n} \text{Poisson}(y_i|\lambda_i)\, \text{gamma}\left(\lambda_i \,\middle|\, \frac{g(\alpha_i, \boldsymbol{\beta}, x_i)^2}{\sigma_p^2}, \frac{g(\alpha_i, \boldsymbol{\beta}, x_i)}{\sigma_p^2}\right) \\
&\times \text{gamma}\left(\alpha_i \,\middle|\, \frac{h(\gamma, \nu, \tau_i)^2}{\varsigma_p^2}, \frac{h(\gamma, \nu, \tau_i)}{\varsigma_p^2}\right) \\
&\times \prod_{j=1}^{3} \text{gamma}\left(u_{ij} \,\middle|\, \frac{\tau_i^2}{\sigma_{o,i}^2}, \frac{\tau_i}{\sigma_{o,i}^2}\right) \text{gamma}(\tau_i|.001, .001) \\
&\times \text{inverse gamma}(\sigma_{o,i}^2|.001, .001)\, \text{gamma}(\nu|.001, .001) \\
&\times \text{gamma}(\gamma|.001, .001) \prod_{k=1}^{2} \text{normal}(\beta_k|0, 100) \\
&\times \text{inverse gamma}(\sigma_p^2|.001, .001) \\
&\times \text{inverse gamma}(\varsigma_p^2|.001, .001)
\end{aligned}$$

Figure 6.2.2. Hierarchical model of fecundity of spotted owls. Relationships between random variables are shown with solid arrows; deterministic relationships are shown with dashed arrows. The observation of fecundity of each owl (y_i) is a random variable controlled by its average fecundity (λ_i) and sampling variation resulting from the particular year the owl was sampled. The average fecundity of an individual (λ_i) is modeled as a quadratic function of the owl's age (x_i) with parameters α_i, β_1, β_2. We assume age is known. Variation in the λ_i not captured by the model is represented by σ_p^2. We assume that the parameter α_i, the fecundity of owl i at first reproduction, decreases exponentially with increasing territory size τ_i. Observations of territory size (u_{ij}) arise from a distribution with mean τ_i and observation variance $\sigma_{o,i}^2$. The rate of

we add uncertainty in observations of territory size. The observed territory size is a random variable arising from a distribution governed by the true territory size (τ_i) and measured observation variance ($\sigma^2_{o,i}$).

We must deal with two parts of equation 6.2.14 that might be confusing. First are the sources of uncertainty. We have an explicit parameter for the process variance (σ^2_p), but there doesn't appear to be a parameter controlling variance in the observations. Does that mean we assume there is no observation variance? The answer is no. Remember that the variance of the Poisson distribution is the same as the mean, so the observation variance is implicit in the likelihood—variance that is highly constrained. We also raise a caution here, which we treat in more detail in section 6.3. The fecundity model lacks replication at the level of the individual; that is, we observe only a single fecundity for each owl. This means we would probably not be able to separate observation variance from process variance if the distribution for the likelihood were less constrained than the Poisson. We discuss this important issue more fully in section 6.3.

Box 6.2.2 When Are Predictor Variables Included in the Posterior Distribution?

A common error in writing expressions for the posterior and joint distribution is to include predictor variables, that is, the **x**, on the right-hand side of the conditioning in the posterior distribution, when we assume (rightly or wrongly) that the **x** are measured without error. They are *known*. Hence, they are not random variables and should not be included in the posterior distribution. It is fine that they are arguments to a deterministic function representing an ecological process, but if we include them in the posterior distribution, then the factoring of the joint distribution doesn't work out in a sensible way.

(*continued*)

Figure 6.2.2 (*continued*).
decrease in α_i is controlled by the parameter ν. The maximum possible value of α_i is γ, which occurs at territory size of 0. Variation in the α_i not represented in the exponential model is represented by ς^2_p. The expressions for the posterior and joint distributions of the unobserved and observed are shown at the bottom of the figure. Note the correspondence between the diagram and the expression for the joint distribution. Quantities at the heads of the solid arrows are on the left-hand side of the conditioning symbols. Quantities at the tails are on the right-hand side. Quantities at the tails of solid arrows with no arrow leading into them must have prior distributions with numeric arguments. The quantities at the tails of dashed arrows are treated as known.

(Box 6.2.2 *continued*)

Consider a simple example. We have a deterministic model $g(\boldsymbol{\theta}, x_i)$, the output of which gives the mean of a response (y_i). Variation in y_i occurs because our model omits many influences, which we quantify with process variance σ_p^2. We assume the y_i are measured perfectly, but we nonetheless treat them as random variables because of the uncertainty about the process that our model fails to capture. We drop the $g()$ wrapper to make the factoring more clear. Consider the *wrong* expression for the posterior and joint distributions,

$$\left[\boldsymbol{\theta}, \sigma_p^2 | y_i, x_i\right] \propto \left[y_i, \boldsymbol{\theta}, \sigma_p^2, x_i\right], \tag{6.2.18}$$

$$\left[\boldsymbol{\theta}, \sigma_p^2 | y_i, x_i\right] \propto \left[y_i | \boldsymbol{\theta}, \sigma_p^2, x_i\right] [\boldsymbol{\theta}] \left[\sigma_p^2\right] [x_i], \tag{6.2.19}$$

which is obviously incorrect because we require a prior on the known value of the observation x_i. The *correct* expression is

$$\left[\boldsymbol{\theta}, \sigma_p^2 | y_i\right] \propto \left[y_i, \boldsymbol{\theta}, \sigma_p^2\right], \tag{6.2.20}$$

$$\left[\boldsymbol{\theta}, \sigma_p^2 | y_i\right] \propto \left[y_i | \boldsymbol{\theta}, \sigma_p^2,\right] [\boldsymbol{\theta}] \left[\sigma_p^2\right], \tag{6.2.21}$$

as illustrated in panel **A** of the following diagram. The x_i are implicitly part of these expressions, as shown in the Bayesian network. It would also be correct to write

$$\left[\boldsymbol{\theta}, \sigma_p^2 | y_i\right] \propto \left[y_i | g\left(\boldsymbol{\theta}, x_i\right), \sigma_p^2\right] [\boldsymbol{\theta}] \left[\sigma_p^2\right]$$

to highlight the deterministic model $g(\boldsymbol{\theta}, x_i)$.

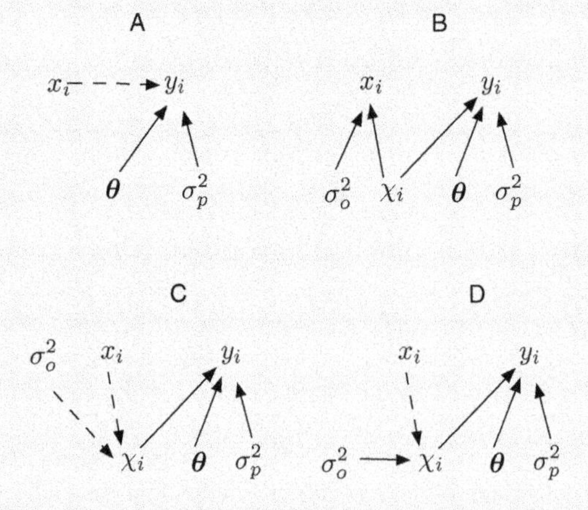

(*continued*)

(Box 6.2.2 *continued*)

Sometimes, we treat the predictor variables as random variables because we wish to model errors in observing them. If we assume that the observations of the predictor variable are imperfect, then we might model them arising from a distribution $[x_i | \chi_i, \sigma_o^2]$, where χ_i is the true, unobserved value of x_i, and σ_o^2 represents uncertainty in the observation process. Our deterministic model is now $g(\boldsymbol{\theta}, \chi_i)$. As shown in panel B, we now have an expression for the posterior that factors correctly:

$$[\boldsymbol{\theta}, \sigma_p^2, \chi_i, \sigma_o^2 | y_i, x_i] \propto [y_i | \boldsymbol{\theta}, \sigma_p^2, \chi_i] [x_i | \chi_i, \sigma_o^2] [\boldsymbol{\theta}] [\sigma_p^2] [\sigma_o^2] [\chi_i]. \quad (6.2.22)$$

One more point bears mentioning. Models for predictor variables that take the form $[\chi_i | x_i, \sigma_o^2]$ are sometimes seen in the scientific literature. These models portray the true, unobserved value of the predictor variable as a random variable determined by the *known* observation and *known* observation variance (panel C). In this case, the expression for the posterior and joint distributions is

$$[\boldsymbol{\theta}, \sigma_p^2, \chi_i | y_i] \propto [y_i | \boldsymbol{\theta}, \sigma_p^2, \chi_i] [\chi_i | x_i, \sigma_o^2] [\boldsymbol{\theta}] [\sigma_p^2]. \quad (6.2.23)$$

Again, the deterministic model is $g(\boldsymbol{\theta}, \chi_i)$. Note that there is no prior on χ_i because it is seen on both sides of a conditional symbol. Also note that x_i and σ_o^2 are no longer seen in the expression for the posterior because they are not random variables. Hence, they do not require a prior. We think a better way to do this would be to treat σ_o^2 as a random variable informed by a strong prior developed in calibration studies (panel D), in which case,

$$[\boldsymbol{\theta}, \sigma_p^2, \sigma_o^2, \chi_i | y_i] \propto [y_i | \boldsymbol{\theta}, \sigma_p^2, \chi_i] [\chi_i | x_i, \sigma_o^2] [\boldsymbol{\theta}] [\sigma_p^2] [\sigma_o^2]. \quad (6.2.24)$$

6.2.2 Multilevel Models: Controls on Nitrous Oxide Emissions from Agricultural Soils

Data in ecological research are often collected at multiple scales or levels of organization in nested designs (fig. 6.2.3). "Group" is a catchall term for the upper level in many different types of nested hierarchies. Groups can logically be formed by populations, locations, species, treatments, life stages, and individual studies. Measurements are taken within groups on individual organisms, plots, species, time periods, and so on. Measurements may also be taken on the groups themselves, that is, covariates that apply at

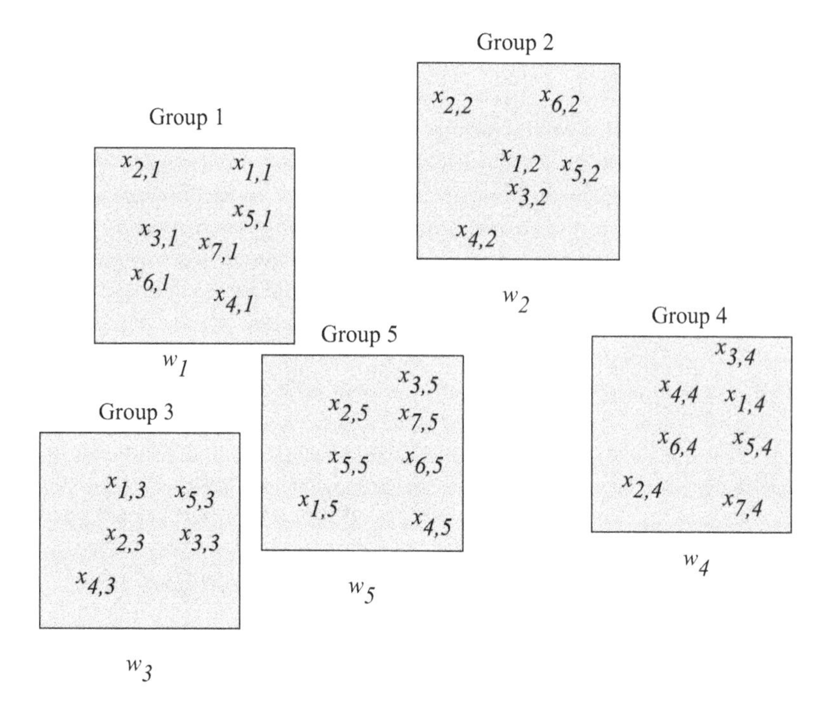

Figure 6.2.3. Observations of states and processes in ecology are often made at different scales of time, space, or level of organization. We can think of the upper level as a "group" with associated observations w_j on the jth group. Observations are also made within each group (x_{ij} paired with a response y_{ij}), where the notation ij means the ith observation in group j. In this illustration there are variable numbers of observations within groups (i.e., the design is unbalanced), and $j = 1, \ldots, 5$ observations for groups. Note that when the number of observations is unbalanced, as it is here, product symbols in likelihoods for observations within groups must have an upper index appropriate for the number of observations (e.g., n_j).

the upper level of organization or spatial scale. Multilevel models represent the way a quantity of interest responds to the combined influence of observations taken at the group level and within the group.

Here, we modify the hierarchical example developed by Qian et al. (2010) (using the data of Carey (2007)) to illustrate a multilevel model. Nitrous oxide (N_2O), a greenhouse gas roughly 300 times more potent than carbon dioxide in forcing atmospheric warming, is emitted when nitrogen is added to the soil in synthetic fertilizers. Carey (2007) conducted a meta-analysis of effects of nitrogen fertilizer addition ($\text{gN} \cdot \text{ha}^{-1} \cdot \text{d}^{-1}$), reviewing 164 studies. In this example, studies occurred at different locations, forming the group level in the hierarchy. Soil carbon content ($gC \cdot g^{-1}$organic matter) was measured as a group-level covariate that was assumed to be measured without error. Replicate observations of N_2O emission, also assumed to be measured without error, were paired with measurements of fertilizer addition ($kgN \cdot ha^{-1}$). The type of fertilizer was also studied, but we chose to omit this effect to simplify the example. A total of 1085 observations were made across all the studies.

We could model the observations of N_2O emission as

$$g\left(\alpha_j, \beta, x_{ij}\right) = \alpha_j + \beta x_{ij}, \tag{6.2.25}$$

$$\left[\alpha_j, \boldsymbol{\beta}, \mu, \sigma_j^2, \varsigma^2 | y_{ij}\right] \propto \left[y_{ij} | g(\alpha_j, \beta, x_{ij}), \sigma_j^2\right] \times \left[\alpha_j | \mu, \varsigma^2\right][\beta][\mu]\left[\sigma_j^2\right]\left[\varsigma^2\right], \tag{6.2.26}$$

where y_{ij} is the ith observation of N_2O emissions in study j, x_{ij} is a paired measurement of fertilizer addition and β is the change in N_2O emissions per unit change in fertilizer addition. The model $g\left(\alpha_j, \beta, x_{ij}\right)$ represents the hypothesis that emissions increase in direct proportion to fertilizer additions. The intercept α_j varies among studies as a random variable drawn from a distribution with parameters μ and ς^2. We explicitly represent variation among studies using the distribution of the α_j. Representing the intercepts this way sets this analysis apart from conventional, single-level regression that could be performed separately for each of the 164 individual sites or by pooling all the data across sites to estimate a single intercept and slope. The σ_j^2 represents the uncertainty about N_2O emissions that comes from sampling variation within a study, and the ς^2 represents the uncertainty that arises as a result of variation among studies. An advantage of this hierarchical approach is known as *borrowing strength*, which means that estimates of the intercepts from locations with small data sets are made more precise by studies with larger data sets (box 6.2.3).

Box 6.2.3 What Does "Borrowing Strength" Mean?

You'll often see the phrase "borrowing strength" in papers that use Bayesian hierarchical models. In this context, borrowing strength refers to the sharing of information among unknowns in Bayesian models. For example, consider the situation in which a researcher measures leaf area index (LAI) for each plant in

(continued)

(Box 6.2.3 *continued*)

a set of five plots. Suppose that the numbers of plants in each plot are 10, 12, 8, 10, and 2. Clearly, the last plot will carry less information about plot-level LAI because of its smaller sample size. A classical Bayesian remedy for this small sample situation is to specify a hierarchical model to help learn about plot-level mean LAI. In this case, let y_{ij} be the LAI measurement for plant i ($i = 1, \ldots, n_j$) on plot j ($j = 1, \ldots, J$), and let the plot-level mean LAI be z_j. A complete Bayesian model could be formulated as

$$y_{ij} \sim \text{normal}(z_j, \sigma_y^2), \tag{6.2.27}$$

$$z_j \sim \text{normal}(\mu, \sigma_z^2), \tag{6.2.28}$$

$$\mu \sim \text{normal}(\mu_0, \sigma_0^2), \tag{6.2.29}$$

$$\sigma_z^2 \sim \text{inverse gamma}(\alpha_z, \beta_z), \tag{6.2.30}$$

$$\sigma_y^2 \sim \text{inverse gamma}(\alpha_y, \beta_y). \tag{6.2.31}$$

Getting ahead of ourselves a bit, the full-conditional distribution for the mean of plot with small sample size z_5, is $[z_5|\cdot]$, where the notation reads "the distribution of z_5 conditional on the data and other parameters that influence its value." (We cover full-conditionals in detail in section 7.3.2.1). Thus, the distribution of z_5 will contain two terms:

$$[z_5|\cdot] \propto \prod_{i=1}^{n_5} \text{normal}(y_{i5}|z_5, \sigma_y^2)\text{normal}(z_5|\mu, \sigma_z^2), \tag{6.2.32}$$

with one term containing the portion of data collected at plot 5 and a second term that depends on the mean of plot-level means and an associated variance component.

In fitting the model, all the data (not just the data for plot 5) will help estimate μ and σ_z^2. We can think of normal($z_5|\mu, \sigma_z^2$) as a "prior" for z_5. The information contained in this second term will lead to a posterior for z_5 that is more precise than the resulting posterior from a model where σ_z^2 is assumed to be known and vague a priori.

In a sense, the information about μ and σ_z^2 from the rest of the plots will "shrink" z_5 for plot 5 to the appropriate distribution of plot-level means. The plots with more data will provide more information about the proper amount of dispersion in the distribution of plot-level means, which, in turn, provides more

(continued)

(Box 6.2.3 *continued*)

information about plots with smaller sample size. Therefore, the variance of z_5 will be smaller than it would have been if we had simply used a vague prior for all z_j.

This concept of borrowing strength is not unique to Bayesian statistics, as it can be interpreted as a random effect in the model. However, the Bayesian perspective of these types of random effects is particularly clear and rigorous. We will return to the general concept of "shrinkage" in later chapters when we describe statistical regularization and its many benefits, including model selection.

We now seek to explain some of the variation among sites using the observation of soil carbon content taken at the group level (similar to the example in sec. 6.2.1, where we modeled owl fecundity as a function of the covariate age of owl). Instead of simply estimating an average value for the intercept in each study (eq. 6.2.26), we model the intercept for each study as a linear function of the observations of soil carbon at the group level; that is, the α_j are predicted using the model $h\left(\kappa, \eta, w_j\right) = \kappa + \eta w_j$ (fig. 6.2.4). Choosing lognormal distributions[11] for the distribution of the data and the intercepts makes sense, because they are continuous and nonnegative:

$$y_{ij} \sim \text{lognormal}(\log(\alpha_j + \beta x_{ij}), \sigma_j^2), \tag{6.2.33}$$

$$\alpha_j \sim \text{lognormal}(\log(\kappa + \eta w_j), \varsigma^2). \tag{6.2.34}$$

The parameters in these expressions might require some review. Let μ be the first parameter of a lognormal distribution. The median of the lognormal distribution equals e^{μ}, so log(median) $= \mu$. We could also model the first parameter using the mean, but the expression is more complicated (app. A.2) Thus, if our deterministic model predicts the median of the posterior distribution, we take its log to obtain the first parameter. The second parameter is the variance of the random variable on the log scale.

We are not limited to modeling the intercept at the group level; we could also allow the slopes to vary among sites or allow both intercepts and slopes to vary. See Gelman and Hill (2009, pg. 376) for details. Moreover, we

[11]Gamma distributions could also be used.

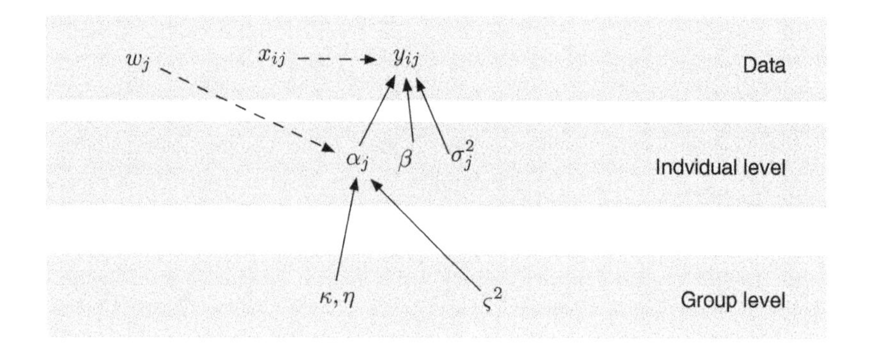

$$\begin{aligned}
g(\alpha_j, \beta, x_{ij}) &= \alpha_j + \beta x_{ij} \\
h(\kappa, \eta, w_j) &= \kappa + \eta w_j \\
\left[\boldsymbol{\alpha}, \beta, \boldsymbol{\sigma^2}, \kappa, \eta, \varsigma^2 | \mathbf{y}\right] &\propto \prod_{j=1}^{164}\prod_{i=1}^{n_j} \text{lognormal}\left(y_{ij} | \log\left(g(\alpha_j, \beta, x_{ij})\right), \sigma_j^2\right) \\
&\times \text{lognormal}\left(\alpha_j | \log\left(h(\kappa, \eta, w_j)\right), \varsigma^2\right) \\
&\times \text{inverse gamma}\left(\sigma_j^2 | .001, .001\right) \\
&\times \text{normal}\left(\beta | 0, 1000\right) \text{gamma}\left(\kappa | .001, .001\right) \\
&\times \text{normal}\left(\eta | 0, 1000\right) \text{inverse gamma}\left(\varsigma^2 | .001, .001\right)
\end{aligned}$$

Figure 6.2.4. Bayesian network and posterior and joint distributions for the meta-analysis of effects of fertilizer on N_2O emissions. The y_{ij} are observations of N_2O emissions accumulated from 164 different studies. The y_{ij} are modeled as a linear function of the level of fertilizer added within a given study (x_{ij}) where j indexes study and the subscript ij indicates the ith observation within study j. The n_j are the number of observations from study j. Intercepts in the individual-level model (α_j) varied among groups (i.e., studies) as a function of soil carbon content (w_j). Uncertainty within individual studies (σ_j^2) was allowed to vary among studies. Lognormal distributions were chosen for y_{ij} and α_j because both quantities must be nonzero. Inverse gamma priors were chosen for the variances $(\sigma_j^2, \varsigma^2)$ because an inverse gamma distribution is a conjugate for the variance of a lognormal distribution assuming the mean is known (table A.3).

emphasize that our choice of linear models to represent the process of N_2O emission is in no way mandatory. We could use any functional form that makes biological sense.[12]

[12] It might make *more* sense, for example, to model the intercept as an asymptotic function of soil carbon, something like $h(\kappa, \eta, w_i) = \kappa w_i/(\eta + w_i)$.

6.2.3 Hidden Processes: Effects of Predation by Treesnakes on Lizard Populations

Ecologists often wish to answer questions about the state of a system that changes over time or space. Many of the states that we strive to understand cannot be observed directly but instead arise from processes that are "hidden" (e.g., Newman et al., 2006; Tavecchia et al., 2009; Liberg et al., 2012; Gimenez et al., 2012). We must make inferences about these unobservable states and hidden processes from the behavior of quantities that we *can* observe. We refer to unobservable states as *latent*. Here, we illustrate an especially valuable use of Bayesian hierarchical models: estimating latent states and how they change over time, over space, and in response to perturbation.

Campbell et al. (2012) used predation by brown treesnakes (*Boiga irregularis*) on lizards on the island of Guam as a model system to test the hypothesis that exothermic predators influence the abundance of their exothermic prey. This research offers an especially useful example because it illustrates how hierarchical models can be used to analyze designed, manipulative experiments. The research team observed lizard abundance on four 1 ha plots, which we index by $m = 1, \ldots, 4$. All treesnakes were removed from two of the plots, and the other two were left as controls with extant levels of snake abundance. Lizards were counted on five transects (indexed by i) within each plot. Counts were repeated seven times on each transect (on different days, indexed by t) within each of six monitoring periods. We are omitting the monitoring period dimension of the experiment and consider only one period, which means that time (t) refers strictly to repeated measures of transects.

The research team needed to estimate the true, unobserved lizard abundance on each transect based on counts along the transect. A key problem in this kind of research is that counting all individuals is virtually impossible, because some lizards that are present inevitably escape detection. The mismatch between what we are able to observe and the true state we want to understand requires building a model of the data, a way to estimate the probability that a lizard is observed (ϕ) on a transect at a given time conditional on its being present. A sensible approach for modeling the data would start with

$$y_{itm} \sim \text{binomial}(z_{im}, \phi), \tag{6.2.35}$$

$$\phi \sim \text{beta}(1, 1), \tag{6.2.36}$$

which simply says that the observations of the number of lizards counted on transect i observed at time t on plot m can be represented as a random variable y_{itm} drawn from a binomial distribution where z_{im} is the true number of individuals on the ith transect of plot m, and ϕ is the probability of observing an individual. Thus, z_{im} represents the number of "trials" on transect i in plot m, that is, the number of lizards that were present and might be found; y_{itm} is the number of "successes," the number of lizards that were found; and ϕ is the probability of a success on a single trial, the probability that we would observe a lizard if it were present (Royle, 2004). By taking replicate observations on the transect and assuming for the moment that z_{im} is known, we can estimate ϕ on the back of a napkin using a beta-binomial conjugate prior relationship (as shown in sec. 5.3). The full expression for the posterior and joint distributions is

$$[\phi|\mathbf{y}] = \prod_{i=1}^{5}\prod_{t=1}^{7}\prod_{m=1}^{4} \text{binomial}\,(y_{itm}|z_{im}, \phi)\ \text{beta}\,(\phi|1, 1) \qquad (6.2.37)$$

Recall that we do not include $\mathbf{z}$ in the posterior distribution because at this point we are assuming it is known.

Our initial model (eq. 6.2.35) requires the assumption that there is a *single* detection probability applying identically to all time periods and all transects within a plot in much the same way that we assumed that all owls had the same average fecundity (sec. 6.2.1). We could improve on the model by allowing each transect and time to have its own detection probability, reasoning that observability is likely to vary over time and space. Transects might differ in the availability of lizard hiding places, and observation times might include different temperatures that affect lizard activity levels, both of which would alter lizard exposure to the observer. Campbell et al. (2012) included this variability in the data model using

$$y_{itm} \sim \text{binomial}(z_{im}, \phi_{itm}), \qquad (6.2.38)$$

$$\text{logit}(\phi_{itm}) = \alpha_0 + \alpha_{1,itm}, \qquad (6.2.39)$$

$$\alpha_0 \sim \text{normal}(0, 100),\ \alpha_{1,itm} \sim \text{normal}(0, \sigma^2_{\alpha_{1,itm}}), \qquad (6.2.40)$$

where inverse logit α_0 is the overall mean probability of detection of lizards on the plot, and $\alpha_{1,itm}$ is a *random effect* (box 6.2.1) of transect and observation time. Thus, $\alpha_{1,itm}$ represents the variation in detection probability that arises from differences among transects and sampling occasions.

It might be useful to think of equation 6.2.38 as an "intercept only" (i.e., α_0) linear model, making it analogous to examples developed earlier that had group-level intercepts (secs. 6.2.2 and 6.2.4). As before, we are allowing the "groups" time and transect to have their own intercepts drawn from a distribution with mean inverse logit(α_0) (recall inverse logit from sec. 2.2.1.3). Variation is created around this mean probability of detection by random variation among transects and observation times—variation that is represented by the parameters $\alpha_{1,itm}$. Notice that we are not modeling how this variation arises, as we might do if we had covariates, say on temperature or vegetation cover. Instead, we simply acknowledge that the variation exists and include it using the random-effect terms.

It might appear at first glance that the model (eqs. 6.2.38–6.2.40) contains only a single source of uncertainty, the random effect $\alpha_{1,itm}$ and the associated parameter controlling its distribution ($\sigma^2_{\alpha_1}$). This appears to contradict the idea that random effects are included in models to capture uncertainty that extends beyond sampling variation. However, recall that a sampling variability is included in the binomial likelihood (eq. 6.2.38), the variance of which is $n\phi\,(1-\phi)$ (sec. 3.4.3.1). So, sampling variability is implicit in the binomial, and hence, there are two sources of uncertainty in this observation model—uncertainty arising from sampling and uncertainty arising because the ability to detect lizards varies among times and transects.

We could achieve the exact same meaning with slightly different notation:

$$y_{itm} \sim \text{binomial}(z_{im}, \phi_{itm}), \tag{6.2.41}$$

$$\text{logit}(\phi_{itm}) \sim \text{normal}(\alpha_o, \sigma_o^2). \tag{6.2.42}$$

In the first case (eqs. 6.2.38–6.2.40) we add a mean 0 random effect ($\alpha_{1,itm}$) to an overall mean. In the second case (eqs. 6.2.41–6.2.42), we model random effects as random draws of logit(ϕ) from a normal distribution with mean α_o and variance σ^2. It is important to see that these are algebraically identical, because both types of notation are widely used in the literature.

A third equally correct alternative for this model would be

$$y_{itm} \sim \text{binomial}(z_{im}, \phi_{itm}), \tag{6.2.43}$$

$$\phi_{itm} \sim \text{beta}(\gamma, \nu). \tag{6.2.44}$$

In this case, we are not transforming ϕ_{itm} so that it takes on values appropriate for the normal distribution ($-\infty$ to $+\infty$). Instead, we are choosing a distribution appropriate for a random variable that can take on values in the interval 0 to 1. The mean of this beta distribution, you will recall from section 3.4.4, is $\gamma/(\gamma+\nu)$ (also see appendix table A.2). We will use this formulation to portray the detection process in the full, hierarchical model, developed next, because we think it is the easiest to understand.

Until now, we assumed that the true number of lizards on a transect was known, which of course, it is not. We now develop a model of processes controlling the true, unobserved lizard abundance on each transect (z_{im}). Thus, z_{im} is now a random variable. The main interest in this study is the variation of numbers of lizards among plots, particularly the variation that is contributed by the snake-removal treatment. So, now we model the means of the four plots (two of which had predators removed) using

$$z_{im} \sim \text{Poisson}(e^{\beta_{0,m}+\beta_1 x_m}), \qquad (6.2.45)$$

$$\beta_{0,m} \sim \text{normal}(0, 100),$$

$$\beta_1 \sim \text{normal}(0, 100), \qquad (6.2.46)$$

where x_m is an indicator variable equal to 1 if plot m had snakes removed and 0 otherwise. It is entirely reasonable to assume that the x_m are measured without error, because treatments were assigned by the investigators. In this model, $e^{\beta_{0,m}}$ is the mean abundance of lizards on plot m in the absence of treatment. Note that the average abundance on transect i within plot m is determined by $e^{\beta_{0,m}}$ as modified by the treatment effect, (e^{β_1}), the proportional change in the mean number of lizards caused by removing treesnakes. Campbell et al. (2012) made repeated observations at the transect scale but modeled log mean abundance (i.e., $\beta_{0,m}$) at the plot scale, recognizing that plots were the experimental units.

The Bayesian network for the relationships between the knowns and unknowns and the expression for the posterior and joint distributions are shown in figure 6.2.5. The process models (equations 6.2.45–6.2.46) and data model (equations 6.2.38–6.2.40) are linked by latent state z_{im}, the true number of lizards on transect i within plot m (fig. 6.2.5). This example shows how hierarchical models can be used in designed experiments to represent variation at multiple scales, to estimate unobservable quantities, to estimate the effect of a treatment, and to properly account for uncertainty arising from multiple sources.

6.2.4 Multilevel Models: Functional Traits of Tropical Trees Mediate Effects of Light and Size on Growth

Relationships between traits of individuals that control physiological and reproductive function are known to shape demographic rates of populations. The relationship between species-functional traits and demography has

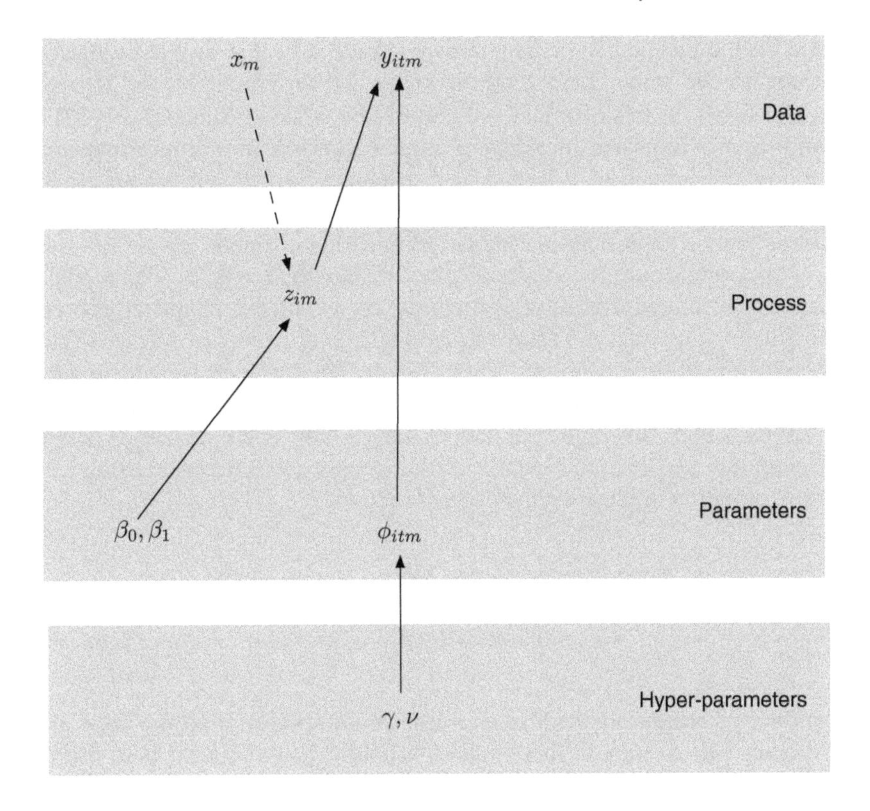

$$
\begin{aligned}
[\boldsymbol{\phi}, \boldsymbol{z}, \boldsymbol{\beta}, \gamma, \nu, |\boldsymbol{y}] \propto \prod_{i=1}^{5} \prod_{t=1}^{7} \prod_{m=1}^{4} & \text{binomial}\left(y_{itm} | z_{im}, \phi_{itm}\right) \\
& \times \text{beta}\left(\phi_{itm} | \gamma, \nu\right) \\
& \times \text{Poisson}\left(z_{im} | e^{\beta_0 + \beta_1 x_m}\right) \\
& \times \text{normal}\left(\beta_0 | 0, 100\right) \text{normal}\left(\beta_1 | 0, 100\right) \\
& \times \text{gamma}\left(\gamma | .001, .001\right) \\
& \times \text{gamma}\left(\nu | .001, .001\right)
\end{aligned}
$$

Figure 6.2.5. Bayesian network for model of effects of snake predation on lizards modified from Campbell et al. (2012). The data (y_{itm}) are the number of lizards observed on transect i at time t on plot m. The true number of lizards on transect i of plot m is z_{im}. The parameter ϕ_{itm} is the probability of detecting a lizard that is truly present. These probabilities arise from a beta distribution with parameters γ and ν. The model $e^{\beta_0+\beta_1 x_m}$ represents the effect of removing predators from plot m on the true number of lizards on a transect i.

provided fundamental insight into the trade-offs that shape life-history strategies (Westoby et al., 2002; Westoby and Wright, 2006; van Kleunen et al., 2010). In this example, we feature the work of Rüger et al. (2012), who used a Bayesian hierarchical model to reveal how functional traits modify the influence of light and size on the growth rate of species of tropical trees. We modified the original analysis in minor ways to improve consistency with principles established in earlier chapters.

The growth rate of individual trees within each species was modeled as a power function of light availability ($x_{1,ij}$) and tree diameter at breast height $\left(x_{2,ij}\right)$, where i indexes individuals, and j indexes species. These were assumed to be measured without error. Thus, species were treated as a group-level variable, and there were measurements of individual growth responses and covariates for individuals within each species. The true growth rate $\left(\lambda_{ij}\right)$ of individual i within species j was modeled using a log transformation to linearize the power function,

$$g\left(\boldsymbol{\beta}_j, \mathbf{x}_{ij}\right) = \beta_{0,j} + \beta_{1,j}\log(x_{1,ij}) + \beta_{2,j}\log(x_{2,ij}),$$

$$\lambda_{ij} \sim \text{lognormal}(g(\boldsymbol{\beta}_j, \mathbf{x}_{ij}), \sigma^2_{p,j}), \tag{6.2.47}$$

where $\sigma^2_{p,j}$ represents the process variance for species j (on the log scale). Another way to look at this model that might be more clear is to use the untransformed power function for growth, $g(\boldsymbol{\beta}_j, \mathbf{x}_{ij}) = e^{\beta_{0,j}} x_{1,ij}^{\beta_{1,j}} x_{2,ij}^{\beta_{2,j}}$, in which case the true growth rate is $\lambda_{ij} \sim \text{lognormal}(\log(g(\boldsymbol{\beta}_j, \mathbf{x}_{ij})), \sigma^2_{p,j})$.

Functional traits of species (wood density, maximum height, leaf area, seed mass, leaf mass per area, and leaf nutrient content) making up the data vector $\mathbf{w}_j$ were used as group-level covariates to model the β coefficients (fig. 6.2.6) in the individual-level model using vectors of group-level parameters, $\boldsymbol{\alpha}$, $\boldsymbol{\gamma}$, and $\boldsymbol{\eta}$:

$$\beta_{0,j} \sim \text{normal}(\alpha_0 + \mathbf{w}'_j\boldsymbol{\alpha}, \varsigma_0^2), \tag{6.2.48}$$

$$\beta_{1,j} \sim \text{normal}(\gamma_0 + \mathbf{w}'_j\boldsymbol{\gamma}, \varsigma_1^2), \tag{6.2.49}$$

$$\beta_{2,j} \sim \text{normal}(\eta_0 + \mathbf{w}'_j\boldsymbol{\eta}, \varsigma_2^2). \tag{6.2.50}$$

Tree growth was observed as the increment in tree diameter measured at two points in time. The process model (eq. 6.2.47) was related hierarchically to observations (y_{ij}) of tree growth (fig. 6.2.6, Data level) with a mixture distribution (sec. 3.4.5) to reflect uncertainty in the observations (Chave et al., 2004; Rüger et al., 2011):

$$\left[y_{ij}|\lambda_{ij},\phi,\sigma_o^2\right] = (1-\phi)\cdot\text{normal}\left(\lambda_{ij},\frac{\sigma_{o,1}^2}{d_{ij}^2}\right)+\phi\cdot\text{normal}\left(\lambda_{ij},\frac{\sigma_{o,2}^2}{d_{ij}^2}\right). \tag{6.2.51}$$

This is a clever approach because it breaks the observation uncertainty into two parts. Small routine errors caused by a slightly different placement of the calipers or tape measure cause size-dependent uncertainty, represented in the variance ($\sigma_{o,1}^2$), which was estimated from a separate calibration data set (Rüger et al., 2011; Chave et al., 2004). The same data set was used to estimate $\sigma_{o,2}^2$, representing large, size-independent errors caused by missing a decimal place or recording a number with the wrong tree. We simplify the example by treating $\sigma_{o,1}^2$ and $\sigma_{o,2}^2$ as known. The estimate of ϕ obtained in calibration studies became an informative prior. The observation variances were divided by the squared number of days (d_{ij}^2) that elapsed between the two diameter measurements of the tree used to calculate the observed annual growth rate (y_{ij}). It was assumed that the magnitude of errors was proportional to the time between measurements.[13]

The full expression for the joint and posterior distributions (fig. 6.2.6) assembles the model of processes governing the true, unobserved growth of individual trees (eq. 6.2.47), the model explaining species effects on the true growth rate using functional traits (eqs. 6.2.48–6.2.50); and a model of the data relating the true growth rate to the observations of growth rate (eq. 6.2.51).

This example illustrates how hierarchical models can be used to deal with two common problems in ecology. We seek to understand how characteristics of species, in this case their functional traits, modify responses of individuals to their local environments. The ecological process we seek to understand is itself hierarchical. We also require a hierarchical model because we are modeling an underlying process (tree growth) based on observations that are not a perfect representation of the process (increments in tree diameter). As in the owl fecundity example, we must separate the uncertainty that arises from imperfect observations from the uncertainty created by the failure of our model to represent the process. This separation of sources of uncertainty is essential to many kinds of ecological analysis.

[13]Be sure you understand why d_{ij}^2 does not appear on the left-hand side of the likelihood. It is because it is measured without error. If this it not clear, see box 6.2.1.

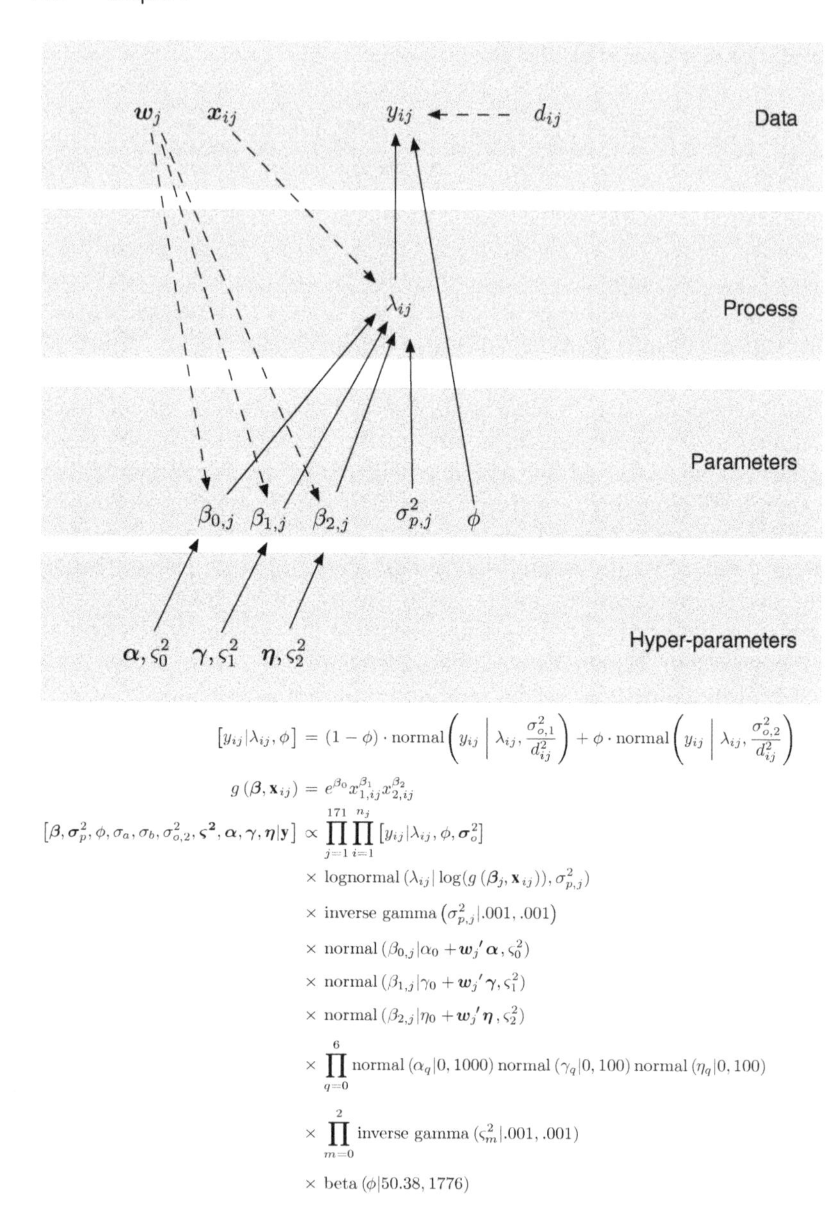

Figure 6.2.6. Hierarchical model of controls on growth of tropical trees (Rüger et al., 2012). Observations (y_{ij}) of the growth rate of an individual (λ_{ij}) were made with error. A mixture model $[y_{ij}|\lambda_{ij}, \boldsymbol{\sigma}^2_o, \phi]$ was used to account for two types of error: mistakes made in measurement of individuals that depended on tree size and

6.2.5 Multiple States, Multiple Types of Data

We now cover an important topic: the use of multiple types of data to estimate parameters and latent states. Ecologists are accustomed to data sets that contain multiple covariates, but the idea of multiple responses may be unfamiliar. In chapter 2 we introduced the idea that models of relatively high dimension may be needed to represent complex ecological relationships, for example, the functioning of ecosystems or the dynamics of populations and communities. In these cases we must use multiple parameters to represent composite forcing and multiple interactions. Models that have many parameters and that predict more than one latent state usually require multiple types of data. Otherwise, their unknowns will not be identifiable.

We first show how to use multiple data sources in a general way before offering a more specific example. Assume we have a deterministic model $g(\boldsymbol{\theta}_p, \mathbf{x})$ that predicts the central tendency of L latent states[14] $(z_1, \ldots, z_L)$. These states might be estimates of fluxes of different molecular forms of carbon and nitrogen from soils; the numbers of individuals of different species in a community and independently obtained estimates of their proportions in the community; temperature, pH, turbidity, and salinity of an estuary. There are w vectors of data on these states, $(\mathbf{y}_1, \ldots, \mathbf{y}_w)$. We have a data model $h(\boldsymbol{\theta}_d, \mathbf{z})$ that relates the observations to the true value of the latent state.

We might reasonably choose a multivariate distribution for the likelihood, something like

$$\mathbf{y}_i \sim \text{multivariate normal}(h(\boldsymbol{\theta}_d, \mathbf{z}_i), \boldsymbol{\Sigma}_y), \tag{6.2.52}$$

$$\mathbf{z}_i \sim \text{multivariate normal}(g(\boldsymbol{\theta}_p, \mathbf{x}_i), \boldsymbol{\Sigma}_z), \tag{6.2.53}$$

[14]To keep our notation compact we are using a static model as an example here, but bear in mind that the model could just as easily be a dynamic simulation model predicting the central tendency of L states $z_{1,t}, \ldots, z_{L,t}$ at time t using model $g(\boldsymbol{\theta}, \mathbf{x}_{t-1}, \mathbf{z}_{t-1})$. Our example will use this type of model.

Figure 6.2.6 (*continued*).
size-independent mistakes resulting from errors like missing a decimal place or recording a number with the wrong tree. Note that $\sigma^2_{o,1}$ and $\sigma^2_{o,2}$ are treated as known in our version of the model, but were random variables in the original paper. The true growth rate was modeled using a power function with individual traits (diameter at breast height and light availability) as predictor variables. The intercept of the power function was modeled as a linear function of six species traits: wood density, maximum height, leaf area, seed mass, leaf mass per area, and leaf nutrient content. The symbols are defined in the text.

were the $\boldsymbol{\Sigma}$ are covariance matrices (box 3.4.4). In this case, we are modeling the observations and the true states as vectors following a multivariate normal distribution. The elements of $\mathbf{y}_i$ and $\mathbf{z}_i$ can be correlated and can have their own individual variance terms. There is nothing wrong with this approach as long as we can plausibly assume that the stochasticity in the latent state and in the observations can be represented using the normal distribution.

The assumption of normality for all states and responses constrains our options. Often, we wish to understand quantities of interest that have different support—some are strictly positive, some are 0 or 1, some range from 0 to 1, some are numbers of individuals in categories.[15] Assuming a variance that is constant with the mean, as is the case with the normal distribution, may not be reasonable. We often require a more flexible approach.

Combining data sets using the independent product rule allows us to choose probability distributions that are appropriate for the support of each random variable, that is, each observation or each unobserved state. It follows that a general expression for the posterior and joint distributions exploiting w data sets is

$$\left[\boldsymbol{\Theta}_d, \boldsymbol{\Theta}_p, \boldsymbol{\sigma}_o^2, \boldsymbol{\sigma}_p^2, \mathbf{z}|\mathbf{y}_1, \ldots, \mathbf{y}_L\right] \propto \prod_{l=1}^{w}\prod_{i=1}^{n_l}\left[y_{li}|h\left(z_{li}\boldsymbol{\theta}_{dl}\right), \sigma_{ol}^2\right]_l \times\left[z_{li}|g\left(\boldsymbol{\theta}_p, \mathbf{x}_i\right), \sigma_p^2\right]\left[\boldsymbol{\theta}_{dl}, \boldsymbol{\theta}_p, \boldsymbol{\sigma}_{ol}^2, \sigma_p^2\right], \tag{6.2.54}$$

where i indexes individual observations and states, and n_1 is the number of observations in data set L. Note that there is a subscript l on the likelihood, which indicates that we can use different distributions as needed for the different data sets, realizing, of course, that we may need some judicious moment matching to transform the means and variances into the proper parameters. The main point here is to recognize the dual products, one product taken over data sets and likelihoods (indexed by l) and the other one taken over individual observations within a data set (indexed by i). Thus, we have used multiple types of data by multiplying the total likelihood of each data set.

We now make equation 6.2.54 more specific by offering a hypothetical example based on age- or stage-structured population modeling (Caswell, 1988), a widely used approach for modeling dynamics of animal and plant populations. Presume we are interested in an organism whose life history can be described by m stages (or age classes) in the state vector $\mathbf{z}_t$, which contains the true number of individuals in each stage at time t. Avoiding

[15] An additional complication arises when the $\mathbf{y}$ vectors differ in length.

species-specific details, assume we have an appropriately composed projection matrix $\mathbf{A}$ containing fertilities and survival probabilities. We notate the survival and fertility parameters in $\mathbf{A}$ collectively as $\boldsymbol{\theta}$, so that our deterministic model is $g(\boldsymbol{\theta}, \mathbf{z}_{t-1}) = \mathbf{A}\mathbf{z}_{t-1}$, and our stochastic model of the process is

$$\log(\mathbf{z}_t) \sim \text{multivariate normal}(\log(g(\boldsymbol{\theta}, \mathbf{z}_{t-1})), \sigma_p^2 \mathbf{I}), \tag{6.2.55}$$

where $\mathbf{I}$ is an $m \times m$ matrix with 1s on the diagonal. The log transform means that we are portraying each of the i stages at time t as a lognormally distributed random variable with median $\mathbf{a}_i' \mathbf{z}_{it-1}$ (the inner product[16] of the ith row of matrix $\mathbf{A}$ with the state vector $\mathbf{z}_{t-1}$) and process variance σ_p^2. We could allow for different process variances for each stage and for their covariance by explicitly specifying a covariance matrix, but we want to keep things simple for this example.

Now, assume we have a data set $\mathbf{y}_1$ containing the total census of individuals in the population at T times. We also have independent data on the number of individuals in each stage in the population $\mathbf{Y}_2$ obtained by sampling a subset of the population and classifying each individual into an appropriate category. Thus, the matrix $\mathbf{Y}_2$ consists of rows containing the classification counts, one row for each of the T time points. We assume for simplicity that there are no gaps in the two time series of data, but missing data could be modeled if necessary. A reasonable expression for the posterior and joint distributions is

$$\begin{aligned}
\left[\boldsymbol{\theta}, \mathbf{z}, \sigma_p^2 | \mathbf{y}_1, \mathbf{Y}_2\right] \propto & \underbrace{\prod_{t=2}^{T} \text{Poisson}\left(y_{1,t} \,\middle|\, \sum_{i=1}^{m} z_{it}\right)}_{\text{likelihood for census data}} \\
& \times \underbrace{\text{multinomial}\left(\mathbf{y}_{2,t} \,\middle|\, \sum_{i=1}^{m} y_{2,it}, \frac{\mathbf{z}_t}{\sum_{i=1}^{m} z_{it}}\right)}_{\text{likelihood for classification data}} \\
& \times \underbrace{\text{multivariate normal}\left(\log(\mathbf{z}_t) | \log\left(g\left(\boldsymbol{\theta}, \mathbf{z}_{t-1}\right)\right), \sigma_p^2 \mathbf{I}\right)}_{\text{process model}} \\
& \times [\boldsymbol{\theta}]\,[\mathbf{z}_1]\left[\sigma_p^2\right].
\end{aligned} \tag{6.2.56}$$

[16]If we have two vectors $\mathbf{u}$ and $\mathbf{v}$ with three elements each, then their inner product is $\mathbf{u}'\mathbf{v} = u_1 v_1 + u_2 v_2 + u_3 v_3$. This is also called the *dot product*.

Be sure you understand the notation $\mathbf{y}_{2,t}$, which is a vector of classification counts, a row from the matrix $\mathbf{Y}_2$ at time t.

There are several points worth emphasizing here. First, look at the likelihood for the census data. The estimate of the true, unobserved population size is the sum over the m stages, which forms our estimate of the mean of the distribution of the random variable, $(y_{1,t})$, the number of individuals counted at time t. Next, focus on the likelihood for the classification data. A vector of proportional contributions of each stage to the total population is $\mathbf{z}_t/\sum_{i=1}^{m} z_{it}$, where elements of this vector are the number in each stage divided by the total. This vector forms the second parameter of the multinomial likelihood. The first argument is simply the total number of individuals classified.

You may be wondering, what happened to the observation variance in equation 6.2.53? Once again, we need to remember the relationship between parameters and moments (sec. 3.4.4). The variance of the Poisson is the same as its single parameter, the mean. This variance reflects the idea that if we censused the population on different days or under different conditions, we would obtain different counts simply because of sampling error.[17] In the case of the multinomial, the observation variance for the estimate of the number of individuals in category $y_{2,it}$ is $n_t p_{it}(1 - p_{it})$, where $n_t = \sum_{i=1}^{m} y_{2,it}$ and $p_{it} = z_{it}/\sum_{i=1}^{m} z_{it}$. So, the observation variance is implicit in the values of the parameters of the multinomial.

Recall that we calculate the total probability of the data conditional on a parameter (i.e., the total likelihood) from the product of the probabilities of individual observations as we are doing in multiplying across the i observations in individual likelihoods in equation 4.1.5. The products of these total likelihoods for the two data sets gives us the combined probability of the two data sets conditional on the values of latent state $\mathbf{z}$. We are not limited to two likelihoods; we might have many. Moreover, although we must assume independence here to keep things simple, we could use nonindependent data sets as long as we properly modeled the dependence among them, a topic that is beyond the scope of this book but that you could tackle after mastering the principles we present.

The final item to notice is that we are modeling the latent state $\mathbf{z}_t$ as a continuous random variable to represent an inherently discrete state—the number of individuals in each age and sex class. Isn't there something wrong here? Consider what we are portraying in the process model—

[17] Actually, it would be wise to replicate the census for each t so that we could explicitly estimate the observation variance. We are ignoring the possibility of bias arising from over- or undercounting to keep this example simple. See section 6.2.3 for an example where we account for bias in the counts.

the *median of the distribution* of the true population state—which we use to construct parameters for use in the Poisson and the multinomial likelihoods. The mean of the Poisson and the probability of membership in the multinomial are *continuous* and should be modeled with a continuous random variable. There is a prior on z_1 because it is an initial condition that we treat as a parameter in the process model.

6.3 When Are Observation and Process Variance Identifiable?

We have spoken frequently about partitioning uncertainty by separately modeling process variance and observation variance. Some conditions make this possible, and others do not. To explain this idea we introduce the statistical term *identifiability*. For inference on a model to be possible, the parameters in the model must be identifiable, which loosely means that it is possible to learn the true value of the model's parameter(s) conditional on an infinite number of observations. In practice, this means that different values of the parameter(s) must generate different probability distributions of the observable variables.

Consider a general hierarchical expression for the posterior and joint distributions of observations and parameters,

$$\left[\boldsymbol{\theta}, \mathbf{z}, \sigma_p^2, \sigma_o^2 | \mathbf{y}\right] \propto \prod_{i=1}^{n} \left[y_i | z_i, \sigma_o^2\right] \left[z_i | g\left(\boldsymbol{\theta}, \mathbf{x}_i\right), \sigma_p^2\right] \left[\boldsymbol{\theta}, \sigma_p^2, \sigma_o^2\right], \quad (6.3.1)$$

where $g(\boldsymbol{\theta}, \mathbf{x}_i)$ is a deterministic model of an ecological process, y_i is an observation of the process, σ_p^2 is the process variance, and σ_o^2 is the observation variance that, in this case, arises purely from sampling.

We can identify σ_p^2 and σ_o^2 if and only if one or more of the following conditions hold:

1. There are replicate observations for each of the unknown states; that is,

$$\left[\boldsymbol{\theta}, \mathbf{z}, \sigma_p^2, \sigma_o^2 | \mathbf{Y}\right] \propto \prod_{i=1}^{n} \prod_{j=1}^{J} \left[y_{ij} | z_i, \sigma_o^2\right] \left[z_i | g(\boldsymbol{\theta}, \mathbf{x}_i), \sigma_p^2\right] \left[\boldsymbol{\theta}, \sigma_p^2, \sigma_o^2\right], \quad (6.3.2)$$

where j indexes multiple observations for each i. Obtaining replicates, of course, requires thoughtful design, which is a great reason for writing the model as part of the design of any research. Examples 6.2.2, 6.2.3, and 6.2.4 all had replicates at the data level.

2. The model has strong and differing "structure" in the data and process models. Structure can mean very different distributions for the process and observation models, as in section 6.2.1, where the Poisson has a highly constrained variance. Differing structure can also mean strong and differing spatial or temporal relationships in the data or in the process, a somewhat advanced topic that is beyond the scope of this book. See Cressie and Wikle (2011) for a full treatment.
3. The model has strongly informative priors on parameters, particularly the variance components. These could come from previous studies, as described in example 6.2.4, again illustrating the value of informative priors (sec. 5.4).

A model in which the two variance components would probably not be identifiable is

$$\left[\boldsymbol{\theta}, \mathbf{z}, \sigma_p^2, \sigma_o^2 | \mathbf{y}\right] \propto \prod_{i=1}^{n} \underbrace{\text{normal}(y_i | z_i, \sigma_o^2)\ \text{normal}(z_i | g(\boldsymbol{\theta}, \mathbf{x}_i), \sigma_p^2)}_{\text{variances not identifiable}} \tag{6.3.3}$$

$$\times \text{inverse gamma}(\sigma_p^2 | .001, .001)\ \text{inverse gamma}(\sigma_o^2 | .001, .001)$$

$$\times [\boldsymbol{\theta}].$$

Note that there is no replication, and the distributions for the process and the data are the same. In this case, there is no information in the data about the source of uncertainty in y_i. It might arise from the ecological process or from the way it is observed.

II Implementation

We have seen how Bayesian models are diagrammed and written as posterior and joint distributions. This is the starting point for all analyses, because writing a model makes an unambiguous statement about the relationships between the unobserved quantities (unknowns) that we seek to understand and the observed quantities (knowns) that we use to gain that understanding. A properly written model provides the starting point for useful conversations between ecologists and statisticians. It is an essential component of papers and proposals that use Bayesian methods. We will expand our instruction on writing models in part III.

In chapter 7, we focus on how we gain insight from properly constructed Bayesian models, step C in the general modeling process sketched in the preface (fig. 0.0.1). We explain an algorithm that makes it possible to estimate parameters and latent quantities. In the next two chapters (8 and 9), we cover inferential procedures used to learn from a single model and from multiple models (fig. 0.0.1 D).

7 Markov Chain Monte Carlo

7.1 Overview

Our goal in this chapter is to explain how to implement Bayesian analyses using the Markov chain Monte Carlo (MCMC) algorithm, a wickedly clever set of methods for Bayesian analysis made popular by the seminal paper of Gelfand and Smith (1990). Many readers may never implement a Bayesian analysis in computer code but would like to understand how modern Bayesian modeling works. Other readers will never write code from scratch to implement MCMC but will instead rely on a flavor of the software (Lunn et al., 2000; Plummer, 2003) that performs MCMC provided that instructions are composed to mimic model statements like those written in previous chapters. Finally, some readers will choose to write their own MCMC algorithms from scratch in languages like R or C. No matter which of these groups describes you, it is vital to understand how the algorithm works to use it reliably and to interpret results that emerge from MCMC.[1]

We start our explanation of MCMC with a heuristic, high-level treatment of the algorithm, describing its operation in simple terms with a minimum of formalism. In this first part, we explain the algorithm so that all readers can gain an intuitive understanding of how to find the posterior distribution by sampling from it.[2] Next, we offer a somewhat more formal

[1]It is our view that you probably shouldn't use the software without programming at least a simple analysis start to finish, but our view is not universal.

[2]We admit this sounds like cheating. Bear with us.

treatment of how MCMC is implemented mathematically. Finally, we discuss implementation of Bayesian models via two routes—by using software and by writing your own algorithm.

7.2 How Does MCMC Work?

Recall from section 5.1 that dividing the joint distribution by the marginal distribution of the data normalizes the joint distribution so that it meets all the requirements of a probability mass function or probability density function (sec. 3.4.1). Recall that this normalization is what allows a Bayesian model to treat all unknowns as random variables arising from a probability distribution, and it is this treatment of unknowns as random variables that sets Bayesian analysis apart from analyses based on likelihood (sec. 5.2). It follows that the denominator of Bayes' theorem, the integral of the joint distribution taken over all the parameters, is a key element of inference based on Bayes' theorem.

However, integrating the joint distribution analytically is impossible except for the most simple problems. It is easy to see why—a model with three parameters requires a three-dimensional integral of a complicated function. This usually can't be done for three variables, let alone 30. What this means is that the sophisticated models like those illustrated in section 6.2 have always been theoretically sound, but until recently, there was no possible way to make use of the theory, because it was impossible to do the integration needed to find the posterior distributions of the unknowns.

MCMC allows us to find the marginal[3] posterior distributions of each of the unknowns while deftly avoiding any formal integration. The essential idea of MCMC is that we can learn about the unknowns by making many random draws from their marginal posterior distributions. The accumulation of these many samples forms the "shape" of the posterior distribution in the same way that taking many samples of data gives shape to a histogram. These many samples allow us to calculate quantities of interest—means, variances, and quantiles. Moreover, we can also easily obtain posterior distributions of any quantity that is *derived* from the unknowns we estimate—an enormous benefit of MCMC.

[3]Recall that marginal distributions give the distribution of a single random variable that is part of a joint distribution (sec. 3.4.2). This is what we desire—inference about each of the unobserved quantities in the joint posterior distribution. How we make that inference from MCMC output will be covered in chapter 8

You are entirely within your rights to wonder, how is it possible to draw samples from something that is unknown? To answer your question, recall that the posterior distribution is not *completely* unknown—it is informed by the known data acting through the likelihood and the known prior, that is, by the proportionality between the joint and the posterior distributions. Bayes' theorem (sec. 5.1) tells us that for data in the vector $\mathbf{y}$ and for any number of unknown parameters and latent quantities in the vector $\boldsymbol{\theta}$,

$$\underbrace{[\boldsymbol{\theta}|\mathbf{y}]}_{\text{posterior}} \propto \underbrace{[\mathbf{y}|\boldsymbol{\theta}]}_{\text{likelihood}} \underbrace{[\boldsymbol{\theta}]}_{\text{prior}} . \tag{7.2.1}$$

The problem that MCMC solves is finding a specific proportionality allowing the posterior distribution $[\boldsymbol{\theta}|\mathbf{y}]$ to be a proper probability density function that integrates to 1.[4]

It may help to think about the problem this way. Recall from the section on likelihood that the difference between the posterior distribution and the likelihood profile[5] is that the area under the likelihood profile is not equal to 1 because it has not been normalized (fig. 5.2.1), whereas the area under the posterior distribution integrates to 1 because it has been normalized. However, given the same likelihood and the same priors, the shapes of the curves are *identical.* We can think of MCMC heuristically as an algorithm that "normalizes the likelihood profile weighted by the prior" (sec. 4.4) using the joint distribution, which is *known* as the product of the likelihood and the prior. In so doing, MCMC chooses values in the simulated posterior distribution in proportion to their probability.

Before explaining how this normalization happens, it would be helpful to explain some terminology. The word "chain" in Markov chain Monte Carlo describes a sequence of values accumulated from random draws from the posterior distribution (fig. 7.2.1). Draws are accumulated in a vector $\boldsymbol{\theta}$, with elements $\theta^{(k)}$, $k = 1, \ldots, K$, where K is the total number of draws. Be sure you understand that the superscript (k) indexes iterations—it is not an exponent. So, in a chain of length K the first element has index $k = 1$, and the last element has $k = K$. The algorithm is called Markov because any element $\theta^{(k)}$ depends only on the value at the previous step in the chain, $\theta^{(k-1)}$, and is independent of all earlier values. The algorithm is called Monte Carlo because the elements of the chain are accumulated from K random draws from a distribution, not unlike numbers picked from

[4]Or, in the less common case of discrete-valued parameters, a probability mass function that sums to 1 over the support of the parameters.

[5]Assuming that priors are vague or that both the Bayesian and likelihood estimate use the same informative priors.

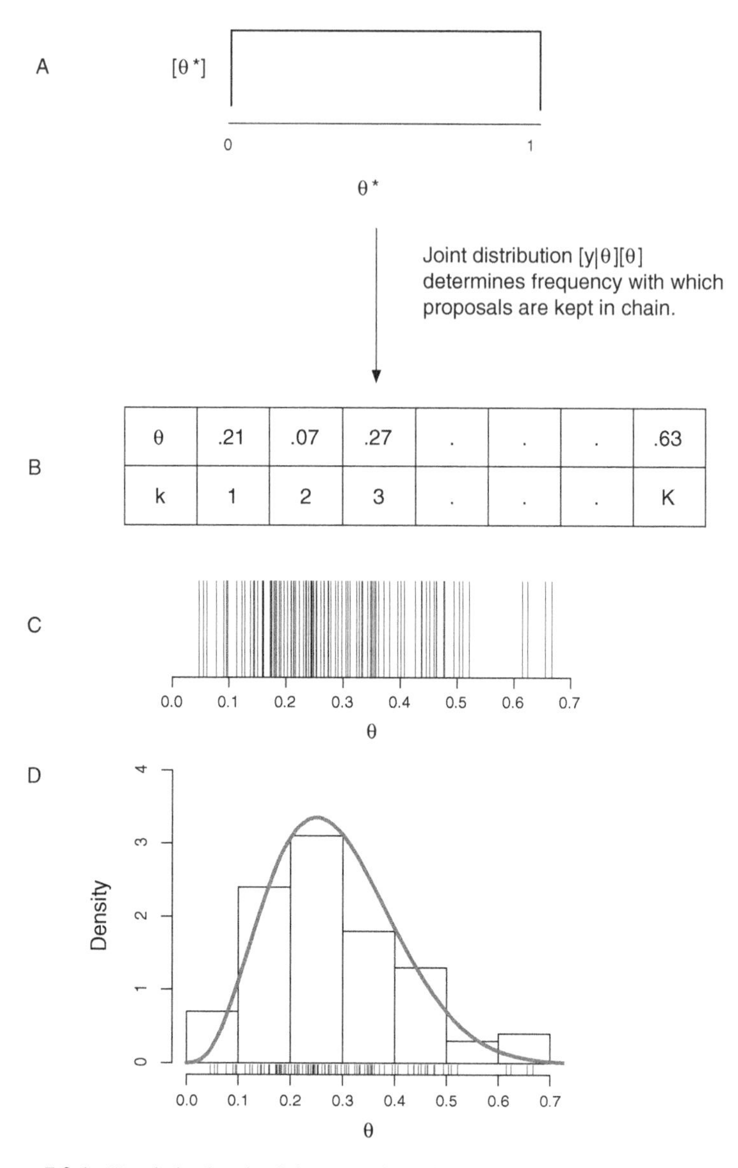

Figure 7.2.1. Heuristic sketch of the operation of the Markov chain Monte Carlo algorithm for a parameter θ with support continuous on 0–1. (**A**) Proposals are made from a distribution, in this example, a uniform on 0–1. (**B**) The MCMC algorithm produces a sequential vector (the chain) containing random draws from the posterior distribution indexed by $k = 1, \ldots, K$. The joint distribution $[\mathbf{y}|\theta]\,[\theta]$ arbitrates the frequency of values of θ that are kept in the vector. (**C**) Each vertical line is a single value of θ kept in the chain. Values of θ with a high likelihood as defined by the joint

a roulette wheel in a casino. For now, simply think of this distribution as a way to generate random numbers that are within the support of the parameter we are trying to estimate.

Here are the steps in the MCMC algorithm (fig. 7.2.1), presented heuristically and with some minor simplifications. We first focus on a single parameter and then expand our treatment to any number of parameters. We have a data set $\mathbf{y}$ relevant to the parameter θ we seek to estimate. We have chosen a prior distribution for θ, that is, $[\theta]$, a distribution with numeric arguments that might be informative based on past research or might be vague if we have no prior knowledge of θ. We choose a value for the starting point in our chain, the value of θ at $k = 1$, notated $\theta^{(1)}$. This choice can be arbitrary, but it is usually best to choose a reasonable value, based on what we know a priori. Next, we choose a proposed, new value of θ, $\theta^{(*)}$ (fig. 7.2.1 A), from a proposal distribution appropriate for the support for θ. Again, for now, you can think of that proposal distribution as being arbitrary. We will refine your thinking soon.

The essence of MCMC is this: We make many sequential choices from the proposal distribution. We use the joint distribution to choose values at each draw so that the frequency of the values of θ accumulated in the chain is proportional to their probability in the posterior distribution (fig. 7.2.1 B,C). This is to say that the joint distribution is used to decide the frequency of values that accumulate in the chain. A large number of values of θ proportional to their membership in $[\theta|\mathbf{y}]$ defines the posterior distribution in the same way that a large number of data points define a distribution in a histogram (fig. 7.2.1 D).

What about models with multiple parameters? How do we use MCMC to estimate a vector of m unobserved quantities? Again the idea is simple. Each of the m unknowns has its own chain (i.e., $\boldsymbol{\theta}_1, \boldsymbol{\theta}_2, \boldsymbol{\theta}_3, \ldots, \boldsymbol{\theta}_m$), and we subjectively assign an initial value to all chains. The MCMC algorithm cycles over each parameter (or latent quantity), treating it as if it were the *only* unknown just as in the preceding single-parameter case. So, at each

---

Figure 7.2.1 (*continued*).
distribution appear in the chain more frequently than values with a low likelihood. (**D**) The vertical hash lines above the x-axis correspond to panel C. Assigning values in the chain to "bins" determines the shape of the posterior distribution (solid line). We can normalize the frequency to approximate density in the same way we empirically normalize a histogram. With a sufficient number of iterations, the bars in the "histogram" can be very narrow. The heights of these narrow bins provide a smooth fit to the posterior distribution. Moments and other quantities of interest are calculated directly from the values in the chain.

step in the chain, we choose a value for each parameter *assuming the other parameters are known*. We then repeat this process for each parameter, effectively turning a multivariate problem into a series of univariate ones solved one at a time.[6]

Again, we have omitted a few details important for implementing the algorithm but not critical to understanding it. The key idea is that we use the joint distribution to arbitrate the frequency of values we keep in the chain. If we have more than one parameter, we have multiple chains, but we use the same process for each one, treating the other parameters as known, updating one parameter at a time using the joint distribution to choose its value. If we do this many times, then the relative frequency of the values of $\boldsymbol{\theta}$ accumulated in the chain is proportional to their probability in the posterior distribution of $[\boldsymbol{\theta}|\mathbf{y}]$.

7.3 Specifics of the MCMC Algorithm

We now add more detail to the heuristic developed in the previous sections. Our detailed description of MCMC is organized into three parts. We start by formally explaining what it means "to sample from the posterior distribution," an idea that, we admit, on first acquaintance might strike you as voodoo. After covering what it means to sample from the posterior distribution for a single parameter, we then show how to decompose a multivariate joint distribution into the univariate distributions needed for sampling multiple parameters. This procedure works for any model, with 2 or 200 parameters. Finally, we touch on some technical issues needed to evaluate the samples obtained from the MCMC algorithm.

7.3.1 Sampling from the Posterior Distribution of a Single Parameter

There are two flavors of sampling from the posterior distribution. In the first flavor, called *accept-reject sampling*, some proposals are kept and others are discarded. In the second flavor, called *Gibbs sampling*, *all* proposals are kept because the proposal distribution is especially well chosen. Understanding how we "keep all proposals" in a Gibbs sampler is best reserved for later, when we discuss multivariate problems

[6]Strictly speaking, these problems do not need to be univariate. We can break problems down into groups of related variables.

(sec. 7.3.4). For now, we consider algorithms for accept-reject sampling, which comprises three steps repeated many times:

1. Propose a new value $\theta^{(*)}$ for the parameter.
2. Compute a probability of accepting the proposed value using a ratio based on the joint distribution evaluated at the proposed value ($\theta^{(*)}$) and the previous value ($\theta^{(k-1)}$).
3. Accept the proposed value (i.e., $\theta^{(k)} = \theta^{(*)}$) with the probability computed in step 2. If you do not accept the proposed value, then retain the previous value (i.e., $\theta^{(k)} = \theta^{(k-1)}$).

How we implement these steps is described in the next three sections.

7.3.1.1 The Proposal Distribution

We said previously that implementing MCMC begins with a vector that is empty except for its first element, which we assign subjectively. We fill in the remaining elements of the vector by making proposals from a distribution, which we call, unsurprisingly, the *proposal distribution.* The proposal distribution is a key concept in MCMC.

In accept-reject sampling, there are two types of proposal distributions: independent and dependent. The value of the proposal drawn from an independent distribution does not depend in any way on the current value of the chain. Draws from a dependent proposal distribution depend on the current value. We emphasize dependent proposal distributions because these are the most widely used.

The idea of a dependent proposal distribution is simple. Assume that $\theta^{(k)}$ is the current value in the chain, and we want to generate a proposal for the next value in the chain, $\theta^{(k+1)}$. We will call the proposal $\theta^{(*)}$. We draw a sample from

$$\left[\theta^{(*)}|\theta^{(k)}\right], \tag{7.3.1}$$

which simply says we make a random draw from a distribution conditioned on $\theta^{(k)}$. For example, we might make a proposal drawn from a normal distribution (fig. 7.3.1 A) with mean $\theta^{(k)}$ and variance σ^2_{tune},

$$\theta^{(*)} \sim \text{normal}(\theta^{(k)}, \sigma^2_{\text{tune}}). \tag{7.3.2}$$

The variance σ^2_{tune} is a *tuning parameter*, chosen subjectively to set the width of the proposal distribution. The tuning parameter determines the

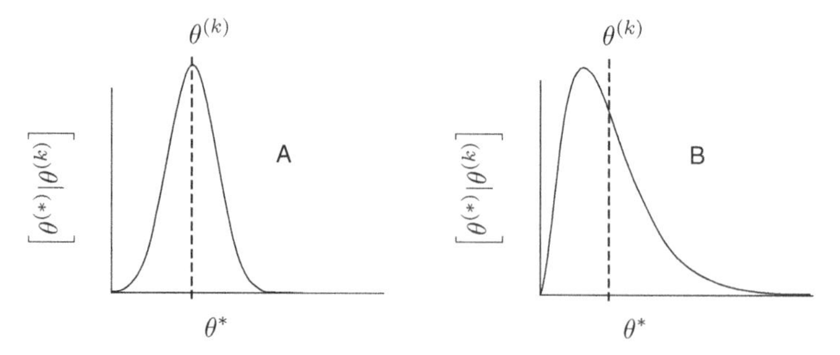

Figure 7.3.1. Example proposal distributions. In both panels, we draw a proposed new value for $\theta^{(*)}$ from a distribution centered on the current value in the chain, $\theta^{(k)}$, shown by the dashed vertical lines. In panel **A**, the proposal distribution is symmetric because $\left[\theta^{(*)}|\theta^{(k)}\right] = \left[\theta^{(k)}|\theta^{(*)}\right]$. In panel **B**, the proposal distribution is not symmetric, because $\left[\theta^{(*)}|\theta^{(k)}\right] \neq \left[\theta^{(k)}|\theta^{(*)}\right]$. It is important to understand that the "appearance of symmetry" does not make a distribution symmetric. Rather, it must have the property that the probability density (or probability for discrete parameters) of the proposal conditional on the current value is equal to the probability density of the current value conditional on the proposal.

frequency that proposals are accepted into the chain.[7] In this example, our proposal distribution is *symmetric*, a concept that is important to understanding the material in the next section. A symmetric proposal distribution has the property

$$\left[\theta^{(*)}|\theta^{(k)}\right] = \left[\theta^{(k)}|\theta^{(*)}\right], \tag{7.3.3}$$

which means that the probability density of the random variable $\theta^{(*)}$ conditional on $\theta^{(k)}$ is equal to the probability density of the random variable $\theta^{(k)}$ conditional on $\theta^{(*)}$.[8] The normal and the uniform distributions are symmetric. For example,[9] assume $\theta^{(k)} = 40$, $\theta^{(*)} = 50$, and the tuning

[7]When the value of σ^2_{tune} is large, proposals are rejected more frequently. Why this is true will become clear shortly.

[8]It is easy to get confused about notation here. Be sure you understand that $[\theta^{(*)}|\theta^{(k)}]$ defines the distribution of $\theta^{(*)}$ conditional on $\theta^{(k)}$. We can use that distribution as a source of random draws of $\theta^{(*)}$ or as a way to calculate the probability density of $\theta^{(*)}$. Both uses are described in this section.

[9]If you try to reproduce these results, be careful that the argument to the function you use is the variance, which is what we used here.

parameter $\sigma^2 = 100$. Then,

$$\left[\theta^{(*)}|\theta^{(k)}\right] = \text{normal}(50|40, 100) = 0.0242 \quad (7.3.4)$$

$$\left[\theta^{(k)}|\theta^{(*)}\right] = \text{normal}(40|50, 100) = 0.0242. \quad (7.3.5)$$

However, if the support of θ did not include all real numbers, but was, say, nonnegative, then a lognormal or gamma distribution would be a better choice for generating proposals:

$$\theta^{(*)} \sim \text{lognormal}(\log(\theta^{(k)}), \sigma^2) \quad (7.3.6)$$

or

$$\theta^{(*)} \sim \text{gamma}\left(\frac{\left(\theta^{(k)}\right)^2}{\sigma^2}, \frac{\theta^{(k)}}{\sigma^2}\right). \quad (7.3.7)$$

In both these cases, the proposal distributions are not symmetric, because

$$\left[\theta^{(*)}|\theta^{(k)}\right] \neq \left[\theta^{(k)}|\theta^{(*)}\right]. \quad (7.3.8)$$

7.3.1.2 Metropolis Updates

We wish to include values in the chain with a frequency proportional to their probability in the posterior distribution. To do this, we update the chain using Metropolis sampling, as follows. The probability density (or probability for discrete parameters) of the current value in the chain conditional on the data is

$$\left[\theta^{(k)}|\mathbf{y}\right] = \frac{[\mathbf{y}|\theta^{(k)}][\theta^{(k)}]}{[\mathbf{y}]}, \quad (7.3.9)$$

which, of course, you recognize as the posterior distribution of $\theta^{(k)}$. Likewise, the probability density of the proposed, new value in the chain conditional on the data is

$$\left[\theta^{(*)}|\mathbf{y}\right] = \frac{[\mathbf{y}|\theta^{(*)}][\theta^{(*)}]}{[\mathbf{y}]}, \quad (7.3.10)$$

that is, the posterior distribution of $\theta^{(*)}$. Solving these expressions is complicated by the need to find the integral $[\mathbf{y}] = \int[\mathbf{y}|\theta][\theta]d\theta$, which is often impossible. However, we need concern ourselves only with the probability of the proposed value of θ *relative* to the probability of the current value to allow us to choose which one is *more* probable. Metropolis sampling finds the more probable value using the ratio

$$R = \frac{[\mathbf{y}|\theta^{(*)}][\theta^{(*)}]/[\mathbf{y}]}{[\mathbf{y}|\theta^{(k)}][\theta^{(k)}]/[\mathbf{y}]} \tag{7.3.11}$$

$$= \frac{[\mathbf{y}|\theta^{(*)}][\theta^{(*)}]}{[\mathbf{y}|\theta^{(k)}][\theta^{(k)}]}. \tag{7.3.12}$$

That $[\mathbf{y}]$ cancels obviates the need for integration. In the Metropolis algorithm, we keep the proposed value whenever $R \geq 1$. When $R < 1$ we keep the proposed value with probability R and keep the current value with probability $1 - R$. In practice, this means we execute the following steps:

1. Draw $\theta^{(*)}$ from a *symmetric* proposal distribution centered on $\theta^{(k)}$.
2. Calculate R based on the joint distributions of the current value $\theta^{(k)}$ and the proposed value $\theta^{(*)}$.
3. Draw a random number u from a uniform distribution defined on $(0, 1)$.
4. If $R > u$, keep $\theta^{(*)}$ as the next value in the chain, that is, $\theta^{(k+1)} = \theta^{(*)}$. Otherwise, retain the current value, $\theta^{(k+1)} = \theta^{(k)}$.

You will also see this algorithm written as follows:

1. Calculate R as

$$R = \min\left(1, \frac{[\mathbf{y}|\theta^{(*)}][\theta^{(*)}]}{[\mathbf{y}|\theta^{(k)}][\theta^{(k)}]}\right). \tag{7.3.13}$$

2. Keep $\theta^{(*)}$ as the next value in the chain with probability R, and keep $\theta^{(k)}$ with probability $1 - R$.

A numerical example of how to calculate the Metropolis R is given in box 7.3. If we execute these steps many times, the chain *converges*, such that the frequency of a given value in the chain is proportional to its density in the posterior distribution. Convergence occurs when adding new proposals to the chain does not change the relative representation of values in the chain. The technical details of convergence are covered in section 7.3.4.

Box 7.3 Calculating the Metropolis Ratio

An example helps clarify the formalism of how to calculate R in the Metropolis and Metropolis-Hastings algorithms. Assume we have data on nitrogen mineralization rate (mg g^{-1} soil day^{-1}), $\mathbf{y} = (0.32,\ 0.32,\ 0.15,\ 0.12,\ 0.21)'$. We want to estimate the posterior distribution of the mean μ and variance σ^2. The posterior and joint distributions are

$$[\mu, \sigma^2|\mathbf{y}] \propto \prod_{i=1}^{5} \text{normal}(y_i|\mu, \sigma^2)\text{normal}(\mu|0, 100) \text{ inverse gamma}(\sigma^2|.001, .001). \tag{7.3.14}$$

We choose a normal likelihood because the observations are continuous and can be positive or negative. We focus attention on estimating μ to illustrate how we calculate the Metropolis ratio R and use it to choose a new value in the chain. We then give an example of calculating R for Metropolis-Hastings (sec. 7.3.1.3). The table gives the output for iterations $k = 5000$ to 5010 for an MCMC algorithm.

k	5000	5001	5002	5003	5004	5005	5006	5007	5008	5009	5010
μ	0.189	0.264	0.176	0.147	0.222	0.276	0.211	0.114	0.18	0.221	
σ^2	0.0161	0.00654	0.00904	0.0131	0.0106	0.0913	0.078	0.124	0.0057	0.0061	

We want to obtain a new value of μ for $k = 5010$. The full-conditional distribution for μ is

$$[\mu|\cdot] = \prod_{i=1}^{5} \text{normal}(y_i|\mu, \sigma^2)\text{normal}(\mu|0, 100). \tag{7.3.15}$$

We sample from this distribution using Metropolis, freely admitting that there are more efficient ways to do this, as described in section 7.3.2.3. We first calculate the total probability of the data multiplied by the prior using the most recent ($k = 5009$) estimate of μ and σ^2. We have

$$\prod_{i=1}^{5} \text{normal}\left(y_i|\mu^{(k=5009)}, \sigma^{2(k=5009)}\right) \text{normal}\left(\mu^{(k=5009)}|0, 100\right)$$

$$= \text{normal}(0.32|0.221,\ 0.0061)$$
$$\times \text{normal}(0.32|0.221,\ 0.0061)$$

(continued)

(Box 7.3 *continued*)

$$\begin{aligned}
&\times \text{normal}(0.15|0.221,\ 0.0061)\\
&\times \text{normal}(0.12|0.221,\ 0.0061)\\
&\times \text{normal}(0.21|0.221,\ 0.0061)\\
&\times \text{normal}(0.221|0,\ 100)\\
&= 7.89. \qquad (7.3.16)
\end{aligned}$$

If you are trying to reproduce this result, be careful about the arguments to the normal probability density function. We are using variance. Many functions in software, such as, `dnorm( )` in R (R Core team, 2013), use the standard deviation. We now propose a new $\mu^{(k=5010)}$ by making a draw centered on $\mu^{(k=5009)}$:

$$\mu^* \sim \text{normal}(0.221, \sigma^2_{\text{tune}} = 0.02) \qquad (7.3.17)$$

$$\mu^* = 0.171. \qquad (7.3.18)$$

Note that the mean of the proposal distribution is $\mu^{(k=5009)}$. The value of the tuning parameter σ^2_{tune} is chosen subjectively to improve algorithm efficiency, but it does not influence inference about the posterior. We now calculate

$$\begin{aligned}
\prod_{i=1}^{5} \text{normal}\left(y_i|\mu^*, \sigma^{2(k=5009)}\right) \text{normal}(\mu^*|0, 100) &= \text{normal}(0.32|0.171,\ 0.0061)\\
&\times \text{normal}(0.32|0.171,\ 0.0061)\\
&\times \text{normal}(0.15|0.171,\ 0.0061)\\
&\times \text{normal}(0.12|0.171,\ 0.0061)\\
&\times \text{normal}(0.21|0.171, 0.0061)\\
&\times \text{normal}(0.171|0,\ 100)\\
&= 2.51. \qquad (7.3.19)
\end{aligned}$$

The Metropolis ratio R is

$$R = \frac{2.51}{7.89} = 0.318. \qquad (7.3.20)$$

The proposal is not as probable as $\mu^{(k=5009)}$ because R is less than 1. Therefore, we keep the proposal (i.e., μ^*) with probability 0.318 and keep the current value

(*continued*)

(Box 7.3 *continued*)

(i.e., $\mu^{(k=5009)}$) with probability $1 - 0.318$. In practice, this means we draw a random number $u \sim \text{uniform}(0, 1)$. If u is less than or equal to 0.318, we keep $\mu^* = 0.171$ as the new element of the chain. If u is greater than 0.318, then we keep $\mu^{(k=5009)}$ as the new element. If R had been greater than 1, we would have kept the proposal.

What if we had used a nonsymmetric proposal distribution (as described in section 7.3.1.3)? For example, what if we used $\mu^* \sim \text{gamma}\left(\frac{0.221^2}{0.02}, \frac{0.221}{0.02}\right)$, obtaining $\mu^* = 0.191$. In this case, we would calculate R as

$$R = \frac{2.51}{7.89} \times \frac{[\mu^{(k=5009)}|\mu^*, \sigma^{2(k=5009)}]}{[\mu^*|\mu^{(k=5009)}, \sigma^{2(k=5009)}]} \tag{7.3.21}$$

$$= \frac{2.51}{7.89} \times \frac{\text{gamma}\left(0.221 \,\middle|\, \frac{0.191^2}{0.0061}, \frac{0.191}{0.0061}\right)}{\text{gamma}\left(0.191 | \frac{0.221^2}{0.0061}, \frac{0.221}{0.0061}\right)} \tag{7.3.22}$$

$$= \frac{2.51}{7.89} \times \frac{4.07}{5.39} = 0.0240. \tag{7.3.23}$$

If we were using a nonsymmetric proposal distribution because the random variable had nonnegative support, then it would probably make sense to rewrite likelihood and the priors as well, but we forgo that change to keep the illustration simple.

Our students are often puzzled by the idea that we retain values of θ that are actually less probable than the proposal. This will sometimes happen in step 4 where we keep the less probable value with probability $1 - R$. Why would we do this? Recall that we are trying to characterize the full posterior distribution, including its tails, which by definition includes improbable values. If we did not occasionally keep less probable values, then the chain could get "stuck" on a very probable value, and we would not be able to find the full shape of the posterior distribution.

The efficiency of the sampling is determined by the number of iterations required to approximate the posterior distribution. Efficient samplers require fewer iterations than inefficient ones. A rule of thumb for efficient sampling is that proposals should be accepted in about 40% of the iterations (Gelman et al., 1996). The analyst can change the acceptance rate by altering the tuning parameter σ^2_{tune} using trial and error.

7.3.1.3 Metropolis-Hastings Updates

Updating the chain using Metropolis sampling requires that proposals come from a symmetric distribution, for example, the normal or the uniform. This requirement is quite limiting. The uniform is usually a poor choice for a proposal distribution (despite our example in figure 7.2.1), because all values are equally likely. As a result, proposals do not exploit information about the most probable values of θ. The normal distribution is not appropriate for generating proposals for parameters and latent quantities with support that does not include all real numbers.

Metropolis-Hastings sampling relaxes the requirement for symmetric proposal distributions, allowing us to use any distribution to propose new values in the chain, by multiplying R by the ratio $[\theta^{(k)}|\theta^{(*)}]/[\theta^{(*)}|\theta^{(k)}]$, that is,

$$R = \frac{[\mathbf{y}|\theta^{(*)}][\theta^{(*)}]}{[\mathbf{y}|\theta^{(k)}][\theta^{(k)}]} \frac{\left[\theta^{(k)}|\theta^{(*)}\right]}{\left[\theta^{(*)}|\theta^{(k)}\right]}. \tag{7.3.24}$$

The only differences between Metropolis and Metropolis-Hastings are the way we calculate R and the kind of proposal distributions we are entitled to use. Everything that applies to the Metropolis algorithm with respect to convergence and efficiency also applies to Metropolis-Hastings.

7.3.2 Sampling from the Posterior Distributions of Multiple Parameters

So far, we have covered the use of MCMC for a single parameter, but we have reiterated that the power of the methods we teach becomes apparent when we must deal with complex problems—problems with multiple parameters and latent quantities and with sources of uncertainty that arise in different ways. We now describe how MCMC equips us to deal with these challenging problems. In essence, we take a complex multivariate problem and break it down it into a series of univariate ones that can be solved one at a time.

Our goal is to find the marginal posterior distributions of parameters and latent quantities that are components of multivariate joint distributions like those illustrated in section 6.2 on hierarchical models. Thus, we need a marginal posterior distribution for each of the unobserved quantities. MCMC enables us to find these distributions using four steps that apply to all problems:

1. Write a mathematical expression for the proportionality between the posterior and the joint distributions.

2. Using the expression in step 1 as a guide, write the *full-conditional distribution* for each of the unobserved quantities.
3. Decide on specific sampling procedures to make draws from the full-conditional distribution of each of the unknowns.
4. Iterate over all the full-conditional distributions in sequence, sampling a single parameter at each iteration, treating the other parameters as known and constant at that iteration.

Writing code for your own MCMC sampler requires all four steps. Alternatively, you may be delighted to know that for many models current MCMC software will accomplish steps 2–4 for you. All you need do is to translate the expression for the joint distribution (step 1) into proper code, which is usually pretty straightforward. We briefly illustrate how to do this in section 7.4. However, your ability to write proper code will be dramatically enhanced by understanding what is going on under the hood when software implements MCMC.

7.3.2.1 Composing Full-Conditional Distributions

Expressions for the posterior and joint distributions contain all the information required to build an MCMC sampler for any problem, because these expressions show how to write the full-conditional distribution for each parameter, effectively transforming a multivariate posterior distribution into a series of univariate posterior distributions. We define full-conditional distributions in the following way. Let $\boldsymbol{\theta}$ be a vector of length q containing all the unobserved quantities we seek to estimate. Let $\boldsymbol{\theta}_{-j}$ be a vector of length $q-1$ including all the elements of $\boldsymbol{\theta}$ except θ_j. The full-conditional distribution of θ_j is $[\theta_j|\boldsymbol{\theta}_{-j}, \mathbf{y}]$. It is the posterior distribution of θ_j conditional on the data and the other parameters. We are often able to simplify $[\theta_j|\boldsymbol{\theta}_{-j}, y]$ by ignoring some of the $\boldsymbol{\theta}_{-j}$ because they contain no information about θ_j.

The full-conditional distribution is most easily explained by example. Consider posterior and joint distributions

$$\left[\alpha, \beta, z, \sigma^2, \varsigma^2|y\right] \propto \underbrace{\left[y|z, \sigma^2\right]}_{\text{a}} \underbrace{\left[z|g\left(\alpha, \beta, x\right), \varsigma^2\right]}_{\text{b}} \underbrace{[\alpha]}_{\text{c}} \underbrace{[\beta]}_{\text{d}} \underbrace{\left[\varsigma^2\right]}_{\text{e}} \underbrace{\left[\sigma^2\right]}_{\text{f}}, \tag{7.3.25}$$

a Bayesian network for which is shown in figure 7.3.2 A. Labels a–f have been added to the distributions to facilitate the subsequent discussion. To add a touch of realism to this example, imagine the deterministic model $g(\alpha, \beta, x)$ that predicts the mean or median of a latent quantity of interest

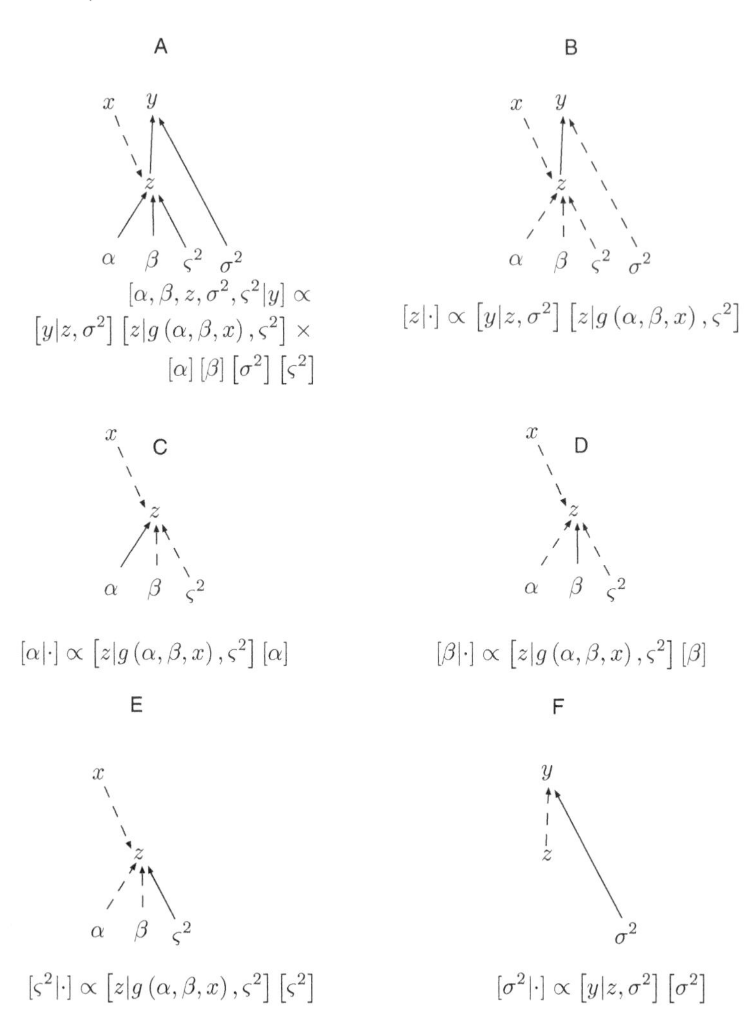

Figure 7.3.2. (**A**) Bayesian network for the posterior and joint distributions for the model given in the text (eq. 7.3.25). Full-conditional distributions are shown in panels **B–F**. The full-conditional distributions are identified by stochastic relationships shown by solid arrows. Quantities at the tails of dashed arrows are assumed to be known and hence are represented as deterministic relationships shown by dashed arrows. The full-conditional distribution and Bayesian network for the latent state z is shown in panel **B**, for α in panel **C**, for β in **D**, ς^2 in **E**, and for σ^2 in **F**. Note that these distributions (**B–F**) are univariate.

(z, e.g., population size, mineralization rate, species richness of a community) as a function of a covariate x with process uncertainty controlled by ς^2. In this example, we are assuming the x is known, as it might be, for example, when we apply a treatment. We have an observation y of the latent state that is made with uncertainty represented by σ^2. Note that we assume a single observation and covariate to keep things simple.[10] Using specific distributions for the parameters might require some moment matching, but for now, we are keeping the distributions generic.

What are the full-conditional distributions of the parameters and latent states in this model? The central idea is to identify the components of a univariate posterior distribution for each parameter from the distributions in which the parameter appears, treating the other parameters and latent states as if they were known constants. Another way to do this is to use the Bayesian network. The conditional posterior distribution for each parameter comprises the nodes and arrows that enter and leave that parameter. Other nodes and arrows are ignored (fig. 7.3.2).

We now illustrate these ideas, beginning with the latent state (z) (fig. 7.3.2 A). The full-conditional distribution for z includes distributions a and b from equation 7.3.25 because they are the only distributions where z appears. Thus,

$$[z|\cdot] \propto \underbrace{\left[y|z,\sigma^2\right]}_{\text{a}} \underbrace{\left[z|g\left(\alpha,\beta,x\right),\varsigma^2\right]}_{\text{b}}. \tag{7.3.26}$$

We are using the notation "·" as a stand-in for "other quantities treated as known," so that the left-hand side of equation 7.3.26 reads "the distribution of z conditional on the data and the other parameters." Be sure that you understand that we are treating all quantities *except* z on the right-hand side of equation 7.3.26 as known.[11] Why can we ignore distributions c–f in the full-conditional for z? Because they contain no information about z that is not contained in distributions a and b.

We compose the full-conditional for α as

$$[\alpha|\cdot] \propto \underbrace{\left[z|g\left(\alpha,\beta,x\right),\varsigma^2\right]}_{\text{b}} \underbrace{[\alpha]}_{\text{e}}. \tag{7.3.27}$$

It contains distributions b and e (equation 7.3.25), because these are the only distributions where α appears (fig. 7.3.2 B). Likewise, we write the

[10] To accommodate a full data set we would simply take products over the multiple observations and latent states in distributions a and b, assuming independence of these random variables.

[11] Which means we could have written equation 7.3.26 as $[z|y,x,\cdot] \propto [y|z,\cdot][z|\cdot]$, a form that is elegant but opaque. We chose to include the α, β, σ^2, and ς^2 for the sake of clarity.

full-conditional for β using distributions b and d:

$$[\beta|\cdot] \propto \underbrace{\left[z|g(\alpha,\beta,x),\varsigma^2\right]}_{\text{b}} \underbrace{[\beta]}_{\text{d}}. \tag{7.3.28}$$

Conditional posteriors for the remaining parameters, ς^2 and σ^2, are shown in figure 7.3.2 parts E and F.

7.3.2.2 Constructing an MCMC Algorithm from the Full-Conditionals

The full-conditional distributions are the basis for constructing an MCMC algorithm in the same way that the expression for the posterior and joint distributions is the starting point for writing the full-conditionals. We use the full-conditionals to specify how to sample the values for each of the unobserved quantities. Extending our example from section 7.3.2.1, we sample the values in the MCMC algorithm for each parameter in sequence from its full-conditional distribution:

$$\begin{aligned}
[z|\cdot] &\propto \underbrace{\left[y|z,\sigma^2\right]}_{\text{a}} \underbrace{\left[z|g(\alpha,\beta,x),\varsigma^2\right]}_{\text{b}}, \\
[\alpha|\cdot] &\propto \underbrace{\left[z|g(\alpha,\beta,x),\varsigma^2\right]}_{\text{b}} \underbrace{[\alpha]}_{\text{c}}, \\
[\beta|\cdot] &\propto \underbrace{\left[z|g(\alpha,\beta,x),\varsigma^2\right]}_{\text{b}} \underbrace{[\beta]}_{\text{d}}, \\
\left[\sigma^2|\cdot\right] &\propto \underbrace{\left[y|z,\sigma^2\right]}_{\text{a}} \underbrace{\left[\sigma^2\right]}_{\text{e}}, \\
\left[\varsigma^2|\cdot\right] &\propto \underbrace{\left[z|g(\alpha,\beta,x),\varsigma^2\right]}_{\text{b}} \underbrace{\left[\varsigma^2\right]}_{\text{f}}.
\end{aligned} \tag{7.3.29}$$

This might seem like an impenetrable thicket of notation, but the process of writing each line is really simple. We examine the expression for the posterior and joint distributions for each parameter and write the distributions that contain the parameter, omitting the others. It is perhaps important to remember that the priors in these expressions—that is, distributions c, d, e and f—must have numeric arguments that are not shown here, but recall that these arguments are known, as we will soon show (eq. 7.3.30).

Several features are particularly important to understand in equation set 7.3.29. Sampling in sequence matters, but the order of the sequence does not. This simply means that we can start with any of the parameters—reshuffling the order in which we sample them doesn't change the outcome. However, note that as we move from top to bottom in the sequence of full-conditionals, we use the updated version of the parameter on the right-hand side of the proportionality $\propto$, that is, the value at $k+1$ if the parameter was updated earlier in the sequence.

It is also important to understand what is meant by "sample from the conditional posterior distribution." The Metropolis-Hastings algorithm can be used to do this sampling for any parameter, as described in section 7.3.1.3. Metropolis-Hastings must be used for parameters in nonlinear deterministic models.[12] However, sometimes we can update parameters much more rapidly than is possible with any accept-reject algorithm. We will explain how to do this in sect. 7.3.2.3. For now, let's go through the mechanics of designing a sampler to fit the model using Metropolis-Hastings.

The first step is to choose specific probability distributions to replace the generic [ ] notation. This requires that we know some details about the data and the parameters. We assume the data are continuous and nonnegative, suggesting a lognormal or a gamma distribution for the likelihood. The data provide $j = 1, \ldots, J$ replicate observations y_{ij} for each of the $i = 1, \ldots, n$ states z_i. We assume our deterministic model is a Michaelis–Menten; that is, $g(\alpha, \beta, x) = \alpha x/(\beta + x)$ with x's measured without error. We know from experience that the parameter α must fall somewhere in the range 0–500 and that reasonable values for β must be somewhere between 0 and 200. The expression for the posterior and joint distributions is

$$
\begin{aligned}
\left[\alpha, \beta, \mathbf{z}, \sigma^2, \varsigma^2 | \mathbf{Y}\right] \propto \prod_{i=1}^{n}\prod_{j=1}^{J} & \underbrace{\text{lognormal}(y_{ij} \mid \log(z_i), \sigma^2)}_{\text{a}} \\
& \times \underbrace{\text{lognormal}(z_i \mid \log(g(\alpha, \beta, x_i)), \varsigma^2)}_{\text{b}} \\
& \times \underbrace{\text{uniform}(\alpha \mid 0, 500)}_{\text{c}} \underbrace{\text{uniform}(\beta \mid 0, 200)}_{\text{d}} \\
& \times \underbrace{\text{inverse gamma}(\sigma^2 | .001, .001)}_{\text{e}} \\
& \times \underbrace{\text{inverse gamma}(\varsigma^2 | .001, .001)}_{\text{f}}.
\end{aligned}
\qquad (7.3.30)
$$

[12]Assuming they cannot be transformed to a linear form.

Labels a–f show the correspondence between the specific distributions shown here and the general distribution notation in equation 7.3.25 as well as in equation set 7.3.29.

As before, we use the expression for the posterior and the joint distributions (eq. 7.3.30) to compose the full-conditionals and arrange them in a sampling algorithm. There are $n+4$ unobserved quantities: the n values of z_i and the four parameters. We must write a full-conditional for each one and sample from it. This is easier than it sounds, because the full-conditionals for the z_i are virtually identical; only the values of the data change among them. We start by setting up $n+4$ vectors to store the estimates, and we assign an initial value to each vector.

We choose a proposal distribution for each of the unobserved quantities. Lognormal distributions are reasonable proposal distributions for all the parameters, because all are continuous and nonnegative. So, for example, we might draw a proposal for z_1 as $z_1^{(*)} \sim \text{lognormal}(\log(z_1^{(k)}), \sigma_{\text{tune}}^2)$, where σ_{tune}^2 is the tuning parameter for z_i. Understand that the probability density of $z_1^{(*)}$ conditional on $z_1^{(k)}$ is given by $\text{lognormal}(z_1^{(*)} | \log(z_1^{(k)}), \sigma_{\text{tune}}^2)$, and the probability density of $z_1^{(k)}$ conditional on $z_1^{(*)}$ is given by $\text{lognormal}(z_1^{(k)} | \log(z_1^{(*)}), \sigma_{\text{tune}}^2)$.

We start with the values for the unobserved state z, although this choice is arbitrary. For each of the z_i we draw a proposal $z_i^{(*)}$ from $z_i^{(*)} \sim \text{lognormal}(\log(z_i^{(k)}), \sigma_{\text{tune}}^2)$ and compute

$$r = \frac{\left(\prod_{j=1}^{J} \underbrace{\text{lognormal}\left(y_{ij} | \log\left(z_i^{(*)}\right), \sigma^{2(k)}\right)}_{\text{a}} \times \underbrace{\text{lognormal}\left(z_i^{(*)} | \log\left(g\left(\alpha^{(k)}, \beta^{(k)}, x_i\right)\right), \varsigma^{2(k)}\right)}_{\text{b}}\right)}{\left(\prod_{j=1}^{J} \underbrace{\text{lognormal}\left(y_{ij} | \log\left(z_i^{(k)}\right), \sigma^{2(k)}\right)}_{\text{a}} \times \underbrace{\text{lognormal}\left(z_i^{(k)} | \log\left(g\left(\alpha^{(k)}, \beta^{(k)}, x_i\right)\right), \varsigma^{2(k)}\right)}_{\text{b}}\right)}, \tag{7.3.31}$$

$$R = \min\left(1, r \times \frac{\text{lognormal}\left(z_i^{(k)} | \log\left(z_i^{(*)}\right), \sigma_{\text{tune}}^2\right)}{\text{lognormal}\left(z_i^{(*)} | \log\left(z_i^{(k)}\right), \sigma_{\text{tune}}^2\right)}\right). \tag{7.3.32}$$

Be sure you see that we are using the ratio (lognormal($z_i^{(k)}|$ $\log(z_i^{(*)}),\ \sigma_{z_i}^2$))/(lognormal($z_i^{(*)}|\log(z_i^{(k)}),\ \sigma_{z_i}^2$)) in the final calculation of R because this multiplication is required for an asymmetric proposal distribution appropriate for the support of z, as described in section 7.3.1.3. We then assign $z_i^{(k+1)} = z^{(*)}$ with probability R and $z_i^{(k+1)} = z_i^{(k)}$ with probability $1 - R$. It is important to understand that we are using the current value in the chain for all quantities except the proposed new value $z^{(*)}$. What this means in practice, of course, is that the numerator and denominator in the expression for r evaluate to scalars, because they are just total likelihoods multiplied by a prior where all the arguments have numeric values.

Next, we update the value for α. We draw a proposal $\alpha^{(*)}$ and calculate

$$r = \frac{\underbrace{\prod_{i=1}^{n} \text{lognormal}\left(z_i^{(k+1)} | \log\left(g\left(\alpha^{(*)}, \beta^{(k)}, x_i\right)\right), \varsigma^{2(k)}\right)}_{\text{b}} \underbrace{\text{uniform}\left(\alpha^{(*)}|50, 500\right)}_{\text{c}}}{\underbrace{\prod_{i=1}^{n} \text{lognormal}\left(z_i^{(k+1)} | \log\left(g\left(\alpha^{(k)}, \beta^{(k)}, x_i\right)\right), \varsigma^{2(k)}\right)}_{\text{b}} \underbrace{\text{uniform}\left(\alpha^{(k)}|50, 500\right)}_{\text{c}}}, \tag{7.3.33}$$

$$R = \min\left(1, r\frac{\text{lognormal}\left(\alpha^{(k)}|\log\left(\alpha^{(*)}\right), \sigma_\alpha^2\right)}{\text{lognormal}\left(\alpha^{(*)}|\log\left(\alpha^{(k)}\right), \sigma_\alpha^2\right)}\right), \tag{7.3.34}$$

assigning $\alpha^{(k+1)} = \alpha^{(*)}$ with probability R and $\alpha^{(k+1)} = \alpha^{(k)}$ with probability $1 - R$. Note that we are using the updated values for the latent state $z_i^{(k+1)}$ in the calculation of r because the choice of a new value for the z_i preceded the calculation for α. Recall that σ_α^2 is the tuning parameter of the proposal distribution for α.

These illustrations should clarify what we mean by "sample from the full-conditional distributions." Extending the examples across all the parameters would be tedious and would preclude using them as a useful exercise. We urge you to write the sampling algorithm for β, σ^2, and ς^2.

You might reasonably guess that these several calculations—calculate R, generate a random number to choose between the current value and a proposed one, test for the choice—are computationally intensive. Things would proceed much more nimbly if we could simply make a proposal and keep it, forgoing all the computations needed to choose between the proposed value and the current one. We now describe a way to do that.

7.3.2.3 Gibbs Updates

Recall the concept of *conjugate distributions* (section 5.3, appendix table A.3). These are cases in which the posterior distribution $[\theta|y]$ has the same form as the prior, $[\theta]$, and the prior $[\theta]$ is called a conjugate of the likelihood $[y|\theta]$. We said that one of the reasons that conjugate distributions are important is that they can be used to dramatically accelerate MCMC sampling, as described next.

We can directly calculate the parameters of the posterior distribution of a parameter of interest whenever the prior for the parameter is a conjugate of the likelihood for the parameter. When this is the case, we can use a conjugate formula to calculate the parameters of the posterior (sec. 5.3, app. table A.3). Recall the set of full-conditionals we composed earlier (equation set 7.3.29). Forget any specifics about these distributions added using equation 7.3.30, so that we have generic distributions again (i.e., equation set 7.3.29), which we reproduce here for convenience:

$$\left[\alpha, \beta, z, \sigma^2, \varsigma^2|y\right] \propto \underbrace{\left[y|z, \sigma^2\right]}_{\text{a}} \underbrace{\left[z|g\left(\alpha, \beta, x\right), \varsigma^2\right]}_{\text{b}} \underbrace{[\alpha]}_{\text{c}} \underbrace{[\beta]}_{\text{d}} \underbrace{\left[\sigma^2\right]}_{\text{e}} \underbrace{\left[\varsigma^2\right]}_{\text{f}}. \tag{7.3.35}$$

Now, imagine that all the pairs of distributions in equation set 7.3.29 are conjugates—a is conjugate with b and f; b is conjugate with c, d, and e. Recall that when distributions are conjugate, then it is possible to calculate the parameters of the posterior *directly*, as we described in section 5.3. So, for example, if distributions a and b are conjugates, then it is possible to calculate the parameters of the posterior distribution of z conditional on the values of the other parameters and the data using a formula like those tabulated in appendix table A.3. We notate these directly calculated posterior parameters as γ_z, η_z for the latent quantity z; $\gamma_\alpha, \eta_\alpha$ for the parameter α; γ_β, η_β for the parameter β, and so on. The algorithm for updating each parameter proceeds as follows:

1. Calculate the parameters $(\gamma_z^{(k+1)}, \eta_z^{(k+1)})$ of the full-conditional distribution of $z^{(k+1)}$, using y, $z^{(k)}$, $g\left(\alpha^{(k)}, \beta^{(k)}, x\right)$, and $\varsigma^{(k)}$. Make a draw of $z^{(k+1)}$ from $\left[z^{(k+1)}|\gamma_z^{(k+1)}, \eta_z^{(k+1)}\right]$ and store it in the chain.[13]
2. Calculate the parameters $(\gamma_\alpha^{(k+1)}, \eta_\alpha^{(k+1)})$ of the full-conditional distribution of $\alpha^{(k+1)}$ using $z^{(k+1)}$, $g(\alpha^{(k)}, \beta^{(k)}, x)$, and $\varsigma^{(k)}$. Make a draw of $\alpha^{(k+1)}$ from $[\alpha^{(k+1)}|\gamma_\alpha^{(k+1)}, \eta_\alpha^{(k+1)}]$ and store it in the chain.

[13]Be sure you understand what "make a draw from" means. We have a distribution defined by the parameters $\gamma_z^{(k+1)}$, $\eta_z^{(k+1)}$. We choose a random value $z^{(k+1)}$ from that distribution. Values of $z^{(k+1)}$ that have a low probability or probability density given $\gamma_z^{(k+1)}$, $\eta_z^{(k+1)}$ will occur less frequently than those that have a high probability or probability density.

3. Calculate the parameters $(\gamma_\beta^{(k+1)}, \eta_\beta^{(k+1)})$ of the full-conditional distribution of $\beta^{(k+1)}$ using $z^{(k+1)}$, $g(\alpha^{(k+1)}, \beta^{(k)}, x)$, and $\varsigma^{(k)}$. Make a draw of $\beta^{(k+1)}$ from $[\beta^{(k+1)}|\gamma_\beta^{(k+1)}, \eta_\beta^{(k+1)}]$ and store it in the chain.

We leave steps 4 and 5 for parameters ς^2 and σ^2 as an exercise.

Each of the preceding steps is called a *Gibbs update*. Notice that there is no accept or reject; we are simply making very smart proposals and accepting all of them. The absence of the computational overhead associated with accept-reject samplers like those described previously[14] means that sampling proceeds much more rapidly.

We can illustrate Gibbs updates in a more concrete way by showing how we would use them to estimate the posterior distribution of the mean of a normally distributed random variable. We call this mean θ. Recall that draws of the random variable y_i from the normal distribution with mean θ arise as

$$y_i \sim \text{normal}(\theta, \varsigma^2). \tag{7.3.36}$$

Of course, we can think of y_i as an observation on some ecological process. For this example, we begin by assuming that the variance of the observations (ς^2) is *known*. It is important to understand that "knowing" ς^2 is not the same as calculating it as the variance of a sample data set. Rather, we are treating it here as a fully observed quantity, as if we had calculated it from *all* the potential observations. In the following discussion, it is particularly important to keep in mind that ς^2 is the variance of the distribution of the *observations* (y_i), not the variance of the distribution of the mean of the observations (θ). We have prior information about θ;

$$\theta \sim \text{normal}(\mu_0, \sigma_0^2). \tag{7.3.37}$$

This information might be informative or vague. Remember that μ_0 and σ_0^2 are numeric arguments; they are known. We have a data set $\mathbf{y}$ with n observations.

Given this information, we wish to estimate the full-conditional distribution of θ. If we assume that the variance in the likelihood (ς^2) is known, then,

$$[\theta|\cdot] \propto \prod_{i=1}^{n} \text{normal}(y_i|\theta, \varsigma^2)\text{normal}\left(\theta|\mu_0, \sigma_0^2\right). \tag{7.3.38}$$

[14]Actually, Gibbs updates can be viewed as a special case of Metropolis-Hastings where all proposals are accepted.

We define μ_1 and σ_1^2 as the parameters of the full-conditional distribution of θ; that is,

$$[\theta|\cdot] = \text{normal}(\mu_1, \sigma_1^2). \tag{7.3.39}$$

Note that σ_1^2 is the updated variance of the distribution of the *mean*, not the variance of the distribution of the *observations*, which of course is ς^2. Equation 7.3.38 shows that we have a normal likelihood for the mean with known variance and a normal prior on the mean, which are conjugates. When this is the case, we can calculate the parameters of the full-conditional distribution of θ directly using the formulas in (app. table A.3):

$$\mu_1 = \frac{\left(\frac{\mu_0}{\sigma_0^2} + \frac{\sum_{i=1}^n y_i}{\varsigma^2}\right)}{\left(\frac{1}{\sigma_0^2} + \frac{n}{\varsigma^2}\right)}, \tag{7.3.40}$$

$$\sigma_1^2 = \left(\frac{1}{\sigma_0^2} + \frac{n}{\varsigma^2}\right)^{-1}. \tag{7.3.41}$$

Notice that every quantity on the right-hand side of these equations is known; μ_0 and σ_0^2 are known as priors. The y_i are observations in hand, and we are assuming (for now) that ς^2 is known. Thus, we have all we need to know to make a draw from the distribution of θ using equation 7.3.39, because we know μ_1 and σ_1^2. So, why wouldn't we just use equations 7.3.40 and 7.3.41 to estimate the parameters of the posterior of θ and be done? Because we must assume that ς^2 is known, which is virtually never the case. Somehow we must learn about ς^2 to estimate θ.

So, what about ς^2? Again, the observations arise from $y_i \sim \text{normal}(\theta, \varsigma^2)$, and we seek to understand the full-conditional[15] distribution of ς^2. If we assume that θ is known, then

$$\left[\varsigma^2|\cdot\right] \propto \prod_{i=1}^{n} \text{normal}(y_i|\theta, \varsigma^2)\text{inverse gamma}(\varsigma^2|\alpha_0, \beta_0). \tag{7.3.42}$$

We define the parameters of the full-conditional distribution of ς^2 as α_1 and β_1, so that

$$\left[\varsigma^2|\cdot\right] = \text{inverse gamma}(\alpha_1, \beta_1). \tag{7.3.43}$$

[15]The distribution is conditional because we must know θ.

We have a normal likelihood with a known mean and unknown variance and an inverse gamma prior on the variance. When this is true we can calculate the parameters of the full-conditional distribution of ς^2 using appendix table A.3:

$$\alpha_1 = \alpha_0 + \frac{n}{2}, \tag{7.3.44}$$

$$\beta_1 = \beta_0 + \frac{\sum_{i=1}^{n}(y_i - \theta)^2}{2}. \tag{7.3.45}$$

Again, remember that β_0 and α_0 are known arguments to priors, so in practice they would be numeric. It follows that all quantities on the right-hand side of equations 7.3.44 and 7.3.45 are known.

It might seem that we have tied ourselves in a knot. We need to know ς^2 to estimate θ, and we need to know θ to estimate ς^2. This is just the kind of problem that MCMC can solve, because at each step in the chain we treat all the parameters save one as known. Equations 7.3.40–7.3.45 give us all we need to construct a very fast sampler for θ and ς^2. The algorithm is as follows:

1. Use the current value of $\varsigma^{2(k)}$ to calculate $\mu_1^{(k+1)}$ and $\sigma_1^{2(k+1)}$ from equations 7.3.40 and 7.3.41. Make a draw from $\theta^{(k+1)} \sim \text{normal}(\mu_1^{(k+1)}, \sigma_1^{2(k+1)})$ and store it in the chain.
2. Use the updated value of $\theta^{(k+1)}$ to calculate $\alpha_1^{(k+1)}$ and $\beta_1^{(k+1)}$ using equations 7.3.44 and 7.3.45. Make a draw from $\varsigma^{2(k+1)} \sim \text{inverse gamma}(\alpha_1^{(k+1)}, \beta_1^{(k+1)})$ and store it in the chain.
3. Repeat steps 1 and 2.

A sufficient number of repetitions usually converge on the posterior distributions of θ and ς^2 much more quickly than if we used Metropolis-Hastings. However, the estimates would be the same.

7.3.3 Hybrid Samplers

Most ecological models of interest will have several parameters and latent states. Some of the full-conditional distributions will be conjugate; others will not. Thus, the preceding algorithms will often include different updating algorithms for sampling from the posterior distribution. Where full-conditionals are conjugate, we use Gibbs; when they are not, we use Metropolis-Hastings.

7.3.4 Evaluating the Output of MCMC

It is important to understand that MCMC is an algorithm, not a mathematical expression. Unlike a properly composed mathematical equation, MCMC can produce answers that are dramatically wrong. However, we can safeguard against this outcome by evaluating the MCMC output to be sure it provides reliable inference about posterior distributions of parameters and latent quantities. Evaluating the output of MCMC is *not* the same as evaluating a model. We cover model evaluation in section 8.1. Our objective here is to provide a high level understanding of the properties of MCMC output that assure reliable inference, without going into excessive technical detail. We begin, as usual, with a general conceptual treatment and then turn to specific procedures that can be used to evaluate MCMC output.

7.3.4.1 Concepts

The most important characteristic of reliable MCMC output is *convergence* (fig. 7.3.3). A heuristic understanding of convergence depends on remembering what we are trying to accomplish: our objective is to draw samples from the posterior distribution of a parameter (or latent quantity) to accurately characterize that distribution—its moments and shape. We say the chain has converged when we have accumulated a sufficient number of samples to achieve that objective (fig. 7.3.3, lower panels). Adding more samples to a converged chain will not meaningfully change the density of the posterior distribution, and hence, will not change the approximations of its moments. We say these chains are *stationary*, because more iterations will not "move" the distribution—it will maintain its shape and location.

Figure 7.3.3. Illustration of convergence using a Gibbs sampler for the variance (ς^2) of a normal distribution (sec. 7.3.2.3) with three chains (shown in medium, light, and dark gray). Left panels are trace plots with the current value of ς^2, plotted against iteration number. Right panels are density plots, $[\varsigma^2|\mathbf{y}]$ plotted as a function of the values of ς^2 accumulated in the chains. We know the true distribution of ς^2 (shown in black, right panels) because we used it to simulate 10 data points from normal($\mu = 100$, $\varsigma^2 = 25$). Notice in the left panels how the chains "mix" as the number of iterations increases. Notice the convergence of the simulated distributions (medium, light, and dark gray) on the true distribution (black) as the number of iterations increases. The lower two panels show chains that have converged. The upper two panels show chains that have not converged.

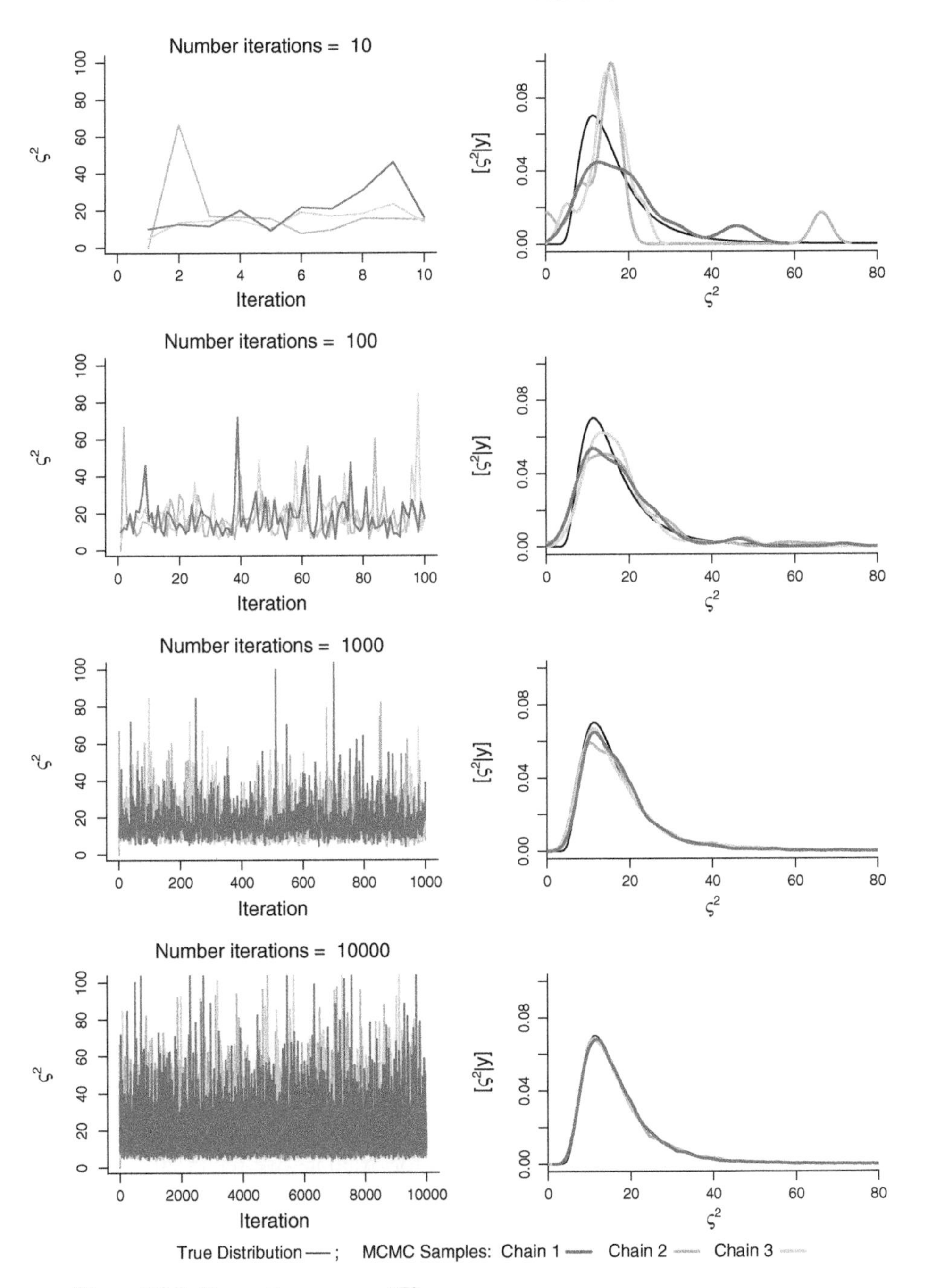

Figure 7.3.3. For caption see page 170.

A related concept is called *burn-in*. Recall that we may start a chain with an initial value that is entirely arbitrary.[16] As a result, the values retained in early iterations of the chain may have very low probability in the posterior we seek, despite being frequently represented in the chain (e.g., figure 7.3.3 top right panel, blue and red densities). The chain will converge on the posterior distribution given a proper model and MCMC algorithm, but we can decrease the number of iterations required by throwing away early values. *After* a sufficient number of iterations, the values retained in the chain occur in proportion to their probability in the posterior distribution. That "sufficient number of iterations" is called burn-in. It is necessary to base inferences on the values in the chain after burn-in. Usually, we implement burn-in informally by looking at trace plots and visually finding a break point where the chain appears stationary.

Two things can cause failure to converge. The cause most easily fixed occurs when a sufficient number of iterations have not been accumulated. However, a chain also can fail to converge because there is not enough information in the data to estimate a parameter or because a model is poorly specified. In these cases, we say that the parameter is not identifiable or perhaps only weakly identifiable. A particularly common cause of nonidentifiability occurs when parameters can trade off such that the different values of parameters—one high and the other low—produce the same likelihood as when their values are reversed. However, we are now venturing into model evaluation, which we cover in greater detail in the next chapter.

Two additional concepts are important, namely, *mixing* and *thinning*. A well-mixed chain implies that the posterior is being explored efficiently. A chain with better mixing can be shorter and still provide accurate estimates of posterior quantities (e.g., posterior mean, variance, or credible intervals) with fewer samples. Mixing is related to *autocorrelation*, or the degree of dependence of one sample on the previous sample. Chains with poor mixing are often highly autocorrelated. For well-specified models, longer chains minimize the influence of poor mixing on inference. Sometimes, mixing can be improved by adjusting the proposal distribution in Metropolis-Hastings, but in Gibbs sampling there's nothing to change other than the model. Poor mixing also can be an indicator of a poorly specified model (i.e., a model with nonidentifiable components).

Thinning refers to the number of samples drawn relative to the number that are stored. So, if we thin by five, for example, we retain 20% of the samples that are drawn. Because Markov chains are inherently correlated,

[16]In practice, it is a good idea to choose reasonable starting values, but this is not a requirement of MCMC if you are sufficiently patient. Chains will eventually converge as long as starting values are within the support of the random variable.

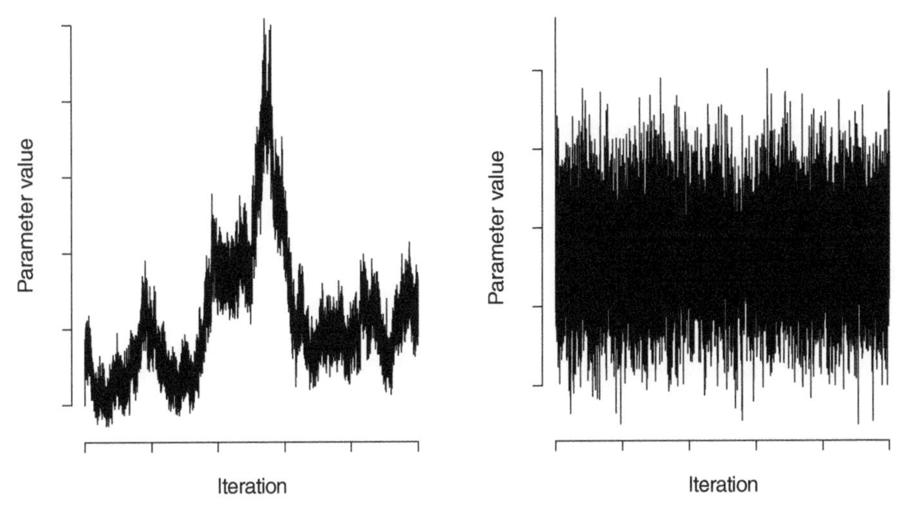

Figure 7.3.4. A trace plot of a chain that has not converged will "wander" among different values of the parameter, so that the band of values is not level and well mixed (left panel). Compare with the trace plot of the converged chain in the right panel.

the tradition has been to "decorrelate" them by thinning. However, inappropriate amounts of thinning may reduce the chain sample size so much in practice that posterior quantities are better approximated without thinning (Link and Eaton, 2012).

7.3.4.2 Methods for Assuring Convergence

Visual Inspection of Trace Plots. Trace plots show the values of an estimated parameter graphed against the number of iterations (fig. 7.3.3, left column). They are useful for obtaining a preliminary, visual assessment of convergence of one or more chains. Trace plots indicate convergence when values are well mixed, showing a level band without sinuosity (fig. 7.3.4).

Gelman-Rubin Diagnostics. Convergence is more formally evaluated by applying diagnostic tests to the MCMC output. The most popular of these was developed by Gelman and Rubin (1992). Their approach is easily conceptualized. When chains have not converged, the variance in parameter values among chains exceeds the variance within chains, as illustrated in the topmost panels of figure 7.3.3, but when chains have converged, the within-chain variance is the same as the variance among chains. The Gelman-Rubin diagnostic requires that we initialize more than one chain with values that are overdispersed relative to the mean of the posterior distribution.[17] Thus, we choose initial values that are well above the mean and well below it.[18] Assuming $j = 1, \ldots, m$ chains with overdispersed starting values and $k = 1, \ldots, K$ retained iterations in each of the m chains, the average within-chain variance for estimates of the parameter θ is

$$w = \frac{1}{m} \sum_{j=1}^{m} \text{var}(\theta_j), \tag{7.3.46}$$

where $\text{var}(\theta_j) = \frac{1}{K-1} \sum_{k=1}^{K} (\theta_{kj} - \bar{\theta}_j)^2$, and $\bar{\theta}_j$ is the mean of chain j. Thus, w is simply the mean of the variances of the j chains. The among-chains variance (b) is

$$b = \frac{K}{m-1} \sum_{j=1}^{m} \left(\bar{\theta}_j - \bar{\bar{\theta}}\right)^2, \tag{7.3.47}$$

where $\bar{\bar{\theta}} = \frac{1}{m} \sum_{j=1}^{m} \bar{\theta}_j$. This is the variance of the means of the chains multiplied by K, because each chain contains K samples. It follows logically that the variance of the stationary distribution of θ is a weighted mean:

$$\sigma_\theta^2 = \left(1 - \frac{1}{K}\right) w + \frac{1}{K} b. \tag{7.3.48}$$

We can now define the potential scale reduction factor,

$$\hat{r} = \sqrt{\frac{\sigma_\theta^2}{w}}. \tag{7.3.49}$$

If $\hat{r}$ is large, then the variance of the stationary distribution is large relative to the variance of the individual chains, and we must increase the number of samples in the chain. If we calculate $\hat{r}$ for each iteration of the chain, we can compute its 95% quantile of the distribution of $\hat{r}$. It is customary to iterate until this 95% quantile is less than 1.1 for all parameters being estimated.

[17] We usually begin implementing an MCMC algorithm with a single chain, because a single chain takes $1/n$ as much time to run as n chains do. Once things seem to be working properly based on visual assessments of convergence, we add chains, usually using three, to assure convergence has been obtained using formal diagnostics.

[18] So, how do you do choose these, if you haven't estimated the mean? The best way is to do a preliminary run with a single chain and assess convergence using trace plots. Then, run multiple chains with overdispersed starting values.

Modern statistical software contains functions that calculate scale reduction factors and their quantiles.[19]

Other Diagnostics. Table 7.1 lists diagnostics that can be useful in evaluating convergence of MCMC output. No single test is definitive, but each can provide insight. The Geweke and Heidelberg-Welch tests can sometimes reveal lack of convergence that is not obvious from visual inspection of trace plots—for example, when trace plots seem entirely stationary, but the moments of parameters could be estimated more precisely given more iterations. The Raftery-Lewis diagnostic is useful in planning the number of iterations that are needed for convergence. Different convergence diagnostics protect against different potential problems (table 7.1).

7.3.5 Verifying MCMC Algorithms with Simulated Data

Verification is a term used in ecosystem modeling to describe the process of checking that the model in computer code does what is expected. This usually means checking to make sure the differential equations representing flows or matter or energy among state variables have been faithfully translated into computer code used for numerical estimation of ecosystem dynamics.

There is an apt analogy in writing MCMC samplers—we need to assure that the algorithm properly recovers the values of "known" parameters and latent states, that is, that the computer code implements the mathematical model properly. This might seem like a formidable problem, because if we knew the values of parameters and latent states, there would be no need for MCMC in the first place. Of course, we will never know the true generating parameters of real data, but we can know their values for data that we *simulate.*[20]

A properly constructed expression for the joint and posterior distribution tells us all we need to know to simulate a data set (or data sets) that would arise from the stochastic and deterministic relationships the expression embodies. We simply view each distribution as a random number generator rather than as a probability mass or probability density function. So, for

[19]See the `gelman.diag()`in the `coda` package (Plummer et al., 2010) in R (R Core Team, 2013).

[20]Charmingly called "fake data" by Gelman and Hill (2009).

TABLE 7.1
Summary of Diagnostics Useful for Evaluating Convergence of Markov Chain Monte Carlo Chains.

Diagnostic	*Requirements*	*Notes*
Gelman-Rubin*	T chains with overdispersed initial values	Compares variance within and among chains. Function in R coda package is `gelman.diag()`.
Heidelberg-Welch†	One chain	Hypothesis test for stationarity using Cramer–von Mises statistic and precision of estimate of mean of posterior distribution using half-width test. Function in R coda package is `heidel.diag()`.
Geweke‡	One chain	Test for equality of the means of the first and last half of a Markov chain. If the samples are drawn from the stationary distribution of the chain, the two means are equal. Function in R coda package is `geweke.diag()`.
Raftery-Lewis§	One chain	Computes the number of iterations needed to estimate the quantile q within an accuracy of $\pm r$ with probability p. Useful for estimating number of iterations for a full implementation of MCMC based on a preliminary one. Function in R coda package is `raftery.diag()`.

*Gelman and Rubin (1992), †Heidelberger and Welch (1983), ‡Geweke (1992), §Raftery and Lewis (1995)

example, the general expression for a hierarchical model (eq. 6.2.1),

$$\left[\boldsymbol{\theta}_o, \boldsymbol{\theta}_p, \sigma_p^2, \sigma_d^2, \mathbf{z}|\mathbf{y}\right] \propto \prod_{i=1}^{n} \left[y_i|\boldsymbol{\theta}_o, z_i\right] \left[z_i|\boldsymbol{\theta}_p, x_i\right] \left[\boldsymbol{\theta}_p\right] [\boldsymbol{\theta}_o], \quad (7.3.50)$$

provides a recipe for simulating a data set $\mathbf{y}^{sim}$ as follows. We specify values for parameters for $\boldsymbol{\theta}_p$ and $\boldsymbol{\theta}_o$ pretending they are known. Recall that, these vectors of parameters include terms controlling stochasticity, and we must choose specific distributions to replace the general notation shown here. We make a draw of z_i from $[z_i|\boldsymbol{\theta}_p, x_i]$ followed by a draw of y_i^{sim}, from

$[y_i|\theta_o, z_i]$. We repeat this process for $i = 1, \ldots, n$. The x_i can be real or imagined data. This produces a vector of n elements ($\mathbf{y}^{sim}$). We can then use the simulated data to test if our MCMC algorithm can recover the parameters used to simulate the data.

It can be unsettling to find that the values of parameters recovered from a single data set are not exactly the same as the parameters you used to simulate a single data set. Does that mean you have a problem with your algorithm? Why would this discrepancy occur? Remember that $\mathbf{y}^{sim}$ is a *single* realization of a stochastic process, so you would not expect the parameters recovered from a single, random realization to be the same as the generating parameters, particularly if the number of data points is small or the stochastic elements of the parameter vectors are large. Thus, you will need to simulate multiple data sets and take the mean of the parameters estimated from each to be sure that your MCMC algorithm can properly recover the parameters used to generate the data. "Fake data simulation" as a way to check MCMC code is covered in detail by Kéry (2010) and Gelman and Hill (2009).

7.4 MCMC in Practice

7.4.1 Implementing MCMC Using BUGS Software

In 1989, the Medical Research Council Biostatistics Unit at Cambridge University developed software (BUGS, Bayesian Inference Using Gibbs Sampling) that implements a proper MCMC algorithm given an expression for the likelihoods and priors of a Bayesian model. There have been several elaborations on this original software (Plummer, 2002; Stan Development Team, 2014). These packages are now widely used by researchers all over the world. The procedure we recommend for implementing MCMC using your choice of software requires two steps:

1. Write a proper mathematical expression for the posterior distribution and joint distributions of your model (e.g., eq. 7.3.30). If you like, use a Bayesian network diagram as a guide.
2. Use the expression for the joint distribution to write BUGS code specifying each of its component distributions and the necessary product terms.

We show the relationship between the mathematical expression for the joint and posterior distributions and some generic BUGS code in box 7.4.

There are many good books that teach you how to write BUGS programs; we particularly like McCarthy (2007); Royle and Dorazio (2008);

Link and Barker (2010); Kéry (2010); Kéry and Schaub (2012). Although our focus is not coding, we nonetheless include a bit of code here to make a point. It is relatively easy to write proper BUGS code *if* you have a properly composed model. The converse is not true. We have seen many examples of models written in code that were not consistent with the correct expressions for the posterior distribution. Learn to write the math first. Discuss the math with your colleagues in statistics. Then, the coding should be straightforward and more likely to be solid. Moreover, you will have a compact expression representing your model that can be a compelling part of papers and proposals.

Box 7.4 The Relationship between Math and Code for a Hierarchical Model

We use equation 7.3.30 to illustrate the relationship between math and code. We assume our model is a Michaelis-Menten equation, that is, $g(\alpha, \beta, x_i) = \alpha x_i/(\beta + x_i)$, and the predictor variables are measured without error. There are n observations, each replicated j times We have

$$[\alpha, \beta, \mathbf{z}, \sigma^2, \varsigma | \mathbf{Y}] \propto \prod_{i=1}^{n}\prod_{j=1}^{J} \underbrace{\text{lognormal}(y_{ij} | \log(z_i), \sigma^2)}_{\text{a}}$$
$$\times \underbrace{\text{lognormal}(z_i | \log(g(\alpha, \beta, x_i)), \varsigma^2)}_{\text{b}} \underbrace{\text{uniform}(\alpha | 0, 500)}_{\text{c}}$$
$$\times \underbrace{\text{uniform}(\beta | 0, 200)}_{\text{d}} \underbrace{\text{inverse gamma}(\sigma^2 | .001, .001)}_{\text{e}}$$
$$\times \underbrace{\text{inverse gamma}(\varsigma^2 | .001, .001)}_{\text{f}}.$$

The code implementing this model is as follows.

```
model{
for(i in 1:n){# product expression over i, n must be
   given as data
    #definition of function g()
```

(*continued*)

(Box 7.4 *continued*)

```
    med[i] <- alpha*x[i]/(beta+x[i])
    #part b, process model
    z[i] ~ dlnorm(log(med[i]),prec.varsigma)
    for(j in 1:J){ #product expression over j,
      J must be given as data
        #part a, data model
        y[i,j] ~ dlnorm(log(z[i]),prec.sigma)
    } # end of j product
} #end of i product
#priors
alpha ~ dunif(0,500) #part c
beta ~ dunif(0,200) #part d
prec.sigma ~ dgamma(.001,.001) #part e
prec.varsigma ~ dgamma(.001,.001) #part f
} #end of model
```

There is an important difference between the code and the math in parts e and f, illustrating that sometimes we need to change distributions to accommodate the constraints of software. In most implementations of the BUGS language (Plummer, 2002; Lunn et al., 2000), the lognormal density function (`dlnorm()`) requires the precision ($1/\sigma^2$) as its second argument. The gamma distribution is the conjugate for the normal precision; the inverse gamma is the conjugate for the normal variance. Using conjugates is not required by MCMC software, but using them may accelerate convergence.

Note that the uniform priors on α and β imply some knowledge of plausible values for these parameters, even though we chose them to have minimum influence on their posterior distributions. It is always good to examine the sensitivity of posteriors to priors, as described in section 5.4.1.

7.4.2 Implementing MCMC Algorithms by Coding Your Own Sampler

Implementing your own MCMC sampler by writing code in a popular programming language like R or C is easier than you might imagine

(Clark, 2007; King and Gimenez, 2009). The general steps are as follows.

1. Write a proper mathematical expression for the posterior distribution and joint distributions of your model (e.g., eq. 7.3.30). If you like, use a Bayesian network diagram as a guide.
2. Using the expression written in step 1, write full-conditional distributions for each unobserved quantity (sec. 7.3.2.1).
3. Identify full-conditionals that can be computed analytically using conjugate relationships (sec. 7.3.2.3) and those that must be computed using accept-reject methods like Metropolis-Hastings (sec. 7.3.1.3).
4. Write functions to sample from the conditional posteriors.
5. Set up storage structures, usually lists or arrays, to store multiple chains for each parameter and latent quantity to be estimated.
6. Write a looping structure to sample from each full-conditional distribution, accumulating values as the loop iterates.

Again, the starting point for coding is a properly composed mathematical expression for the posterior and joint distributions.

8 Inference from a Single Model

In this chapter, we show how to make inferences using MCMC samples, the final step in the modeling process we outlined in the preface (fig. 0.0.1D). We begin our treatment of inference by assuming that we are analyzing a single model. We seek to estimate parameters, latent states, and derived quantities based on that model and the data. These estimates are conditioned on the single model we analyze. We will relax this requirement in the next chapter when we cover inference from multiple models.

8.1 Model Checking

We start this chapter with a truly crucial topic, checking our model. Model checking forms the gateway to inference whether we are using one model or many. We learned in section 7.3.4 that reliable inference from the MCMC algorithm requires that chains have converged. We learned tests for convergence. But we are not home free if our chains pass those tests. We must perform another level of checking before we proceed with inference.

The insights from Bayesian (or likelihood-based) analyses are reliable only if the requirements of statistical theory are satisfied. Every time we compose a likelihood to link models with data we must choose an appropriate deterministic model to represent our ideas about a process. We must also choose probability distributions to represent uncertainty. How do we know our choices are good ones?

Introductory statistics classes covering linear regression teach that residuals must be normally distributed and that this assumption should always be checked. The model underpinning a simple linear regression is

$$y_i = \beta_0 + \beta_1 x_i + \epsilon_i, \tag{8.1.1}$$

$$\epsilon_i \sim \text{normal}(0, \sigma^2). \tag{8.1.2}$$

The ϵ_i are the errors, and if they are not normally distributed, then we cannot rely on statistical theory to back us up when we make inferences using a regression model. We are flying blind, proceeding bravely on our own. Examining the residuals in classic linear regression is an example of model checking.

We learned earlier (chapters 2 and 3) that we are not shackled to classical regression but that we can use any deterministic model

$$\mu_i = g(\boldsymbol{\theta}, x_i) \tag{8.1.3}$$

and any probability distribution

$$\left[y_i | \mu_i, \sigma^2\right] \tag{8.1.4}$$

to learn about ecological processes. Combining the proper deterministic model with the proper probability distribution is a *far* more general approach than linear modeling assuming normally distributed errors and illustrates the great flexibility we have in composing models. We can represent ecological processes with deterministic models that can take any functional form, $g(\boldsymbol{\theta}, x_i)$, linear or nonlinear. Our model may contain any number of parameters $\boldsymbol{\theta}$. We can determine the probability of observations conditional on our model being true using many different statistical distributions to form the likelihood[1] $\left[y_i | \mu_i, \sigma^2\right]$.

This flexibility comes with an obligation. We are obliged to check whether our choices of deterministic models and likelihoods are reasonable. Of course, we should always choose probability distributions for our models by first considering the support of the random variable we wish to model. Violations of statistical theory are assured, for example, if we choose a normal distribution to represent discrete, nonnegative observations. Choosing the proper support is a good starting point—necessary but not sufficient. We must also answer a simple question before proceeding with inference: can our model (eqs. 8.1.3, 8.1.4) give rise to new observations that properly resemble the original data?

[1]Again, we are taking the liberty of treating μ_i as a parameter controlling the central tendency and σ^2 as a parameter controlling the dispersion of a generic distribution $\left[y_i | \mu_i, \sigma^2\right]$ realizing that we may need to use moment matching (sec. 3.4.4) to compose arguments for a specific distribution.

8.1.1 Posterior Predictive Distributions

Understanding model checking requires understanding how new data can be predicted from a Bayesian model. We introduce this topic next. Later (sec. 8.4), we elaborate on making predictions as a type of inference in its own right.

One of the most intellectually satisfying features of the Bayesian approach is that it treats all unobserved quantities in the same way. Until now, we have limited our treatment of unobserved quantities to model parameters and latent states. But there are many other types. Notable among them are predictions of future states and missing data as well as the topic we treat here, new data. New data are "observations" that could arise from the same system we are trying to model. They are predictions of what we would expect to observe conditional on the observed data.

Consider the following general expression of the posterior distribution of two unknowns, a vector of model parameters $\boldsymbol{\theta}$ and a prediction of a single observation, y^{new}:

$$\left[\boldsymbol{\theta}, y^{new}|\mathbf{y}\right] \propto \left[\mathbf{y}|\boldsymbol{\theta}, y^{new}\right]\left[\boldsymbol{\theta}, y^{new}\right], \tag{8.1.5}$$

an expression that should be familiar by now (if not, review chapter 5). The first term on the right-hand side, the likelihood, can be simplified, because $\mathbf{y}$ and y^{new} are conditionally independent in this case.[2] The second term (i.e., $[\boldsymbol{\theta}, y^{new}]$) can be factored to $\left[y^{new}|\boldsymbol{\theta}\right][\boldsymbol{\theta}]$ using the definition of conditional probability (sec. 3.2). It follows that

$$\left[\boldsymbol{\theta}, y^{new}|\mathbf{y}\right] \propto [\mathbf{y}|\boldsymbol{\theta}]\left[y^{new}|\boldsymbol{\theta}\right][\boldsymbol{\theta}]. \tag{8.1.6}$$

We can write 8.1.6 as

$$\left[\boldsymbol{\theta}, y^{new}|\mathbf{y}\right] \propto \left[y^{new}|\boldsymbol{\theta}\right][\boldsymbol{\theta}|\mathbf{y}], \tag{8.1.7}$$

because $[\mathbf{y}|\boldsymbol{\theta}][\boldsymbol{\theta}]$ is proportional to the posterior distribution $[\boldsymbol{\theta}|\mathbf{y}]$. If our sole interest is to predict y^{new}, omitting inference about $\boldsymbol{\theta}$, we can find its marginal posterior distribution by integrating over $\boldsymbol{\theta}$ ("integrating it out" of the left-hand side):

$$\left[y^{new}|\mathbf{y}\right] = \int \left[y^{new}|\boldsymbol{\theta}\right][\boldsymbol{\theta}|\mathbf{y}]\,d\boldsymbol{\theta}. \tag{8.1.8}$$

[2]This means that $[\mathbf{y}|\boldsymbol{\theta}] = \left[\mathbf{y}|\boldsymbol{\theta}, y^{new}\right]$, which is to say that given $\boldsymbol{\theta}$, the value of $\mathbf{y}$ does not depend on y^{new}.

This is the posterior predictive distribution of a new observation, y^{new}. It is important to note that this is a *distribution*, not a fixed quantity.[3] The distribution of y^{new} reflects the uncertainty pertaining to parameters in our model and the ability of the model to make accurate predictions of the data.

Those of us who lack formal statistical training may have difficulty understanding what this integral (eq. 8.1.8) means. Thinking about how we use MCMC to approximate it helps us gain intuition about it (box 8.1). In practice, we can use the value of the model parameters at each iteration $\left(\boldsymbol{\theta}^{(k)}\right)$ of the MCMC algorithm to calculate a new prediction conditional on the parameters. The parameters, in turn, are conditioned on the observed data ($\mathbf{y}$). When we integrate over $\boldsymbol{\theta}$, we are obtaining the probability density of y^{new} at each of the infinite posterior values of $\boldsymbol{\theta}$. Samples from MCMC approximate this distribution by computing $y^{new(k)}$ from $\boldsymbol{\theta}^{(k)}$ at each iteration of the algorithm. This process is known formally as *composition sampling* (Tanner, 1996).

Box 8.1 Marginal Posterior Distributions

There are several integrals throughout the text that give the analytical expressions for marginal posterior distributions and posterior predictive distributions. These may be challenging to interpret. It has helped our students to think about these by seeing a simple example of Monte Carlo integration, the method we use to approximate the moments of the marginal posterior distributions of parameters and states.

Assume we have the 15 observations shown in figure 8.1.1 A, and we fit the simple regression model

$$g(\boldsymbol{\beta}, x_i) = \beta_0 + \beta_1 x_i, \tag{8.1.9}$$

$$y_i \sim \text{normal}\left(g(\boldsymbol{\beta}, x_i), \sigma^2\right), \tag{8.1.10}$$

$$\left[\boldsymbol{\beta}, \sigma^2 | \mathbf{y}\right] \propto \prod_{i=1}^{15} \text{normal}\left(y_i | g(\boldsymbol{\beta}, x_i), \sigma^2\right) \text{normal}(\beta_0|0, 1000) \, \text{normal}(\beta_1|0, 1000) \times \text{inverse gamma}\left(\sigma^2 | .001, .001\right)$$

(continued)

[3]Don't be confused with the marginal probability of the data, $\int [\mathbf{y}|\theta][\theta]\, d\theta$, which is a distribution before the data are observed and a constant after the data are observed.

(Box 8.1 *continued*)

to those observations. We also obtain a derived quantity for the mean of the response when the predictor variable equals x_4, $\mu_4 = \beta_0 + \beta_1 x_4$, and we make a prediction of a new observation $y_4^{new} \sim \text{normal}\left(\mu_4, \sigma^2\right)$.

The following table illustrates converged MCMC output including the first and last five samples for parameters and derived quantities. There were a total of 10,000 iterations; k gives the iteration number. The first 5000 iterations were discarded for burn-in. The quantity μ_4 is the model prediction of the mean of the response when the predictor variable value is x_4. The quantity y_4^{new} is the prediction of a new observation for a predictor variable value equal to x_4.

k	5001	5002	5003	5004	5005	...	9996	9997	9998	9999	10000
β_0	9.84	10.9	10.6	10.7	11.9	...	12.5	9.84	11.1	10.9	11.7
β_1	1.38	1.22	1.25	1.12	1.14	...	1.04	1.23	1.27	1.06	1.14
σ^2	10.8	6.32	4.96	4.57	5.76	...	12.4	9.77	8.19	6.88	13.1
μ_4	13.3	14	13.8	13.6	14.7	...	15.1	12.9	14.3	13.6	14.6
y_4^{new}	9.29	14.3	9.44	15.3	18.5	...	11.4	16.8	12.8	14.5	15.8

The tabular output and the corresponding histograms in figure 8.1.1 show how we "integrate out" all but one parameter to obtain its marginal posterior distribution. For example, the marginal posterior distribution of β_0 is

$$[\beta_0|\mathbf{y}] = \int\int \left[\beta_0, \beta_1, \sigma^2|\mathbf{y}\right] d\beta_1 d\sigma^2. \tag{8.1.11}$$

The distribution represented by this integral is approximated using the elements of row 1 of the table. Assembling these values in a normalized histogram (fig. 8.1.1 B) shows the shape of the distribution in the same way that the heights of bars of an increasingly narrow width can be used to approximate any continuous function and its integral. (To see this, you might take a look at fig. 7.2.1 D). Note that we are not "ignoring" the other parameters, because their values in each column are part of the joint distribution that includes β_0.

The more iterations we obtain, the closer the approximation becomes, because we can divide the range of β_0 into increasingly narrower bars in the histogram. If you think about the definition of a definite integral, increasing the number of iterations is like making the "width of the bars" used to approximate a function infinitesimally small by increasing their number to infinity and their

(*continued*)

(Box 8.1 *continued*)

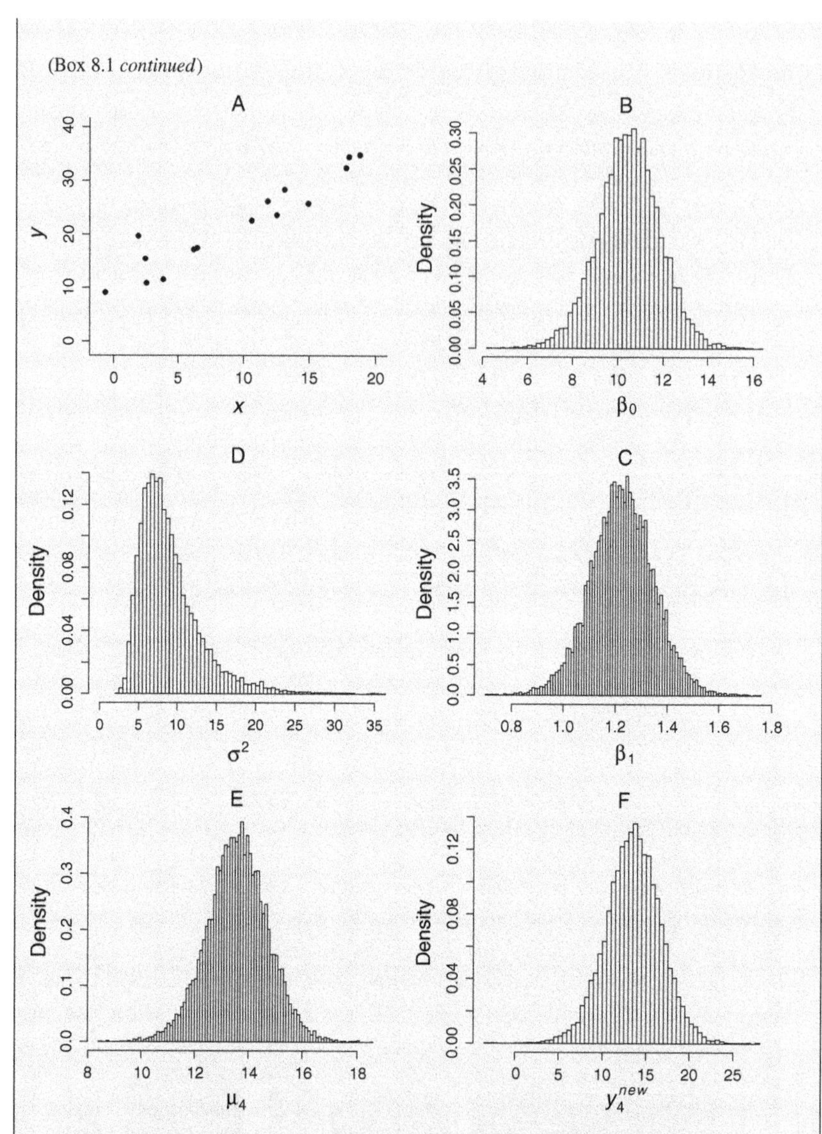

Figure 8.1.1. (**A**). Data used to fit the regression model (equation 8.1.9). (**B–F**) Normalized histograms plotted from MCMC output in individual rows of the preceding table. These histograms approximate the marginal posterior distribution of each of the parameters $(\beta_0, \beta_1, \sigma^2)$ and predictions (μ_4, y_4^{new}).

(continued)

(Box 8.1 *continued*)

width to zero. The integral from a to b for the function f is defined as $\lim_{\substack{\Delta x \to 0 \\ n \to \infty}} \sum_{i=1}^{n} f(x_i)\,\Delta x = \int_a^b f(x)dx$, where $f(x_i)$ gives the height of bars used to approximate the continuous function $f(x)$, and Δx gives their width. As the width of the bars goes to 0, and the number of bars goes to ∞, the approximation of the area under the curve $\sum_{i=1}^{n} f(x_i)\,\Delta x$ approaches the analytical value $\int_a^b f(x)dx$.

We can also use the row for β_0 to approximate moments of its marginal posterior distribution using Monte Carlo integration. For example, the mean is given analytically by the integral

$$\mathrm{E}(\beta_o|\mathbf{y}) = \int \beta_0\,[\beta_0|\mathbf{y}]\,d\beta_0, \tag{8.1.12}$$

which is approximated, simply enough, using

$$\mathrm{E}(\beta_0|\mathbf{y}) \approx \frac{1}{K}\sum_{k=1}^{K} \beta_0^{(k)}, \tag{8.1.13}$$

where K is the number of MCMC iterations after burn-in—in this case, 5000—and k indexes a single iteration. Similarly, the variance is $\text{var}(\beta_0|\mathbf{y}) \approx (\sum_{k=1}^{K}(\beta_0^{(k)} - \frac{1}{K}\sum_{k=1}^{K}\beta_0^{(k)})^2)/K$, the MCMC sample variance of the elements of the β_0 row. We can obtain other statistics of interest (e.g., medians, coefficients of variation, quantiles, highest posterior density intervals) by applying the appropriate function to the row. MCMC software makes most of these estimates automatically, but it is good to know how to do this because you may wish to write your own MCMC algorithm. You should know what is going on under the hood in MCMC software to use it properly and because you may want to apply nonstandard functions to the output from MCMC software.

The other rows in the table can be used to approximate the marginal posterior distributions (fig. 8.1.1 C–F) and their moments for other parameters and predictions. The quantity y_4^{new} may require additional explanation. In this case we have a *posterior predictive distribution*,

$$\left[y^{new}|\mathbf{y}\right] = \int \left[y^{new}|\boldsymbol{\theta}\right]\,[\boldsymbol{\theta}|\mathbf{y}]\,\mathrm{d}\boldsymbol{\theta}. \tag{8.1.14}$$

(*continued*)

(Box 8.1 *continued*)

where $\boldsymbol{\theta} = (\beta_0, \beta_1, \sigma^2)$. Thus, the full integral over all three parameters is

$$[y_4^{new}|\mathbf{y}] = \int\int\int [y_4^{new}|\beta_0, \beta_1, \sigma^2] [\beta_0, \beta_1, \sigma^2|\mathbf{y}] \, d\beta_0 d\beta_1 d\sigma^2, \tag{8.1.15}$$

which we admit looks frightening. But, again, what this means can easily be seen in the MCMC output and its histogram. We are "integrating out" the parameters to obtain the posterior predictive distribution of y_4^{new} by focusing solely on the elements in its row. This "integrating out" yields the univariate, marginal distribution of y_4^{new}, as shown in figure 8.1.1 F.

8.1.2 Bayesian p-Values

Recall the question that motivated treatment of the posterior predictive distribution: Can our model (eq. 8.1.4) give rise to new observations that properly resemble the original data?" Bayesian analysis implemented using MCMC provides a straightforward way to answer this question using a procedure called the *posterior predictive check* (Gelman et al., 2004). Recall the traditional definition of a p-value: the probability of observing a more extreme test statistic than the one calculated from the observed data. Posterior predictive checks use Bayesian p-values (P_B) to test whether the distribution of the data that would arise from our model is in some way more extreme than the distribution of the observed data. When this is the case, we made a bad choice of the deterministic model or the likelihood, motivating us to change our model.

P_B is calculated from the posterior predictive distribution of the new data and the distribution of the observed data. We first define a test statistic $T(\mathbf{y}, \boldsymbol{\theta})$ that describes the distribution of the data. $T(\mathbf{y}, \boldsymbol{\theta})$ is a summary of the data. It could be the mean, variance, the coefficient of variation, the kurtosis, the maximum, or the minimum of the observed data set, or it might be an "omnibus" statistic like a squared discrepancy or a chi-square value (Gelman et al., 2004; Kéry, 2010). Next, we estimate the probability that the test statistic calculated from "new" data arising from our model ($\mathbf{y}^{new}$) is more extreme than the test statistic calculated from the observed data ($\mathbf{y}$):

$$P_B = \Pr\left(T(\mathbf{y}^{new}, \boldsymbol{\theta}) \geq T(\mathbf{y}, \boldsymbol{\theta})|\mathbf{y}\right). \tag{8.1.16}$$

We condition on the actual data $\mathbf{y}$ in equation 8.1.16 because we desire a posterior probability of observing a more extreme statistic.[4]

To estimate P_B, we simulate a new data set $\left(\mathbf{y}^{new(k)}\right)$ at each iteration $(k = 1, \ldots, K)$ of the converged chain by sampling from the likelihood. Sampling from the likelihood means that we use it to make random draws of new "data" for each of the data points. For example, assume we have a deterministic model with parameters $\boldsymbol{\theta}$ and covariate x_i, $g(\boldsymbol{\theta}, x_i)$. At each iteration of the MCMC algorithm, the likelihood of an observation y_i conditional on our model and a term for variance[5] is

$$\left[y_i | g(\boldsymbol{\theta}^{(k)}, x_i), \sigma^{2(k)}\right]. \tag{8.1.18}$$

We can use the likelihood to make a random draw of a new observation[6] y_i^{new} from the distribution:

$$\left[y_i^{new} | g(\boldsymbol{\theta}^{(k)}, x_i), \sigma^{2(k)}\right]. \tag{8.1.19}$$

If there are n observations in the original data set, sequentially making these draws from the likelihood n times produces the new data set $\mathbf{y}^{new(k)}$. We create a new data set at *each* of these iterations based on the current value of $\boldsymbol{\theta}^{(k)}$. We then calculate the test statistic $T^{(k)}$ for the new data and the observed data and set an indicator variable $I^{(k)} = 1$ if $T(\mathbf{y}^{new(k)}, \boldsymbol{\theta}) \geq T(\mathbf{y}, \boldsymbol{\theta})$ and $I = 0$ otherwise. We use $I^{(k)}$ to calculate $P_B = \sum_{k=1}^{K} I^{(k)}/K$, the proportion of times that the test statistic based on the new data exceeds the test statistic based on the observed data.

For example, assume we are interested in checking the goodness of fit of the coefficient of variation, $T(\mathbf{y}, \boldsymbol{\theta}) = \text{CV}(\mathbf{y})$, that is, the mean of $\mathbf{y}$ divided by the standard deviation of $\mathbf{y}$. We compute the proportion of times that the coefficient of variation of $\mathbf{y}^{new}$ exceeds the coefficient of variation of $\mathbf{y}$

[4]To compute P_B we need to solve the seemingly formidable integral

$$P_B = \Pr\left(T(\mathbf{y}^{new}, \boldsymbol{\theta}) \geq T(\mathbf{y}, \boldsymbol{\theta}) | \mathbf{y}\right) = \int\int I_{\{T(\mathbf{y}^{new}, \boldsymbol{\theta}) \geq T(\mathbf{y}, \boldsymbol{\theta}) | \mathbf{y})\}} [\mathbf{y}^{new} | \boldsymbol{\theta}][\boldsymbol{\theta} | \mathbf{y}] d\mathbf{y}^{new} d\boldsymbol{\theta}, \tag{8.1.17}$$

where the I is an indicator variable that is equal to 1 when the condition in the subscript is true and zero otherwise. The simple numerical procedure we describe in the text approximates this integral.

[5]Often transformed by moment matching.

[6]It is critical to understand that equation 8.1.18 represents the probability or probability density of y_i. When we make a draw of y_i from that distribution, its frequency will be determined by the quantities $g(\boldsymbol{\theta}^{(k)}, x_i), \sigma^2$.

over all MCMC iterations.[7] This proportion is P_B. If P_B is very large (i.e., close to 1) or very small (i.e., close to 0), we conclude that our model fails to adequately represent the distribution of the data. Calculating P_B requires no more than a few lines of code in an MCMC algorithm (e.g., Gelman and Hill, 2009; Kéry, 2010).

When we observe an extreme P_B, say $< .10$ or $> .90$, we have evidence that our chosen model does not adequately represent the data. We must change the model so that there is no obvious lack of fit. We might choose a different distribution for the likelihood, we might alter our deterministic model by adjusting the number of parameters or changing its functional form. We would not proceed to make inferences until we could be confident that the distribution of data arising from our model did not differ from the distribution of the original data more than we would expect from chance alone.

8.2 Marginal Posterior Distributions

Many researchers use software that implements the MCMC algorithm (e.g., JAGS, Plummer, 2003; or WinBUGS, Lunn et al., 2000). These packages allow users to easily produce crisp density plots of posterior distributions of parameters and lovely tables of statistics summarizing those distributions—means, medians, standard deviations, quantiles, and so on. It is a good idea to know where these plots and tables come from. Moreover, it is useful to know how to summarize the converged chains that required sweaty labor to produce when programming an MCMC algorithm.

Assume we have MCMC output from a model that has been checked (as in sec. 8.1), and we wish to use it to make inferences. An especially useful property of MCMC is marginalization. Recall that the purpose of MCMC is to help us gain insight about the joint distribution of the unobserved quantities (parameters and latent states) conditional on the observed ones (the data). We usually want to make statements about these quantities one at a time rather than jointly. We discussed in section 3.4 that we can learn about a single random variable that is part of a joint distribution by marginalizing over all the random variables except the one we seek to understand.

The marginalization property of MCMC allows us to do this for each parameter and latent state in MCMC output, as illustrated in the box 8.1.

[7]This is easily accomplished in software using the `step( )` function. See Kéry (2010) for examples.

The table rows hold values for different parameters, and columns hold their values at each iteration in the MCMC chain. The marginalization property of MCMC allows us to examine the posterior distributions of each parameter and latent state by treating its "row" as a vector without paying any attention to the other rows. The frequency of values in that vector defines the posterior distribution of the parameter (box 8.1). The posterior distribution of a single parameter or latent state can be shown in scientific papers and proposals by simply plotting a normalized histogram of its values from the converged chain. Alternatively, we can use a kernel density estimator to fit a smooth curve.[8] This is what is going on under the hood when we output smooth curves of the posterior distribution using popular MCMC software.

How do we summarize the marginal posterior distribution statistically? The marginalization property means that we can calculate summary statistics directly from the chain for each parameter (i.e., from its "row" in box 8.1) to estimate the moments of the distribution. So, for example, we approximate the mean of θ, that is, its expected value $\mathrm{E}(\theta|\mathbf{y})$, using

$$\mathrm{E}(\theta|\mathbf{y}) \approx \frac{1}{K}\sum_{k=1}^{K}\theta^{(k)}, \tag{8.2.1}$$

where K is the total number of iterations in the converged chain and $\theta^{(k)}$ is the sample of θ at the kth iteration. This is called *Monte Carlo integration* (box 8.1). Recall from section 3.4.1 that equation 8.2.1 is an approximation of the integral $\mathrm{E}(\theta|\mathbf{y}) = \int \theta\,[\theta|\mathbf{y}]\,d\theta$. The posterior variance is similarly approximated by

$$\mathrm{var}(\theta|\mathbf{y}) \approx \frac{\sum_{k=1}^{K}\left(\theta^{(k)} - \mathrm{E}(\theta|\mathbf{y})\right)^2}{K}. \tag{8.2.2}$$

Again, we see this is an approximation of the second central moment,

$$\mathrm{E}\left((\theta - \mathrm{E}(\theta|\mathbf{y}))^2\right) = \int \left((\theta - \mathrm{E}(\theta|\mathbf{y}))^2\right)[\theta|\mathbf{y}]\,d\theta \tag{8.2.3}$$

[8]The advantage of a smooth curve is that the posterior can be overlaid on the prior, which is good practice. This overlay can be hard to see if a histogram is used for the MCMC output, particularly when the prior is diffuse. Functions for kernel density estimators are widely available in statistical software—for example, see the `density()` function in R.

(sec. 3.4.1). We can also calculate functions of the mean and variance. For example, the standard deviation of θ conditional on the data is

$$\mathrm{sd}(\theta|\mathbf{y}) \approx \sqrt{\mathrm{var}\,(\theta|\mathbf{y})}, \tag{8.2.4}$$

and the coefficient of variation is

$$\mathrm{cv}(\theta|\mathbf{y}) \approx \frac{\sqrt{\mathrm{var}\,(\theta|\mathbf{y})}}{\mathrm{E}\,(\theta|\mathbf{y})}. \tag{8.2.5}$$

Don't allow these formulas to make life complicated. What this means in practice is that you calculate the mean, variance, standard deviation, or coefficient of variation over all the θ in the MCMC output (i.e., the row for any parameter in box 8.1).

How do we show uncertainty associated with the point estimate of a parameter? We can define a $1-\alpha$ Bayesian *credible interval* on θ as the interval between L and U such that

$$\Pr(L < \theta < U) = \int_L^U [\theta|\mathbf{y}]\, d\theta = 1 - \alpha\,, \tag{8.2.6}$$

which says that the probability that the true value of the random variable θ falls between L and U is $1-\alpha$. Credible intervals can be calculated in different ways. We cover the two most widely used ones here.

An equal-tailed interval gives the points L and U such that $\Pr(\theta < L) = \alpha/2$, and $\Pr(\theta > U) = \alpha/2$ (fig. 8.2.1). This means that

$$\int_{-\infty}^{L} [\theta|\mathbf{y}]\, d\theta = \int_{U}^{+\infty} [\theta|\mathbf{y}]\, d\theta = \frac{\alpha}{2}. \tag{8.2.7}$$

Given a mathematical expression for $[\theta|\mathbf{y}]$, we find L and U analytically from its cumulative distribution function (sec. 3.4.1.4). We approximate L and U empirically from the quantiles of the converged MCMC samples: L is the $\alpha/2$ quantile, and U is the $1-(\alpha/2)$ quantile. They are not necessarily symmetric, which precludes use of a conventional interval expression like $\mu \pm w/2$, where w is the width of the interval.

Equal-tailed intervals are easily understood and widely used in the ecological literature. Some Bayesians prefer a different approach that does not require equal areas in tails. A highest posterior density interval (HPDI) for the parameter θ is defined as a subset H of all the possible values of θ,

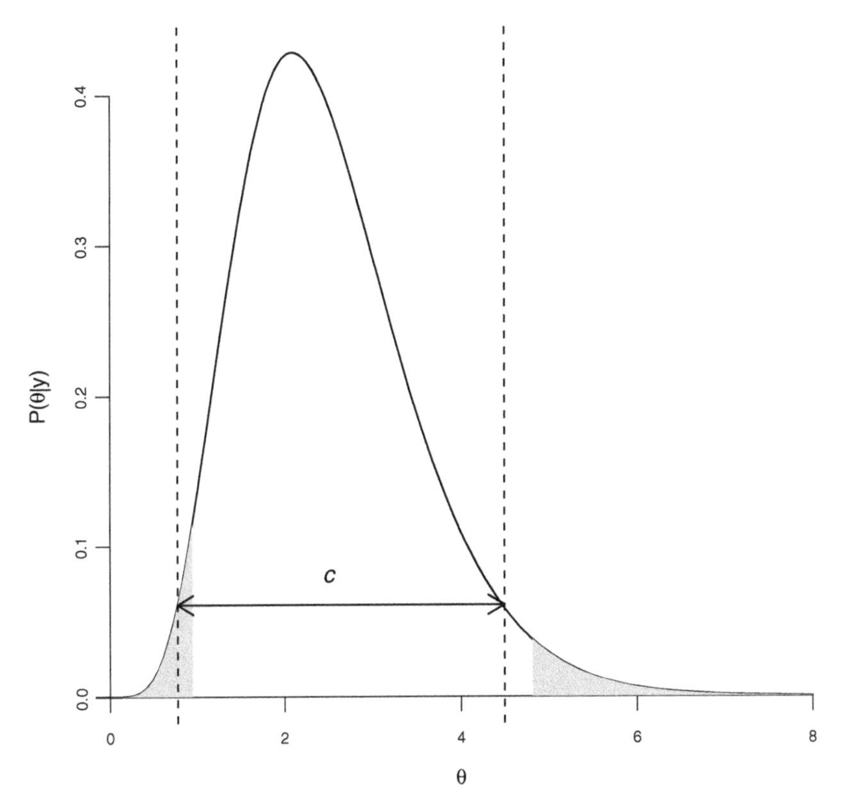

Figure 8.2.1. Illustration of equal-tailed intervals and highest posterior density intervals (HPDIs) for $\alpha = 0.05$. The distribution is gamma with mean = 2.5 and variance = 1. The equal-tailed intervals give the upper (0.975) and lower (0.025) quantiles of the distribution. The shaded areas define the values of θ outside the equal-tailed interval. The HPDI is defined by the c arrow. This is the longest horizontal line that can be placed within the distribution such that the area between the vertical dashed lines and beneath the distribution curve equals $1 - \alpha$. Note that the HPDI (0.74, 4.48) is substantially narrower than the equal-tailed interval (0.94,4.81), because the distribution is skewed. For symmetric distributions, the values would be the same.

such that

$$H = \{\theta : \Pr(\theta|\mathbf{y}) \geq c\}, \tag{8.2.8}$$

where c is the largest number, so that

$$\int_{\theta:[\theta|\mathbf{y}]\geq c} [\theta|\mathbf{y}]\, d\theta = 1 - \alpha. \tag{8.2.9}$$

This integral can be understood as follows. The quantity c represents a horizontal line intersecting the posterior distribution (fig. 8.2.1). The points where the line intersects the distribution curve are the lower and upper limits of the HDPI, L and U. We seek to find the *longest* line between L and U for which the area under the curve and between intersections of the line with the curve equals $1 - \alpha$. Starting at the peak of the distribution, we move the line downward until the area defined by its intersection with the curve equals $1 - \alpha$ (fig. 8.2.1). Although this sounds like a formidable task, functions are available[9] for estimating HPDIs from vectors extracted from MCMC chains, at least for univariate distributions.

There are three situations in which HPDIs are preferable to equal-tailed intervals. HPDIs are particularly useful for asymmetric posterior distributions, in which case an HPDI will be narrower than an equal-tailed interval (fig. 8.2.1). HPDIs can also include values in an interval that would be excluded by equal-tailed intervals. For example, it may make sense for an interval to contain 0 for a random variable for distributions with nonnegative support or support on (0, 1). In these cases Bayesian equal-tailed credible intervals can never include 0, but HPDIs can. Finally, HPDIs are called for when the posterior distribution is multimodal, that is, has more than one peak, although in this case the HPDI can be challenging to calculate.

8.3 Derived Quantities

A second, exceedingly useful property of MCMC is *equivariance*. The idea of equivariance is that any quantity that is calculated from a random variable becomes a random variable with its own probability distribution. We have learned how Bayesian analysis supports inference on parameters and latent states. However, we often seek inference on quantities *derived* from parameters and latent states. For example, we might estimate survivals and fertilities in a stage-structured matrix model and seek inference on the dominant eigenvalue of the projection matrix; we might estimate the proportional contribution of species in plots and seek inference using Shannon's diversity index; we might estimate the mean body mass of a species and want to make inference on its metabolic rate using a scaling equation. The equivariance property of MCMC allows us to easily obtain posterior distributions of these derived quantities—a difficult if not impossible task using classical statistics.

Continuing our example from equations 8.1.3 and 8.1.4, assume we calculate a derived quantity $\varphi^{(k)}$ at each of the K iterations of an MCMC

[9]For example, `HPDinterval()` in the `coda` package (Plummer et al., 2010) or the `HPDregion()` in the `emdbook` package (Bolker, 2013) in R (R Core Team, 2013).

algorithm using a deterministic function, $\varphi^{(k)} = h\left(\boldsymbol{\theta}^{(k)}, z^{(k)}\right)$. We can summarize the distribution of the $\varphi^{(k)}$ from MCMC output in the same way we summarize the distributions of each θ (sec. 8.2).

A particularly useful application of derived quantities is to examine the size of effects of treatments, locations, groups, and so on—an application analogous to the use of contrasts in an analysis of variance. Imagine that we have a model that estimates θ_j for $j = 1, \ldots, J$ levels of a treatment. We wish to know the magnitude of the difference between levels 1 and 2, that is, the posterior distribution of the difference $\delta_{1,2}$ attributable to the treatment. We can do this simply by calculating a derived quantity at each MCMC iteration,

$$\delta_{1,2}^{(k)} = \theta_1^{(k)} - \theta_2^{(k)}, \tag{8.3.1}$$

and using the converged chain for $\delta_{1,2}$ to make an inference. For example, we use the chain as a basis for statements like "The probability that treatment 1 exceeds treatment 2 is p," where p is the proportion of MCMC samples where $\delta_{1,2}$ is positive. This, of course, is our estimate of the area of the posterior distribution of the $\delta_{1,2}$ that is greater than 0. It is also a Bayesian p-value.

Though one of the most commonly desired quantities involves a sum or difference of parameters, it is important to note that we can use any mathematical function $h(\boldsymbol{\theta})$ as a derived quantity. For example, recall the matrix population model we used to illustrate the use of multiple sources of data (eq. 6.2.56). Often, these models are used to learn about the population growth rate λ and stable stage structure $\boldsymbol{\omega}$, which for a linear projection matrix, can be derived as the dominant eigenvalue and associated normalized eigenvector of the matrix. In the preceding example, we could compute the dominant eigenvalue and eigenvector of the matrix $\mathbf{A}$ at each MCMC iteration and use the vector of these accumulated computations across all the iterations as a basis for inference about growth rate and stable age structure. If we did this for more than one population, we could compute the difference in λ for each iteration and in so doing, make inferences about the probability of differences in growth rates among populations.

In all the preceding examples the posterior distribution of derived quantities can be calculated directly from the MCMC output. All that is required is that we calculate the derived quantity at each iteration based on the values of relevant parameters at that iteration. Alternatively, we can use the MCMC output for the relevant parameters as input for a function for a derived quantity of interest *after* the model has been fit. This allows us to use complex numeric functions that would be difficult to embed in MCMC software, for example, the eigenvalue of a large matrix.

The derived function can involve model parameters, latent processes, and/or data (i.e., $h(\mathbf{y}, \mathbf{z}, \boldsymbol{\theta})$). However, when obtaining the desired inference about this derived quantity using MCMC samples, it is important to follow the general rules of Monte Carlo integration; that is, when we approximate integrals using sums based on random samples from the distribution of interest, we must ensure that the samples for each quantity being combined in the function are "aligned" such that they occurred at the same step in the MCMC algorithm, as illustrated in box 8.1. This point is subtle but absolutely critical, because the samples arise from an MCMC algorithm in such a way that they are correlated with each other, and it is this correlation that provides the proper dependence in the joint posterior distribution.

For example, if we are interested in the posterior mean of the derived quantity $\theta_1^{y_i}/\theta_2$, where y_i is a certain observation in our data set, then we could approximate the necessary integral using

$$\mathrm{E}(\theta_1^{y_i}/\theta_2|\mathbf{y}) = \int\int(\theta_1^{y_i}/\theta_2)[\theta_1, \theta_2|\mathbf{y}]d\theta_1 d\theta_2 \tag{8.3.2}$$

$$\approx \frac{\sum_{k=1}^{K}(\theta_1^{(k)})^{y_i}/\theta_2^{(k)}}{K} \tag{8.3.3}$$

if $\theta_1^{(k)}$ and $\theta_2^{(k)}$ are samples arising from the same iteration of the MCMC algorithm, for $k = 1, \ldots, K$.

8.4 Predictions of Unobserved Quantities

A distinguishing feature of Bayesian inference is that unobserved quantities (e.g., parameters, latent state variables, and "future data") are treated as random variables in a Bayesian statistical model. Consequently, inference pertaining to them could more accurately be described as "prediction" rather than "estimation." This interpretation differs from that in non-Bayesian inference, where "unknown observables" (i.e., future data) are treated as random variables, but "unknown unobservables" (i.e., parameters) are often treated as fixed quantities. Latent state variables, if stochastic, are treated as random quantities in both Bayesian and non-Bayesian approaches. Therefore, inference pertaining to them could best be summarized as prediction, especially when inference is desired at different times or locations than when or where the data were collected.

We usually build models to make predictions. Predictions provide the foundation for evaluating the ability of our model to adequately represent the distribution of the data (sec. 8.1). We use predictions to evaluate competing models in the next chapter. But predictions are also one of the most useful applications of a model beyond their value in model checking. We often want to predict what we would *expect to observe* conditional on what we *have observed.* We covered the principles allowing us to make these predictions in section 8.1 when we described posterior predictive distributions. Here we elaborate on the utility of these distributions.

Model predictions are most often used by ecologists in two cases. In the first, most simple case, we have a deterministic model $g(\boldsymbol{\theta}, \boldsymbol{x})$ that we fit to data $\mathbf{y}$, $\left[y_i | g(\boldsymbol{\theta}, x_i), \sigma^2\right]$. We wish to know the posterior predictive distribution of an unobserved response variable $\tilde{y}_i$ based on a newly observed or postulated value of the predictor variable ($\tilde{x}_i$), that is, a prediction of a new observation $\tilde{y}_i$ conditional on the data in hand ($\mathbf{y}$).

A critical choice at this point is whether to seek the posterior predictive distribution of the *mean* of the new observation $\left(\left[\text{E}(\tilde{y}_i)|\mathbf{y}\right]\right)$ or the posterior predictive distribution of an *individual observation* $\left(\left[\tilde{y}_i|\mathbf{y}\right]\right)$. This concept might be familiar from training in classical linear regression, which teaches the difference between the confidence intervals on a single observation of y_i at a given x_i versus confidence intervals on the mean of y_i at a given x_i (fig. 8.4.1). To approximate the posterior predictive distribution of the mean of $\tilde{y}_i$ $\left(\int \left[\text{E}(\tilde{y}_i)|\boldsymbol{\theta}\right] [\boldsymbol{\theta}|\mathbf{y}]\, d\boldsymbol{\theta}\right)$ and its quantiles (fig. 8.4.1, dashed lines) we calculate a fixed quantity $\mu_i = g\left(\boldsymbol{\theta}^{(k)}, \tilde{x}_i\right)$ at each MCMC iteration. This makes sense, because our model predicts the mean (or perhaps some other central tendency) of the independent variable at each value of the dependent variable. We make inference on the mean of the $\tilde{y}_i$ by summarizing the samples of μ_i following the procedures outlined in section 8.2. Using MCMC software, this means simply including the output of the deterministic model, $\mu_i = g(\boldsymbol{\theta}, x_i)$, as a line in the code.

To obtain the posterior predictive distribution of an individual observation $\left(\int \left[\tilde{y}_i|\boldsymbol{\theta}\right] [\boldsymbol{\theta}|\mathbf{y}]\, d\boldsymbol{\theta}\right)$, we calculate $g\left(\boldsymbol{\theta}^{(k)}, \tilde{x}_i\right)$ at each MCMC iteration and then make a draw from the likelihood,

$$\left[\tilde{y}_i^{(k)} | g(\boldsymbol{\theta}^{(k)}, \tilde{x}_i), \sigma^{2(k)}\right],$$

which you will recognize as the same series of steps we used to simulate new data for posterior predictive checks (eq. 8.1.19), except that we are using a newly observed or postulated $\tilde{x}_i$ rather than an x_i from the original data set. Again, we assume conditional independence of the data given θ. The iterative composition sampling scheme effectively computes the necessary integral, and Monte Carlo predictive inference can be obtained using $\tilde{y}_i^{(k)}$ for the $k = 1, \ldots, K$ converged MCMC samples. For example, a posterior predictive mean for $\tilde{y}_i$ and quantiles (fig. 8.4.1, dotted lines) can

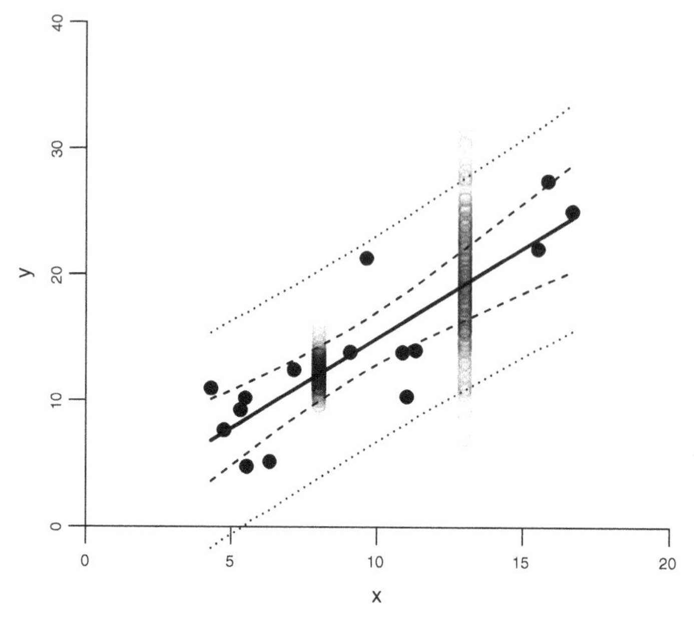

Figure 8.4.1. Example of predictions from a Bayesian regression model $g(\boldsymbol{\beta}, x_i) = \beta_0 + \beta_1 x_i$, $y_i \sim \text{normal}\big(g(\boldsymbol{\beta}, x_i), \sigma^2\big)$. The dotted lines give the 0.025 and 0.975 quantiles of the posterior distribution of the prediction of a new observation at a given x. The dashed lines give 0.025 and 0.975 quantiles for the posterior distribution of the prediction of the mean of the new observations at a given x. The left set of open circles represents 500 draws from the converged MCMC output approximating the posterior predictive distribution of the mean of y at $x = 8$. The right set of open circles represents 500 draws from the converged MCMC output approximating the posterior predictive distribution of the y^{new} at $x = 13$.

be approximated using Monte Carlo integration by

$$\mathrm{E}(\tilde{y}_i|\mathbf{y}) \approx \frac{\sum_{k=1}^{K} \tilde{y}_i^{(k)}}{K}, \tag{8.4.1}$$

as illustrated in box 8.1. This ability to use MCMC sample moments to approximate posterior moments is a key advantage of making inference (sec. 8.2).

The other case of prediction is frequently seen in models that portray how a state z_t changes over time. We discussed the procedure for obtaining predictions from a Bayesian perspective above and in section 8.1.1. However, the posterior predictive distribution we described is useful for learning about "unknown observables" (e.g., missing data). We are often interested in predicting the "process" rather than the data. The process is typically a latent component of a larger hierarchical model that we do not get to observe directly, or at least not completely. That is, we may be interested in gaining an understanding of the process at different times or locations than when or where we observed it. For example, consider a dynamic hierarchical model

$$\mu_t = g\left(\boldsymbol{\theta}, z_{t-1}\right) \tag{8.4.2}$$

$$\left[z_t|\mu_t, \sigma_p^2\right] \tag{8.4.3}$$

$$\left[y_t|z_t, \sigma_o^2\right], \tag{8.4.4}$$

where y_t is an observation of the true state z_t, and $\boldsymbol{\theta}$ is a vector of parameters in our deterministic model. The quantities σ_p^2 and σ_o^2 require some review: σ_p^2 is the process variance that accounts for the uncertainty caused by the failure of our model to perfectly represent the true state, and σ_o^2 is the observation variance that accounts for the uncertainty caused by the difference between the true state and our observation of the true state.[10] The model is hierarchical because z_t appears on both sides of the conditioning. The dynamics in this model are known as first-order "Markovian" because the process depends on itself in time only through its most recent value.

We want to know the probability distribution of the true state at some future point in time conditional on the observations of the state in the past. If the observed data extend from $t = 1, ..., T$, we seek to make inferences about $z_{t>T}$. These inferences are *forecasts* (sensu Clark, 2003b; Cressie and Wikle, 2011) because they are predictions of the future state accompanied by proper estimates of uncertainty. Predictive inference is commonly desired for the data (y_{T+1}) and/or latent process (z_{T+1}), where the subscript $T + 1$ indicates one time interval into the future, or at least a time extending beyond the period of data collection. In the case of the inference on the data (y_{T+1}), we seek the usual posterior predictive distribution, as previously described:

$$[y_{T+1}|y_1, \ldots, y_T] = \int \cdots \int [y_{T+1}|z_{T+1}, \sigma_o^2][z_1, \ldots, z_T, z_{T+1}, \theta_1, \ldots, \theta_m, \sigma_p^2|y_1, \ldots, y_T] \times d\theta_1, \ldots, d\theta_m, d\sigma_o^2 d\sigma_p^2 dz_1 \cdots dz_{T+1}, \tag{8.4.5}$$

[10]We emphasize that we use this formulation for clarity; z_t and μ_t control central tendency, and σ_p^2 and σ_o^2 control dispersion. Depending on the specific distribution we choose, these may need to be functionally matched to the appropriate parameters, as described in section 3.4.4, or by modeling the natural parameters in the distribution directly. We are not implying that these are normal distributions.

which looks baroquely complicated but in practice is quite simple to obtain using samples from MCMC.

Recall that this procedure is called *composition sampling* (Tanner, 1996). It works as follows. Inside the usual MCMC algorithm for fitting the model, simply sample $z_{T+1}^{(k)}$ from $[z_{T+1}|g(\boldsymbol{\theta}^{(k)}, z_T^{(k)}), \sigma_p^{2(k)}]$ given the current values for $\boldsymbol{\theta}^{(k)}$, $\sigma_p^{2(k)}$, $\sigma^{2(k)}$, and $z_T^{(k)}$, then sample $y_{T+1}^{(k)}$ from $[y_t|z_{T+1}^{(k)}, \sigma_o^{2(k)}]$. The iterative composition sampling scheme effectively computes the necessary integral, and Monte Carlo predictive inference can be obtained using $y_{T+1}^{(k)}$ for the $k = 1, \ldots, K$ converged MCMC samples. For example, a posterior predictive mean for y_{T+1} can be approximated using Monte Carlo integration by

$$E(y_{T+1}|y_1, \ldots, y_T) \approx \frac{\sum_{k=1}^{K} y_{T+1}^{(k)}}{K}. \tag{8.4.6}$$

This ability to use MCMC sample moments to approximate posterior moments is a key advantage to making inference.

The procedure just described provides a way to obtain predictive inference concerning the data, but how can we apply similar methods to learn about the process at future times? As it turns out, the predictive distribution for the process is even simpler, and if we have already obtained the MCMC samples for the posterior predictive distribution of the data, then we are done. Technically, we need to find the predictive process distribution

$$\begin{aligned}
&[z_{T+1}|y_1, \ldots, y_T] = \\
&\quad \int \int \cdots \int [z_{T+1}|z_T, \boldsymbol{\theta}, \sigma_p^2][z_1, \ldots, z_T, \theta_1, \ldots, \theta_m, \sigma_p^2|y_1, \ldots, y_T] \\
&\qquad \times d\theta_1, \ldots, d\theta_m, d\sigma_p^2 dz_1 \cdots dz_T,
\end{aligned} \tag{8.4.7}$$

which computationally is merely sampling $z_{T+1}^{(k)}$ from $[z_{T+1}|g(\boldsymbol{\theta}^{(k)}, z_T^{(k)}), \sigma_p^{2(k)}]$ given the current values for $\boldsymbol{\theta}^{(k)}$ and $z_T^{(k)}$. We can then use the MCMC samples $z_{T+1}^{(k)}$ to approximate posterior moments, as before. So, if we have already obtained the posterior predictive samples for the future data, then we have the posterior predictive samples for the future process and we can obtain our desired inference using them directly (sec. 8.2). For example, the mean of the predictive process distribution could be approximated using Monte Carlo integration by

$$E(z_{T+1}|y_1, \ldots, y_T) \approx \frac{\sum_{k=1}^{K} z_{T+1}^{(k)}}{K}. \tag{8.4.8}$$

What about the value of future states q intervals of time beyond the time of the last observation? We simply extend $T+1$ to $T+2, \ldots, T+q$ and iteratively sample from $[z_{T+q}|g(\boldsymbol{\theta}^{(k)}, z_{T+q-1}^{(k)}), \sigma_p^{2(k)}]$. Using MCMC software, this simply means extending the number of sequential outputs of the process model from T to $T+q$.

8.5 Return to the Wildebeest

The wildebeest example from chapter 1 offers a useful example of inference from a single model, so we take it up again here. We have three purposes: to illustrate choices on specific distributions needed to implement the model, to show how informative priors can be useful, and to illustrate some of the inferential procedures we have described in this chapter—posterior predictive checks, marginal posterior distributions, estimates of derived quantities, and forecasting.

8.5.1 Posterior and Joint Distribution

It would be helpful to review the wildebeest model where we left off (fig. 1.2.1). We have described the relationships between the observed and the unobserved quantities for a single data point, but we have not yet chosen specific distributions to represent stochasticity or developed a model for the full data set. An expression for the posterior and joint distributions, modified slightly from figure 1.2.1 to accommodate the data is

$$\left[\mathbf{N}, \boldsymbol{\beta}, \sigma_p^2|\mathbf{y}\right] \propto \prod_{t\in\mathcal{M}} \text{normal}(y_t|N_t, \sigma_{\text{data},t}^2) \tag{8.5.1}$$

$$\times \prod_{t=2}^{53} \text{lognormal}(N_t|\log(g(\boldsymbol{\beta}, N_{t-1}, x_t)), \sigma_p^2) \tag{8.5.2}$$

$$\times \text{normal}(\beta_0|0.234, 0.0185) \prod_{i=1}^{3} \text{normal}(\beta_i|0, 100) \tag{8.5.3}$$

$$\times \text{inverse gamma}(\sigma_p^2|0.01, 0.01) \tag{8.5.4}$$

$$\times \text{normal}(N_1|y_1, \sigma_{\text{data},1}^2). \tag{8.5.5}$$

There is much to explain here, starting with the data model (eq. 8.5.1). We did not have access to the raw density data on individual photographs but instead had a population mean of the total number of animals y_t for the entire Serengeti ecosystem for year t calculated from densities on individual photographs, and a variance for the total population based on the sample of photographs $(\sigma^2_{\text{data},t})$. It is important to understand that we are using $\sigma^2_{\text{data},t}$ as if it were known. We chose a normal distribution because y_t is based on the mean of animals on hundreds of photographs during year t. Sums of things are normally distributed when the number of things is large. We could have used a gamma or a lognormal distribution just as well, but the normal has some advantages of conjugacy that we will describe shortly. Notice the expressions $t \in \mathcal{M}$; the set $\mathcal{M}$ contains all the years for which data are not missing. There are 19 years of data beginning in 1961 and ending in 2008, which means there are 29 years lacking observations. We use the indexing here to match the predictions of the process model, which are made *every* year, with the years for which we have data.

The process model (eq. 8.5.2) uses a lognormal distribution for the latent state (N_t), a somewhat customary choice for dynamic ecological models, as explained in section 3.4.3.2. Note that σ^2_p is on the log scale. Now, look at the product symbol that ranges from 2 to 53. What is going on here? Why 2? Why 53? The lower bound on the product is necessary because the deterministic model depends on N_{t-1}, so if we started at 1, it would not be defined. We estimate the initial condition (i.e., N_1) using an informative prior based on the first year's data (eq. 8.5.5) so that it is treated as a parameter when $t = 2$ in the product. We chose the upper index (53) because we wanted 48 instances of N_t spanning the years 1961–2008, and five years of forecasts (2009–2013) extending five years after the last data were collected. Notice that we treat missing data (i.e., the years not contained in $\mathcal{M}$) and forecasts identically, emphasizing a lovely feature of Bayesian modeling: all unobserved quantities are treated in exactly the same way. Predictions of true states in the past on occasions with missing data are treated in precisely the same way as unobserved states in the future, and with identical consequences—uncertainty about N_t increases when we fail to observe y_t.[11]

Now, look at the priors for $\boldsymbol{\beta}$, starting with β_0. Earlier (sec. 5.4.2), we advocated using informative priors whenever they make sense. This often requires more scholarship and thought than a simple automatic choice of vague priors, but as we show here the effort can be worthwhile. Recall (from sec. 1.2) that β_0 is the intrinsic rate of increase, the instantaneous per capita rate of change in the population when the population size is 0 and rainfall is average; that is, $\beta_0 \approx r_{max}$. What do we know a priori about r_{max} and hence, β_0?

[11]Strictly speaking, these would be forecasts only if we were doing this in year 2008, but we take some liberty here to make an important point.

There are well established scaling relationships for many parameters needed to describe ecological and physiological processes (reviewed by Peters, 1983; Pennycuick, 1992; Marquet et al., 2005). Many of these relationships are believed to have a basis in physical laws (Brown et al., 2002). We would be foolish to ignore what is known about r_{max} from these scaling relationships in a model that contains r_{max}! Sinclair (2003) developed an allometric scaling equation for r_{max} for large vertebrate herbivores, $r_{max} = 1.37M^{-0.315}$ and a relationship for the standard deviation of the population growth rate, $sd(r_t) = 0.805M^{-0.316}$, where M is body mass in kilograms. We are safe using a rough estimate of average body mass of wildebeest of 275 kg, because the species is not strongly sexually dimorphic, which means that we should a priori expect $r_{max} \sim \text{normal}\left(0.234, 0.136^2\right)$. Other prior distributions were left vague to allow us to focus on the effect of informing β_0, but we nonetheless point out that the scaling of equilibrium population density could also be used to inform β_1 (Silva and Downing, 1995; Polishchuk and Tseitlin, 1999; Haskell et al., 2002; Dobson et al., 2003).

8.5.2 Convergence and Model Checking

We implemented the model in JAGS (Plummer, 2003), obtaining two chains for 200,000 iterations each. We evaluated convergence by visual inspection of trace plots and by the diagnostics of Gelman and Rubin (1992) and Heidelberger and Welch (1983) (sec. 7.3.4.2). We conducted posterior predictive checks (sec. 8.1.2) using the test statistics $T(\mathbf{y}, \boldsymbol{\theta})^{(k)} = \sum_t (y_t - N_t^{(k)})^2/y_t$ and $T(\mathbf{y}^{new}, \boldsymbol{\theta})^{(k)} = \sum_t (y_t^{new} - N_t^{(k)})/y_t^{new2}$, where the $y_t^{new(k)}$ were simulated from $y_t^{new(k)} \sim \text{normal}(N_t^{(k)}, \sigma^2_{\text{data},t})$. Recall that (k) indexes the kth iteration in the MCMC algorithm. The Bayesian p-value—the proportion of the total number of iterations for which $T(\mathbf{y}^{new}, \boldsymbol{\theta})^{(k)} \geq T(\mathbf{y}, \boldsymbol{\theta})^{(k)}$—was 0.85, indicating no evidence of lack of fit (fig. 8.5.1).

8.5.3 Marginal Posterior Distributions

8.5.3.1 Parameters and Derived Quantities

We used informative and vague priors on β_0 to illustrate the effect of including information in the prior. The median of the distribution of β_0 with an informative prior was 0.243 (95% equal-tailed Bayesian credible interval, BCI = (0.132, 0.375)), which was remarkably close to the theoretical prediction based on scaling (0.236) (fig. 8.5.2 A). This might lead you to believe that the prior contained all the information, because

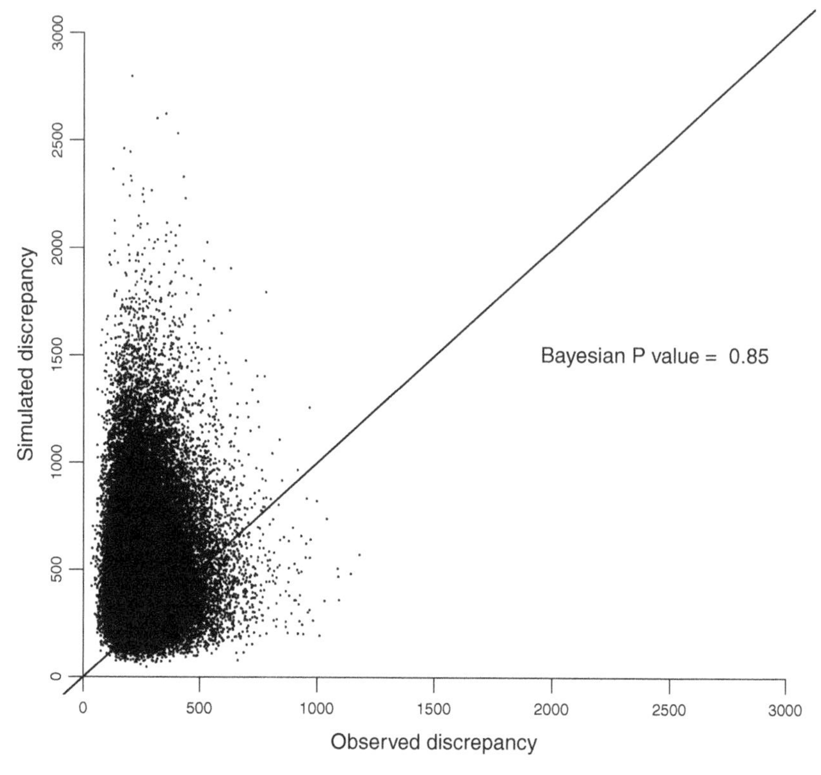

Figure 8.5.1. Results of posterior predictive check. Each data point is $T(\mathbf{y}^{new}, \boldsymbol{\theta})^{(k)}$ plotted against $T(\mathbf{y}, \boldsymbol{\theta})^{(k)}$. The diagonal line indicates $T(\mathbf{y}^{new}, \boldsymbol{\theta})^{(k)} = T(\mathbf{y}, \boldsymbol{\theta})^{(k)}$. The Bayesian p-value is the proportion of the points that fall above the line.

the central tendency in the posterior was not different from the mean of the prior. Two results show this was not the case. First, the estimate of the median of β_0 with a vague prior was 0.25, which also very close to the theoretical value (fig. 8.5.2 A). Moreover, it is easy to see that the marginal posterior distribution of β_0 was substantially narrower than the prior, reflecting the information contributed by the data.

This is a tidy result. That both posteriors are centered on the prior tends to confirm theory, and that the posterior given an informed prior is narrower than the prior confirms the value of the data. That the posterior based on the informed prior shrinks relative to the posterior based on the uninformative one shows the value of the prior. All the posteriors from a joint distribution with an informative prior on β_0 were narrower, even if slightly so, than their posteriors when the prior on β_0 was vague (fig. 8.5.2 A–E). This illustrates

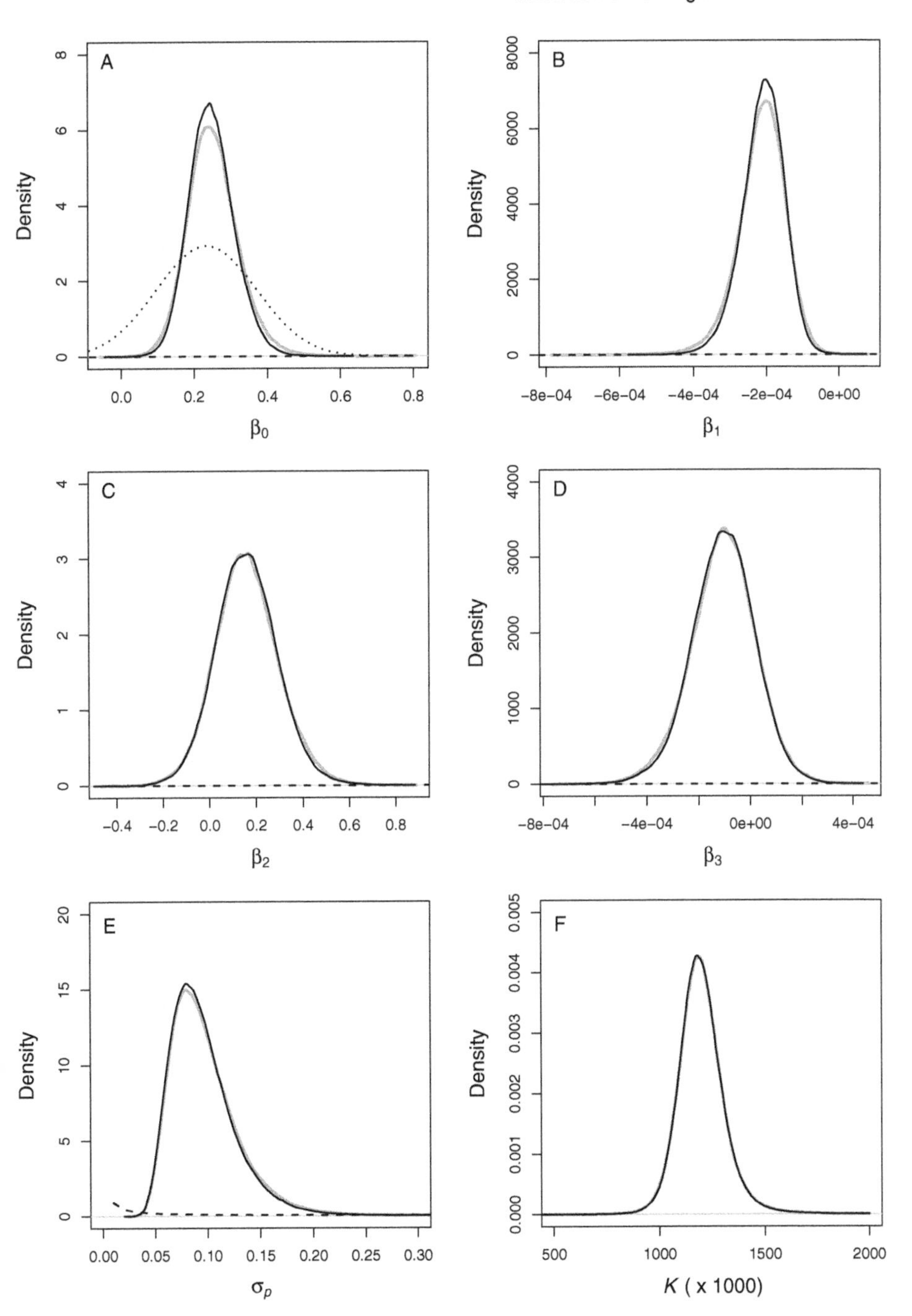

Figure 8.5.2. For caption see following page.

that informing the prior on a single parameter can shrink the estimates of *all* parameters, although it will not always do so, and some of the shrinkage may be minor.

Examining the marginal posterior for β_1 (fig. 8.5.2 B) reveals a clear negative feedback of population size on population growth rate. We can be 99.9% certain[12] that the value of β_1 does not include 0 and that the sign is negative. The median change in per capita growth rate for each 1000 wildebeest in the population was −0.000204 (BCI = −0.00033, −0.000103). We can also be 88% confident that the effect of rainfall was nonzero (fig. 8.5.2 C). We are less sure about the interaction between rainfall and population size. Although the sign of the median of β_3 was negative, there was a 20% chance that its true value was actually positive (fig. 8.5.2 D). Thus, there is not strong support in the data for an interaction between density and rainfall.

We included a derived quantity,[13] the estimated equilibrium density of the wildebeest population, in the MCMC algorithm, $K = -\beta_0/\beta_1$. The estimate of the median of the posterior distribution was 1.193 million wildebeest (BCI = 1.006, 1.439) (fig. 8.5.2 F).

8.5.3.2 Latent States

Posterior distributions of the latent state N_t are reflected in the credible intervals shown in figure 8.5.3. There are several points that need emphasis here. Notice that the credible intervals expand whenever there are missing data—data missing in the past or "missing" in the future. Also see that the estimates of the true state are slightly more precise when we use an informed prior on β_1 relative to an uniformed one. Finally, note that the fit of the model medians in Figure 8.5.3 is not smooth but, rather, tends to jump a bit between time steps. If we assumed that process variance was zero—such that our model was a perfect representation of nature, and the only

[12]This probability is calculated as the proportion of β_1 in the converged MCMC output that were less than 0. It is analogous to the value of the cumulative distribution function.

[13]The mathematical basis for this equation is developed in section 1.2.

Figure 8.5.2. Marginal posterior distributions of parameters and derived quantities in the Serengeti wildebeest model. Posterior distributions obtained using a model with an informative prior on β_0 and vague priors on all other parameters are shown with black lines. Posterior distribution obtained using a model with vague priors on all parameters, including, β_0 are shown with gray lines. Vague priors are shown with dashed lines. The informative prior on β_0 are shown with a dotted line in panel A. There is no prior on K (panel F) because it is a quantity derived from β_0 and β_1.

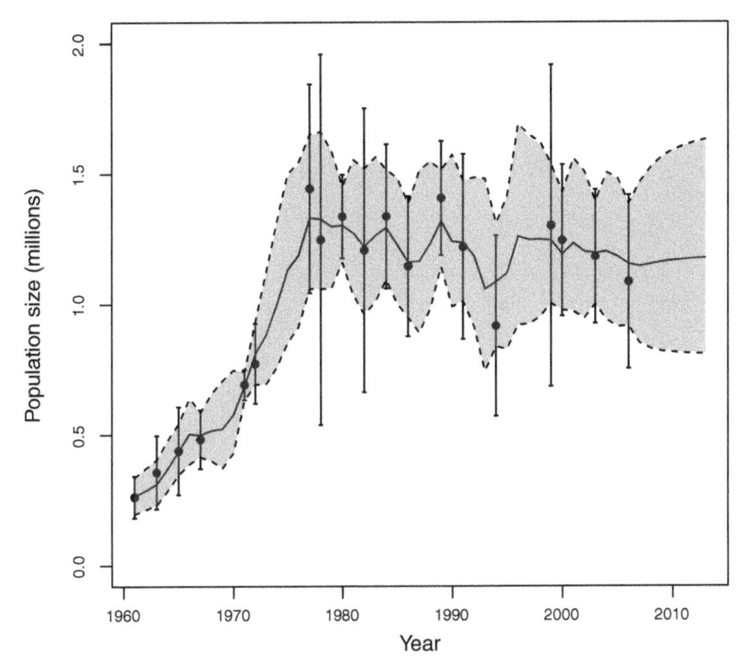

Figure 8.5.3. Estimate of the number of animals in the Serengeti wildebeest population. Dots are the mean observed number of animals with ±2 standard deviations as error bars. The solid line is the median of the marginal posterior distribution of the true, unobserved population size. The shaded area shows .025 and .975 quantiles of the posterior distributions of population size estimated with informative priors. Dotted lines given .025 and .975 quantiles for distributions estimated with vague priors. Shaded area and solid line after 2008 are forecasts based on the data obtained up to 2008.

error arose from sampling—then the curve would be smooth if plotted at the average rainfall. Alternatively, if we assumed that observation variance was 0, such that the true state was exactly the same as the observed state, *then the medians would go through every data point.* That the model responds to every point individually (i.e., is not smooth) but does not go through every point reflects the combined influence of sampling and process variance. It may be counterintuitive that a model like this one that *does* hit every point, or close to it, is probably not a very good model. What this says is that process variance is high relative to observation variance. The predictive value of such a model will be low.

9 Inference from Multiple Models

Specifying models and combining them with observations forms the essence of science. Often, we can learn a great deal about ecology using a single model of a process, as we described in (chapter 8). When would we use a single model as a basis for inference? The simple answer is that one model is appropriate whenever we are satisfied making inferences conditional on that model. This will often be the case when the parameters in our model of an ecological process have a mechanistic basis that has been well established in previous work and when our goal is to estimate those parameters and evaluate their importance to the process we seek to understand. A single model can allow us to meet that goal (Gelman and Rubin, 1995; Ver Hoef, 2015). A single model is usually appropriate for experiments with treatments and controls.

But sometimes we will need to use more than one model to gain insight, as we outlined in our description of the general modeling process (fig. 0.0.1 D). Research in ecology often begins by rounding up the suspects, that is, by assembling competing views of how a process operates. We represent these views as alternative models. Data allow us to evaluate the ability of alternative models to make predictions. We keep the models that succeed in these predictions and use them for inference. We abandon those that fail. In this chapter we describe how to evaluate alternative models with data.

There are two broad ways to formally use multiple models: model selection and model averaging. In both cases we have a set of L alternative models differing in the number of parameters they contain, in their functional forms, or both. Let's call this set of models $\mathcal{M} = \{M_1, \ldots, M_l, \ldots, M_L\}$. We assume that all these models have been chosen thoughtfully by the researcher and have been checked as described in section 8.1. In model selection, we wish to evaluate these models based on some criterion. In model averaging, we want to make inference using all models in the set, or a subset of them, using a weighted average of posterior distributions of individual models. We use the data to estimate the weight (via a posterior probability) of each model in this average. Gelman et al. (2014) developed the case that model selection and averaging are distinct in the Bayesian approach. We organize this chapter by first treating model selection before turning to model averaging.

This chapter is somewhat more technical that earlier ones because the material we treat here is diffuse in the literature, and we want to ensure that you have a solid overview of Bayesian model selection. A more detailed, comprehensive treatment can be found in Hooten and Hobbs (2015), which provides the basis for the material presented here. This is the first chapter that is not required to understand subsequent ones, so if you wish to return to model selection later, then you might read this chapter to get a feel for the material we cover, delaying study until you need it for your work.

9.1 Model Selection

We assume there are at least two models we would like to compare. That is, we want to evaluate each model relative to some predefined characteristic. Predictive ability is by far the most commonly sought model characteristic in the literature on model selection. We emphasize it here.

9.1.1 Out-of-Sample Validation

If we are interested in prediction as our main criterion for model utility, then it makes most sense to evaluate the model in terms of *real* predictive ability. The gold standard in model selection is *out-of-sample validation*, a procedure that assesses the ability of a model to accurately predict data not used to fit the model, generally referred to as *out-of-sample data*. We desire a model that closely predicts out-of-sample data, where "closeness" is measured using a score function. Ideally, this assessment uses a data set that, by design, was obtained solely for the purpose of model selection, a data set withheld from model fitting, which we call $\mathbf{y}_{\text{oos}}$. When we have such data, we can use the mean square prediction error (MSPE) as a score function

$$\text{MSPE} = \sum_{i=1}^{n} \frac{(y_{\text{oos},i} - \hat{y}_{\text{oos},i})^2}{n}, \tag{9.1.1}$$

where n is the number of observations in the out-of-sample set, and $\hat{y}_{\text{oos},i}$ is the model prediction of the ith observation in the out-of-sample set $y_{\text{oos},i}$. Low values of MSPE indicate models with greater predictive ability.

The practice of evaluating models based only on point estimates of parameters or predictions does not naturally incorporate uncertainty. An outstanding advantage of Bayesian inference is the ability to account for various sources of uncertainty, and we can exploit that advantage in model selection using out-of-sample validation. To understand how this method works we must understand prediction from the Bayesian perspective (recall sect. 8.4). In general, data that have not been observed are viewed as random variables, and we treat them like all other unobserved quantities: we seek their posterior distribution. The posterior distribution for predictions is called the *posterior predictive distribution* and can be found using the integral

$$[\mathbf{y}_{\text{oos}}|\mathbf{y}] = \int [\mathbf{y}_{\text{oos}}|\mathbf{y}, \boldsymbol{\theta}][\boldsymbol{\theta}|\mathbf{y}]d\boldsymbol{\theta}. \tag{9.1.2}$$

This integral[1] represents the marginal posterior distribution of the predictions of the out-of-sample data conditional on the data used to fit a model containing parameters $\boldsymbol{\theta}$. Note that we are integrating out the $\boldsymbol{\theta}$ (recall sect. 3.4.2), leaving a marginal distribution. There is potential for confusion here, because you are probably thinking, "I thought the out-of-sample data were known?" Equation 9.1.2 is analogous to the marginal distribution of the data (eq. 5.1.11); it represents a distribution before $\mathbf{y}_{\text{oos}}$ are observed, and a scalar afterward. With both data sets ($\mathbf{y}_{\text{oos}}$, $\mathbf{y}$) in hand, we use equation 9.1.2 to compute the probability (or probability density) of the out-sample-data given the model predictions. The quantity $[\mathbf{y}_{\text{oos}}|\mathbf{y}]$ will be large for models that have high predictive capability.

The log predictive density $\log[\mathbf{y}_{\text{oos}}|\mathbf{y}]$ is a scoring function that is appropriate for Bayesian model validation (Gelman et al., 2014), and is appropriate for the model being fit and the data being predicted. It can be

[1]Understand that there are as many integrations as parameters. We use 1 integral symbol to simplify notation.

approximated by computing

$$\log[\mathbf{y}_{\text{oos}}|\mathbf{y}] \approx \log\left(\frac{\sum_{k=1}^{K}[\mathbf{y}_{\text{oos}}|\mathbf{y},\boldsymbol{\theta}^{(k)}]}{K}\right) \tag{9.1.3}$$

from MCMC output, where k indexes a single iteration, and K is the total number of iterations. What this means in practice is that we compute the total probability (or probability density) of the out-of-sample observations conditioned on the model predictions of those observations at each MCMC iteration, $[\mathbf{y}_{\text{oos}}|\mathbf{y},\boldsymbol{\theta}^{(k)}]$. The log of the mean of $[\mathbf{y}_{\text{oos}}|\mathbf{y},\boldsymbol{\theta}^{(k)}]$ across all iterations is our predictive score. We can then use this score to rank all models in the set $\mathcal{M}$ and find the one that yields the best predictions. Higher values indicate more predictive models.

9.1.2 Cross-Validation

We often lack an additional data set that was collected solely for model selection. Instead, we have a single data set that somehow must be used to estimate model parameters and select among alternative models. The problem in this case is that we cannot use the same data to assess model predictive ability and to estimate model parameters. Therefore, we must leave out a subset of the data ($\mathbf{y}_m$) from the fitting procedure so that the model is unaware of it. We then use the fitted model to predict the $\mathbf{y}_m$ data points. The problem with choosing a single subset of the data to leave out is that we can assess predictive ability only for those specific observations. It is entirely possible to get different results for model selection depending on the choice of particular subset. Thus, it is common to leave out all the data, but only in small subsets sequentially.

In M-fold cross-validation[2] we group the data evenly (or approximately evenly) into M groups and then use each set of left-out data $\mathbf{y}_m$ to compare with the model predictions based on the remaining data ($\mathbf{y}_{-m}$). We then iterate through all groups of data $\mathbf{y}_m$ for $m = 1,\ldots,M$ and compute component scores, which we sum to yield the full cross-validation score for the whole data set:

$$\sum_{m=1}^{M}\log\left(\frac{\sum_{k=1}^{K}[\mathbf{y}_m|\mathbf{y}_{-m},\boldsymbol{\theta}^{(k)}]}{K}\right). \tag{9.1.4}$$

[2]We depart from the more conventional term K-fold cross-validation to avoid confusion with the notation we have used for iterations in MCMC.

What this means in practice is most easily understood when $M = n$ (n is the sample size), a procedure often referred to as *leave-one-out cross-validation*.[3] We create M data sets, each of which omits a single observation. We then fit our model to each data set and calculate the probability (or probability density) of the left-out observation conditional on the model's prediction of that observation. We compute the mean of that probability (or probability density) across the K MCMC iterations for each of the M left-out data sets and sum the log of those means.

The drawback of M-fold procedures for cross-validation is the time required for computation; it requires that the model be refit potentially many times. For large models and large data sets this can lead to computational burdens that are too great for efficient analysis using MCMC, although advances in parallel computing are likely to eventually alleviate this concern.

9.1.3 Information Criteria

Information criteria are widely used by ecologists to select models, particularly when maximum likelihood methods are used to estimate model parameters. The aim of information criteria is to obtain the same assessment of predictive ability provided by cross-validation without regaining out-of-sample data. The same data are used to estimate model parameters and to select among alternative models, avoiding the potential computational costs of cross-validation.

9.1.3.1 The concept of Statistical Regularization

Information criteria involve specific approaches to model selection that fall under the much broader umbrella of statistical regularization. The term *regularization* comes from the use of a function that regularizes an optimization. The optimization problem might be a likelihood to be maximized or a posterior distribution sampled with MCMC. Consider the general expression

$$\mathscr{L}(\mathbf{y}, \boldsymbol{\theta}) + r(\boldsymbol{\theta}, \gamma), \tag{9.1.5}$$

[3]Leave-one-out cross-validation may be preferable when the sample size is small and there are few observations to use as training data, though the resulting estimate of prediction error loses precision as $M \rightarrow n$.

where $\mathcal{L}(\mathbf{y}, \boldsymbol{\theta})$ represents a *loss function*, a function[4] of quantities we observe ($\mathbf{y}$) and those that are unobserved ($\boldsymbol{\theta}$). The loss function expresses how a quantity of interest (e.g., a likelihood) changes as we change $\boldsymbol{\theta}$. As it is composed here, small values of $\mathcal{L}(\mathbf{y}, \boldsymbol{\theta})$ are best. The function $r(\boldsymbol{\theta}, \boldsymbol{\gamma})$ in equation 9.1.5 opposes the minimization of $\mathcal{L}(\mathbf{y}, \boldsymbol{\theta})$ by adding to it as $\boldsymbol{\theta}$ changes. The regulator function r may also depend on some other variables $\boldsymbol{\gamma}$ that may or may not be related to the loss function or its components. Of course, there are other ways to express the loss and regulator relationship. For example, expressions need not be additive, but the expression in equation 9.1.5 is perhaps the most common. We can then obtain statistical inference by minimizing the joint function (eq. 9.1.5) with respect to $\boldsymbol{\theta}$, and perhaps $\boldsymbol{\gamma}$, if not already known. We use regularization because it can improve inference. It can shrink the variance of estimates or increase the accuracy of predictions or both. As we will soon see, the dominant approaches to model selection are variants of the general idea of regularization.

We have already seen two examples of regularization. We showed earlier that information from prior estimates of parameters can be used to influence the maximum likelihood estimate (sec. 4.4) using an approach known as *penalized likelihood*. We can rearrange equation 4.4.1 to

$$-\log(L(\theta|\mathbf{y})) = -\left(\sum_i \log[y_i|\theta] + \log[\theta|\alpha, \beta]\right), \tag{9.1.6}$$

so that we seek the minimum of equation 9.1.6 to obtain the maximum likelihood estimate of the parameter θ. We can see in this example that $\mathcal{L}(\mathbf{y}, \boldsymbol{\theta}) = \sum_i \log[y_i|\theta]$, and $r(\boldsymbol{\theta}, \boldsymbol{\gamma}) = \log[\theta|\alpha, \beta]$. Intuitively, we see that $\log[\theta|\alpha, \beta]$ acts as a "penalty" that regularizes the effect of the negative log likelihood on inference. If the values of θ depart markedly from the values obtained in previous estimates summarized in $\log[\theta|\alpha, \beta]$, then the penalty gets large, making the overall expression less negative, pushing it away from the minimum.

The second place we have seen a regulator is in the Bayesian prior itself, which, as we discussed (sec. 5.2), is used in virtually the same way in likelihood and Bayesian analysis. It can be shown that the Bayesian posterior provides a rigorous framework for regularization, but here we offer a more intuitive treatment based on fig. 5.1.1. The posterior distribution of the unknown parameter represents a compromise between the likelihood

[4]Do not confuse $\mathcal{L}(\mathbf{y}, \boldsymbol{\theta})$ with a likelihood. We use $\mathcal{L}$ to denote a loss in general, and L specifically to denote a likelihood.

and the regulating effects of the prior on the likelihood as illustrated in fig. 5.1.1, which can also be seen algebraically in the general expression for a posterior distribution:

$$\log[\theta|\mathbf{y}] = \log[\mathbf{y}|\theta] + \log[\theta] + \text{constant}. \tag{9.1.7}$$

The Bayesian perspective makes it clear that we are constraining the model parameters with "prior" information. The prior helps us find a better predictive model, usually with lower variance in estimates of a parameter, as we saw in the wildebeest example (sec. 8.5.3.1). We are often taught that the Bayesian prior should be chosen either objectively, to minimize the influence on the posterior, or retrospectively, to best represent existing prior knowledge about the parameters. However, the only rule for specifying prior information in a Bayesian model is that the same data must not be used in the likelihood and to form the prior. The reason for this rule is that it maintains the acyclic nature in the Bayesian network (sec. 3.3) so that the data depend on the parameters, and the parameters depend on either other parameters or fixed quantities. The Bayesian network must be acyclic to ensure that the probability statements about the unknown quantities are valid, because a properly factored joint distribution must have a corresponding acyclic graph.

9.1.3.2 Likelihood-Based Criteria

Akaike's Information Criterion (AIC). We now explain how information criteria fit into the regularization concept, first in a non-Bayesian context. The general idea behind information criteria is that we choose a scoring function a priori that will be used to "score" each of the models based on the balance of fit to the within-sample data and parsimony (or overall predictive ability; Gneiting (2011)). Akaike's information criterion (AIC) depends on the assumption that parameters are "free," which means they are not constrained by a prior or by a hierarchical relationship to other parameters. Given model (M_l) with p free parameters,

$$\text{AIC} = -2\log[\mathbf{y}|\hat{\boldsymbol{\theta}}, M_l] + 2p, \tag{9.1.8}$$

where $\hat{\boldsymbol{\theta}}$ is the maximum likelihood estimate of the model's parameters. It takes the same form as the regularization expression (eq. 9.1.5), with lower values of AIC implying better predictive ability of the model.

AIC provides the same rankings of models as leave-one-out cross-validation under certain conditions (Hastie and Friedman, 2009). This is an attractive result, because it can dramatically reduce the computational burden in finding a good predictive model. However, the result holds only for linear models and normal likelihood functions. We also must assume that the "true" model is in the model set being considered.[5] AIC, being a function of maximum likelihood estimates, does not have a clear Bayesian interpretation but it seems to perform well empirically in situations where it can be used (Hastie and Friedman, 2009).

Bayesian Information Criterion (BIC). The Bayesian information criterion (BIC; Schwarz 1978) is calculated as follows:

$$\text{BIC} = -2\log[\mathbf{y}|\hat{\boldsymbol{\theta}}, M_l] + \log(n)p \tag{9.1.9}$$

where $\hat{\boldsymbol{\theta}}$ are the maximum likelihood estimates of the model parameters, M_l is a specific model, n is the number of observations, and p is the number of "free" parameters in the model. Again, the general form of regularization is easily recognized. The purpose of BIC differs from the purpose of AIC and many other regularization methods. AIC is an information criterion that seeks to provide a measure of predictive ability, whereas BIC is distinctly focused on model averaging (Link and Barker, 2006; Gelman et al., 2014), which is why we condition on a specific model in equation 9.1.9. We briefly discuss the use of BIC for model averaging next. Here, we simply point out that despite its name, BIC has limited utility for Bayesian analysis. BIC relies on maximum likelihood estimates of parameters and requires a count of parameters, as does AIC. Parameter counts are meaningless when priors are informative or when models are hierarchical, because in these cases at least some parameters are not free.

9.1.3.3 Bayesian Criteria

Deviance Information Criterion (DIC). Spiegelhalter et al. (2002) proposed the deviance information criterion (DIC) as a Bayesian alternative to likelihood-based methods (AIC and BIC). Like the other information criteria, DIC contains a loss function—the deviance ($D = -2\log[\mathbf{y}|\boldsymbol{\theta}]$)—plus a penalty or regulator function. However, the penalty differs from AIC and BIC in that it does not require counting parameters. Recall that even the simplest Bayesian models may contain parameters constrained in some way by informative priors. Hierarchical Bayesian models will have latent state variables that are technically unknown but are also highly constrained by both the likelihood and prior. A truly Bayesian criterion requires estimating

[5]Link and Barker (2010) provide a particularly lucid discussion of what is meant by a "true" model. If we generate data using simulation from a model and then fit a set of alternative models to the simulated data, the model used to simulate the data is the "true" model in the set.

an "effective" number of parameters (p_D), because all parameters are not equally constrained. DIC is calculated as follows:

$$\begin{aligned} \text{DIC} &= -2\log[\mathbf{y}|\text{E}(\boldsymbol{\theta}|\mathbf{y})] + 2p_D &\quad (9.1.10)\\ &= \hat{D} + 2p_D. &\quad (9.1.11)\end{aligned}$$

The effective number of parameters is obtained from the deviance calculated two ways, from the posterior mean of the deviance $\bar{D}$, and the deviance computed at the posterior mean of the parameters $\hat{D}$:

$$p_D = \bar{D} - \hat{D}. \quad (9.1.12)$$

The posterior mean deviance is

$$\begin{aligned} \bar{D} &= \text{E}_{\boldsymbol{\theta}|\mathbf{y}}(-2\log[\mathbf{y}|\boldsymbol{\theta}]) &\quad (9.1.13)\\ &= \int -2\log[\mathbf{y}|\boldsymbol{\theta}][\boldsymbol{\theta}|\mathbf{y}]d\boldsymbol{\theta}. &\quad (9.1.14)\end{aligned}$$

In practice, we compute $\bar{D}$ by calculating the deviance at each iteration of the MCMC algorithm,

$$D^{(k)} = -2\log[\mathbf{y}|\boldsymbol{\theta}^{(k)}], \quad (9.1.15)$$

and finding the mean of D across all the iterations, $\bar{D} = \sum_{k=1}^{K} D^{(k)}/K$. We estimate $\hat{D}$ by calculating the deviance using the means of the posterior distributions of each of the parameters. Contemporary software implementing MCMC has built-in functions to calculate DIC.

In the case of linear regression, with vague priors on the regression coefficients, the effective number of parameters p_D approaches the number of coefficients p. DIC has become popular as a result of its similarity to AIC, its simplicity, its implementation in software, and its ease of calculation using MCMC samples.

Overall, it appears that DIC (eq. 9.1.11) is most appropriate as a model-selection criterion in linear models for which the p_D is much smaller than n. Thus, from a Bayesian perspective, DIC is good for comparing the same classes of models as is AIC. DIC is not appropriate for model selection with mixture models or missing-data models (e.g., Spiegelhalter et al., 2002; Celeux et al., 2006; Plummer, 2008). Zero-inflated models are the largest and most heavily used class of models in wildlife ecology (i.e., capture-recapture and occupancy models, Royle and Dorazio, 2008) and are a form of mixture model. The original version of DIC is unsuitable for comparing zero-inflated models.

Watanabe-Akaike Information Criterion (WAIC). Aside from the aforementioned caveats, DIC is a useful information criterion in the Bayesian modeling context when prediction is paramount. However, DIC does not best represent the actual Bayesian predictive procedure. To arrive at predictions, the Bayesian approach summarizes the posterior predictive distribution (sec. 8.1.1), but the posterior predictive distribution is not needed to compute DIC (eq. 9.1.11). This seems to be a mismatch between the type of inference desired and the tool used to obtain it.

Gneiting (2011) recommended using a scoring function that is true to the desired form of inference to assess predictive performance; in particular, a logarithmic scoring function achieves this (i.e., termed "local" and "proper"). Thus, for Bayesian model comparison based on predictive ability we should seek a statistic that considers the log posterior predictive distribution (sec. 8.1.1) for new data $\mathbf{y}^{new}$:

$$\log[\mathbf{y}^{new}|\mathbf{y}] = \log \int [\mathbf{y}^{new}|\boldsymbol{\theta}][\boldsymbol{\theta}|\mathbf{y}]d\boldsymbol{\theta}. \tag{9.1.16}$$

The quantity in equation 9.1.16 is stochastic because $\mathbf{y}^{new}$ is assumed to be unknown. A common technique in developing most information criteria is then to consider the mean of equation 9.1.16 over $\mathbf{y}^{new}$,

$$\mathrm{E}_{\mathbf{y}^{new}}(\log[\mathbf{y}^{new}|\mathbf{y}]) = \int \log \int [\mathbf{y}^{new}|\boldsymbol{\theta}][\boldsymbol{\theta}|\mathbf{y}]d\boldsymbol{\theta}[\mathbf{y}^{new}]d\mathbf{y}^{new}, \tag{9.1.17}$$

which is impossible to compute directly because the true distribution of the new data $[\mathbf{y}^{new}]$ is unknown. Thus, the log pointwise predictive score can be used to estimate the mean log posterior predictive score as follows:

$$\log \prod_{i=1}^{n} [y_i|\mathbf{y}] = \sum_{i=1}^{n} \log \int [y_i|\boldsymbol{\theta}][\boldsymbol{\theta}|\mathbf{y}]d\boldsymbol{\theta}, \tag{9.1.18}$$

(Richardson, 2002; Celeux et al., 2006; Watanabe, 2010), where MCMC is used to compute the integral (Gelman et al., 2014). In practice this involves computing the probability (or probability density) of each data point at each iteration (k) in the converged MCMC output using the current values for the model parameters at that iteration, that is $[y_i|\boldsymbol{\theta}^{(k)}]$. The means of $[y_i|\boldsymbol{\theta}^{(k)}]$ calculated across the $k = 1, \ldots, K$ MCMC iterations are obtained and summed over each of the $i = 1, \ldots, n$ observations.

There are two issues with the score in equation 9.1.18. The product representation of the pointwise predictive score implies that the data are independent (conditioned on $\boldsymbol{\theta}$). In addition, the score relies completely on the observed data $\mathbf{y}$ rather than on the new data $\mathbf{y}^{new}$. The first issue suggests that the score should not be used with models containing dependence in the data (e.g., spatial and time-series models unless it can be shown that there is no dependence). The latter issue implies that in equation 9.1.18 the predictive score for a given model requires regularization because the within-sample data are being used twice. Gelman et al. (2014) suggest using the effective number of parameters for this regularization, (p_D):

$$p_D = \sum_{i=1}^{n} \text{Var}_{\boldsymbol{\theta}|\mathbf{y}}(\log[y_i|\boldsymbol{\theta}]). \tag{9.1.19}$$

As before, Monte Carlo integration is used to approximate p_D. At each iteration (k) in the MCMC algorithm $\log[y_i|\boldsymbol{\theta}^{(k)}]$ is calculated; it represents the probability (or probability density) of each data point conditional on the current parameter values. The variance of $\log[y_i|\boldsymbol{\theta}^{(k)}]$ over $k = 1, \ldots, K$ MCMC samples in the converged chain is then computed for each for $i = 1, \ldots, n$ observations (y_i). These variances are summed across the n observations to estimate p_D.

The Watanabe-Akaike information criterion can then be defined as -2 times the log pointwise predictive score plus the effective number of parameters,

$$\text{WAIC} = -2\sum_{i=1}^{n} \log \int [y_i|\boldsymbol{\theta}][\boldsymbol{\theta}|\mathbf{y}]d\boldsymbol{\theta} + 2p_D \tag{9.1.20}$$

$$\approx -2\left(\sum_{i=1}^{n} \log\left(\frac{\sum_{k=1}^{K} [y_i|\boldsymbol{\theta}^{(k)}]}{K}\right)\right) + 2p_D, \tag{9.1.21}$$

with both elements in the sum approximated using MCMC samples at no extra computational cost beyond that required for calculating DIC (Watanabe, 2013). The addition of p_D in equation 9.1.20 serves as a bias correction in posterior prediction similar to that of AIC and DIC.

WAIC offers many benefits. It is based on the posterior predictive distribution and is fully Bayesian but yields the same results as DIC in linear Gaussian models with uniform priors. Furthermore, unlike DIC, WAIC is valid in both hierarchical and mixture models (Watanabe, 2013). Also, unlike with DIC, the effective number of parameters is always positive. A parameter in p_D counts as a zero if the learning comes entirely from the prior. To determine the correct proportion of each parameter to count, WAIC requires the data (as in DIC) to compute p_D. This is essential in the Bayesian context, where hierarchical structures with strong interdependencies and informative priors are regularly used.

Overall, WAIC seems very appealing; however, the main disadvantage is substantial depending on the area of application: its calculation relies on an independence assumption of the data given the parameters. This assumption will not be valid in spatial models where dependence among the data is one of the key features being modeled. Ando and Tsay (2010) provided a way to relax the independence assumption, but the resulting criterion requires numerous model fits, which eliminates one of the key practical benefits of WAIC (Gelman, 2014).

Posterior Predictive Loss. Posterior predictive loss (Gelfand and Ghosh, 1998) is a model-selection criterion resembling WAIC in its reliance on posterior predictions of new data. A version of posterior predictive loss (D_{sel}) is calculated as

$$D_{sel} = \sum_{i=1}^{n}(y_i - \mathrm{E}(y_i^{new}|\mathbf{y}))^2 + \sum_{i=1}^{n}\mathrm{Var}(y_i^{new}|\mathbf{y}), \tag{9.1.22}$$

where, as before, y_i^{new} represents a new data point simulated from the model, as described in section 8.1.1. In practice this means calculating a new data set at each iteration of the MCMC chain, just as was done in section 8.1. The mean $\mathrm{E}\left(y_i^{new}|\mathbf{y}\right)$ and the variance $\mathrm{Var}\left(y_i^{new}|\mathbf{y}\right)$ of each of the simulated data points are calculated across all the iterations of the MCMC chain, and these quantities are used to calculate D_{sel} via Monte Carlo integration (box 8.1).

Note the similarity of D_{sel} to the WAIC (eq. 9.1.20) and DIC (eq. 9.1.11), for large n. All three criteria contain two terms in a sum: the first is a goodness-of-fit measure, and the second acts as a penalty or regulator. In this case we can see that the penalty $\sum_{i=1}^{n}\mathrm{Var}(y_i^{new}|\mathbf{y})$ will increase in models with too many parameters, because the prediction variance becomes larger with an increasing number of parameters. D_{sel} appears to be appropriate for many classes of hierarchical models, because it depends directly on the posterior predictive distribution rather than the likelihood and posterior mean of the parameters alone. Also, unlike WAIC, the posterior predictive loss approach is suitable for correlated data models (e.g., spatial and temporal models) because it depends directly on the posterior predictive distribution.

9.1.4 Indicator Variable Selection

All the regularization methods discussed in the previous section are model-based approaches to model selection. They are model based because they contain a formal mechanism that trades off model fit for model parsimony. We saw in section 9.1.3.1 that the Bayesian model itself provides a natural model-reduction mechanism via the prior. We saw that likelihood-based models can achieve the same effect using prior information on parameters. Other methods have been developed that explicitly augment the overall model structure with indicator variables that act as on-off switches to control which parameters appear in the selected model (O'Hara and Sillanpaa, 2009).

Consider the basic linear regression model

$$y_i \sim \text{normal}(\beta_0 + \mathbf{x}_i'\boldsymbol{\beta}, \sigma^2),$$

where the parameter vector $\boldsymbol{\beta} = (\beta_1, \ldots, \beta_j, \ldots, \beta_p)'$ contains the individual coefficients corresponding to p predictor variables of interest. We include binary indicator variables z_j, so that $\beta_j = z_j \cdot \theta_j$ for $j = 1, \ldots, p$. In general, a prior would be specified for each (z_j, θ_j) pair, and the full Bayesian model could then be fitted, yielding inference not only about the coefficients β_j but also the selection indicators z_j. If the posterior mean for the distribution of z_j is large (i.e., is closer to one than zero), it indicates that the jth covariate is important in the model; conversely, when the posterior mean of z_j is close to zero, it effectively removes the jth effect from the model, thereby inducing a certain parsimony.

It would seem logical to use independent priors for z_j and θ_j; for example, we might standardize the predictor variable and specify

$$z_j \sim \text{Bernoulli}(p), \tag{9.1.23}$$

$$\theta_j \sim \text{normal}(0, \tau^2), \tag{9.1.24}$$

for all $j = 1, \ldots, p$. However, an independent prior specification often leads to MCMC samples that fail to converge if the prior for θ_j is too vague (i.e., the prior variance (τ^2), is large), because when $z_j = 0$ in an MCMC algorithm, θ_j will be sampled from its prior, and $z_j = 1$ will rarely occur in later iterations, because the θ_j is likely to be far from the majority of posterior mass.

A potential remedy for this problem is to use joint priors for z_j and θ_j that include explicit dependence between the indicators and coefficients (George and McCulloch, 1993; Carlin and Chib, 1995). The joint prior distribution

$[z_j, \theta_j] = [\theta_j|z_j][z_j]$ is decomposed. In this joint prior specification, the Bernoulli prior for z_j is retained, but the prior for θ_j, conditional on z_j, is written as

$$\theta_j|z_j \sim z_j \cdot \text{normal}(0, \tau^2) + (1 - z_j) \cdot \text{normal}(\mu_{\text{tune}}, \sigma^2_{\text{tune}}), \quad (9.1.25)$$

which has the form of a mixture distribution and is often referred to as a "slab and spike" prior (Miller, 2002). The Gibbs variable selection procedure then involves choosing the tuning parameters μ_{tune} and σ^2_{tune} such that normal($\mu_{\text{tune}}, \sigma^2_{\text{tune}}$) is near the posterior, and the MCMC algorithm exhibits better mixing. Surprisingly, the posterior distribution of β_j is largely insensitive to what would seem to be an informative prior on θ_j and z_j (eq. 9.1.25).

A somewhat different approach to specifying priors is "stochastic search variable selection" (George and McCulloch, 1993). The joint prior distribution of θ_j and z_j is factored to give

$$\theta_j|z_j \sim z_j \cdot \text{normal}(0, c\tau^2) + (1 - z_j) \cdot \text{normal}(0, \tau^2). \quad (9.1.26)$$

In this case, the prior does influence the posterior distribution of the β_j. Both c and τ^2 are tuned such that τ^2 is quite small, providing an effective spike at zero, while $c\tau^2$ is larger, creating a slab around zero. The slab then provides the prior for θ_j when the variable β_j is in the model (i.e., when $z_j = 1$).

9.2 Model Probabilities and Model Averaging

Recall the set of alternative models $\mathcal{M} = \{M_1, \ldots, M_l, \ldots, M_L\}$ introduced at the beginning of this chapter. We assume these are thoughtfully composed to represent alternative views on the operation of an ecological process and that all these models have been checked using the methods outlined in section 8.1. Any models that failed these tests were excluded from the set. Here, we describe how we might combine the strengths of multiple models to improve our inference about the process being studied and enhance our ability to make predictions.

It has been argued (e.g., Kass and Raftery, 1995; Link and Barker, 2006) that Bayesian model averaging (BMA) is the proper way to combine inference from multiple models because BMA provides a valid, probability-based mechanism for considering multiple models in the presence of uncertainty. Hoeting et al. (1999) provided details on BMA, complete with a treatment of its implementation. Here, we offer a more general overview.

An important and often overlooked aspect of model averaging is that BMA was not designed as a method for model selection but, rather, as a method for combining posterior distributions. It is important to understand that this is a different objective than the goal of model selection, where some characteristic is evaluated, usually on its ability to predict out-of-sample data. The methods in section 9.1 on model selection are based heavily on finding models that excel at out-of-sample predictive performance. BMA is intended for combining models using within-sample data.

The key outcome of BMA is the average posterior distribution of a quantity of interest, averaged across different models. We define $g \equiv g(\boldsymbol{\theta}, \mathbf{y})$ to reduce notational clutter. Thus, the posterior distribution we seek is

$$[g|\mathbf{y}] = \sum_{l=1}^{L} [g|\mathbf{y}, M_l] \Pr(M_l|\mathbf{y}), \tag{9.2.1}$$

where $[g|\mathbf{y}, M_l]$ is the posterior distribution of g under individual model M_l, and $\Pr(M_l|\mathbf{y})$ is the posterior probability of model M_l. It is easy to see in equation 9.2.1 that the posterior distribution of unobserved quantities conditional on the data is simply a weighted average of the posterior distributions across all models in the set, where the weights are given by the posterior probability of the model, $\Pr(M_l|\mathbf{y})$. Thus, we have a natural and proper Bayesian framework for multimodel inference as long as we can find the required quantities in equation 9.2.1. BMA performed on a set of models $\mathcal{M}$ yields better inference about g than any one of the models alone (Madigan and Raftery, 1994), thus we have a compelling reason to use it.

Recall from section 5.1 the expression for Bayes' theorem for a single model,

$$[\boldsymbol{\theta}|\mathbf{y}] = \frac{[\mathbf{y}|\boldsymbol{\theta}][\boldsymbol{\theta}]}{[\mathbf{y}]}, \tag{9.2.2}$$

where $[\boldsymbol{\theta}]$ is the prior distribution for the parameters. The denominator $[\mathbf{y}]$, which we typically avoid finding analytically, is the marginal data distribution for the given model. Also recall that another name for $[\mathbf{y}]$ is the prior predictive distribution, because it gives us information about the probability of new data *before* they are collected. It follows that $[\mathbf{y}]$ is a natural model discrimination measure by itself. Good models will have a large probability of the data, and poor models will have a small probability. It seems sensible that $[\mathbf{y}]$ is also fundamental in computing the posterior model probabilities $\Pr(M_l|\mathbf{y})$. To show this, we generalize the notation to include information concerning specifying the model associated with $[\mathbf{y}]$.

Thus, $[\mathbf{y}|M_l]$ is the marginal data distribution for model l. We can write the posterior probability for model l as

$$\Pr(M_l|\mathbf{y}) = \frac{[\mathbf{y}|M_l]\Pr(M_l)}{\sum_{l=1}^{L}[\mathbf{y}|M_l]\Pr(M_l)}, \tag{9.2.3}$$

where $\Pr(M_l)$ is the assumed prior model probability, which is usually specified as $1/L$. Note the close resemblance between this expression and Bayes' theorem for discrete parameters,

$$[\theta|y] = \frac{[y|\theta][\theta]}{\sum_{\theta}[y|\theta][\theta]}. \tag{9.2.4}$$

Both equations are based on a likelihood and a prior arranged in the same way—we are using the exact same theoretical basis to find the probability of a model as we use to find the probability distribution of a parameter. The probability of the model (eq. 9.2.3) provides the foundation for Bayes' factors, which can be used to compare one model with another (box 9.2).

Box 9.2 Bayes Factors

Consider the ratio of posterior probabilities for two models, say M_l and $M_{l'}$. Using a bit of algebra we can show that the ratio (i.e., the posterior odds) is

$$\frac{\Pr(M_l|\mathbf{y})}{\Pr(M_{l'}|\mathbf{y})} = \frac{[\mathbf{y}|M_l]\Pr(M_l)}{\sum_{l=1}^{L}[\mathbf{y}|M_l]\Pr(M_l)} \Big/ \frac{[\mathbf{y}|M_{l'}]\Pr(M_{l'})}{\sum_{l=1}^{L}[\mathbf{y}|M_l]\Pr(M_l)} \tag{9.2.5}$$

$$= \frac{[\mathbf{y}|M_l]}{[\mathbf{y}|M_{l'}]}\frac{\Pr(M_l)}{\Pr(M_{l'})} \tag{9.2.6}$$

$$= B_{l,l'}\frac{\Pr(M_l)}{\Pr(M_{l'})}. \tag{9.2.7}$$

Thus, after the data $\mathbf{y}$ have been observed, the posterior odds comparing two models can be written as a constant multiplier times the ratio of prior model probabilities (i.e., the prior odds). The multiplier

$$B_{l,l'} = \frac{[\mathbf{y}|M_l]}{[\mathbf{y}|M_{l'}]} \tag{9.2.8}$$

(continued)

(Box 9.2 *continued*)

is known as the *Bayes factor* and is a function of only the marginal data distributions from each model (Good, 1983). Thus, the posterior evidence in favor of one model over another is found by updating the prior evidence with the data. The interpretation of a Bayes factor is that model M_l is $B_{l,l'}$ times more probable than model $M_{l'}$ in light of the data. We find this statement sufficiently informative without further embellishment. However, similar to the various rules of thumb for comparing models using information criteria, several rules of thumb have been suggested in the literature for Bayes factors, for example, that $B_{l,l'} > 10$ implies strong evidence in favor of model M_l over model $M_{l'}$ (Jeffreys, 1961).

Despite the fact that Bayes factors are often used for model comparison (e.g., Kass and Raftery, 1995), they are especially relevant for model averaging (e.g., Link and Barker, 2006), because the posterior probability of model M_l can be written as

$$\Pr(M_l|\mathbf{y}) = \frac{B_{l,l'}\Pr(M_l)}{\sum_{l=1}^{L} B_{l,l'}\Pr(M_l)}, \tag{9.2.9}$$

and thus any BMA posterior distribution of interest can be found by computing

$$[g(\boldsymbol{\theta},\mathbf{y})|\mathbf{y}] = \sum_{l=1}^{L}[g(\boldsymbol{\theta},\mathbf{y})|\mathbf{y},M_l]\Pr(M_l|\mathbf{y}), \tag{9.2.10}$$

where $g(\boldsymbol{\theta},\mathbf{y})$ is typically a function of either an unknown parameter ($\boldsymbol{\theta}$) or set of data ($\mathbf{y}$) or both. For example, to find the BMA posterior for a single parameter θ that is common among all models we compute

$$[\theta|\mathbf{y}] = \sum_{l=1}^{L}[\theta|\mathbf{y},M_l]\Pr(M_l|\mathbf{y}) \tag{9.2.11}$$

after first finding each posterior model probability (using the Bayes factors). Of course, if we seek only to compare various models without performing BMA, then we need only to find the Bayes factors for those sets of models of interest. An important note, however, is that the Bayes factors will not necessarily indicate better-predicting models but, rather, models that weigh more heavily in Bayesian model averaging. A further advantage of the posterior model probabilities $\Pr(M_l|\mathbf{y})$ (and hence Bayes factors) is that they are guaranteed to find the data-generating model if it is in the set of models being compared (as the sample size gets large). Of course, in practice the data-generating mechanism is never known, but the fact that it can be recovered if known is comforting.

The use of equal prior model probabilities explicitly assumes that there may be no reason to prefer one model over another but, we can specify unequal probabilities if different models are assumed to be more influential a priori. To obtain the necessary marginal data distribution for model l we must integrate over the parameters in the joint distribution of the data $\mathbf{y}$, the model M_l, and the parameters $\boldsymbol{\theta}$, so that

$$[\mathbf{y}|M_l] = \int [\mathbf{y}|\boldsymbol{\theta}, M_l][\boldsymbol{\theta}]d\boldsymbol{\theta}. \tag{9.2.12}$$

Note that for a single model we can eliminate M_l, so equation 9.2.12 is the same expression that typically appears in the denominator of Bayes' theorem for continuous parameters (eq. 5.1.8).

We need only to compute the posterior model weights to find the averaged posterior distribution (eq. 9.2.1). Solving the integral in the marginal data distribution (eq. 9.2.12) is often impossible analytically, which is why most Bayesian studies use MCMC to obtain it. The sum in the denominator of the posterior model probability (eq. 9.2.3) can also become intractable as the number of models L grows.

BMA is attractive because it has a firm basis in statistical theory, but it is often challenging to implement in practice. Various methods exist for calculating the necessary quantities in Bayesian model averaging (e.g., Congdon, 2006), and we describe a widely used approach next. However, we first mention how model weights should *not* be calculated. For many Bayesian models, DIC can be used for ranking models and finding those that should predict better than others, just as AIC would. DIC addresses the issue of model complexity and in many cases yields results quite similar to those of AIC for similar likelihood-based models, but can it be used for Bayesian model averaging? If we follow the same approach used to calculate AIC weights (Burnham and Anderson, 2002), we would be tempted to use $w_j = e^{-\Delta\text{DIC}_j/2} / \sum_l e^{-\Delta\text{DIC}_l/2}$, where ΔDIC_j represents the difference between DIC for model j and the minimum DIC across all models in $\mathcal{M}$. Do these weights w_j approximate posterior model probabilities? Although this approach is seen in the literature,[6] it has no theoretical basis.

The current algorithm of choice for estimating the probabilities of models and Bayes factors is *reversible-jump MCMC* (RJMCMC). We define the parameters in the model as $\boldsymbol{\theta}_l$, noting that the lengths p_l of these parameter vectors $\boldsymbol{\theta}_l$ may vary among the models in set $\mathcal{M}$. In RJMCMC, we treat the

[6]See, for example, Link and Barker (2010) and Stauffer (2008).

model index l as a random variable to be modeled along with the set of all possible parameters $\boldsymbol{\theta}$. The posterior distribution of interest, then, is

$$[\boldsymbol{\theta}, l|\mathbf{y}] \propto [y|\boldsymbol{\theta}_l, l][\boldsymbol{\theta}_l|l][l], \tag{9.2.13}$$

where $[\boldsymbol{\theta}_l|l]$, is the prior distribution for the parameters for model M_l, and $[l]$ is the prior distribution for model M_l itself.[7] The beauty of this specification is that multimodel inference becomes model-based much like indicator variable selection (sec. 9.1.4).

The use of MCMC to implement this model (eq. 9.2.13) involves the usual steps: specify initial values for unknowns and then cycle through the unknowns, updating each one sequentially. However, difficulties arise when sampling the model index l, and hence its associated parameters $\boldsymbol{\theta}_l$, because the set of parameters being sampled changes depending on which model is being sampled. Thus, the potentially different model parameter sets must be accounted for when accepting a Metropolis-Hastings proposal for the parameters in an MCMC algorithm.

RJMCMC approaches have become a popular option for computing Bayes' factors and Bayesian model probabilities (e.g., Johnson and Hoeting, 2011). When prior model probabilities are assumed to be equal, the Bayes' factor ($B_{l,l'}$) can be computed simply by calculating the quotient of summed number of visits to each model (M_l and $M_{l'}$) in the RJMCMC algorithm (Hastie and Green, 2012). RJMCMC can be challenging (or impossible) to implement for complicated models. For details on a variation of RJMCMC that improves its tractability, see Link and Barker (2006), Link and Barker (2010), and Barker and Link (2013).

9.3 Which Method to Use?

It may come as a surprise to ecological researchers accustomed to using likelihood methods and AIC that there are many valid approaches to multimodel inference. There is no consensus in the statistical literature on which method is "best," nor should there be. Each of the approaches we have outlined has strengths and shortcomings, and it is the responsibility of the researcher to make an informed choice of the approach to use. To facilitate that choice, we offer some general considerations here.

[7]This is the first time you have encountered a prior on a model. The distribution $[l]$ simply specifies your prior knowledge of the probability of the model in the same way that $[\theta]$ specifies prior knowledge about the probability of a parameter.

The first factor to consider whether multimodel inference is needed or whether a single model will suffice. Often, it is perfectly justifiable to make inferences by conditioning on a single model. Gelman and Rubin (1995) and Ver Hoef (2015) offer thoughtful treatments of the decision to use one or more than one model to gain insight. It also is possible to use a single model for inference after winnowing a set using model checking (e.g., Gelman and Shalizi 2013; Gelman and Hill 2009).

However, if it is clear that multimodel inference is needed, then the following guidelines might prove helpful. The first choice is between model averaging and model selection. If inference is to be combined across several models in a set, then Bayesian model probabilities are the proper way to obtain a posterior distribution averaged across the set. Model probabilities can be computed using RJMCMC. They can also be approximated using BIC, but only under certain circumstances. Despite its name, the BIC approximation is not terribly useful in a fully Bayesian analysis, because BIC requires maximizing the likelihood and counting parameters. Hoeting et al. (1999) summarized methods for approximating model probabilities that enjoy a formal foundation in statistical theory. It is perhaps important to note that most information criteria (e.g., AIC, DIC, WAIC) lack this foundation for the purposes of Bayesian model averaging.

If model selection as opposed to model averaging is chosen as the approach to analysis, then out-of-sample validation is the method of choice when data have been withheld from model fitting to allow them to be used for model selection. When the same data must be used for both purposes—fitting and selection—then some sort of M-fold scheme for cross-validation remains a very strong choice for model selection if its implementation does not impose excessive computational demands.

When cross-validation is computationally infeasible, then information criteria are especially useful. DIC has become popular as a result of its convenience—it is easily obtained using software for MCMC. Results from DIC resemble those from AIC, and both tend to favor highly parameterized models when the sample size is large. The biggest problem with DIC occurs when the posterior mean of the parameters fails to describe the central tendency of the posterior distribution. As a result, DIC is potentially inappropriate for mixture models widely used in ecology.[8] Furthermore, use of DIC should be limited to cases in which the number of effective parameters is much smaller than the number of observations, which is often not the case in hierarchical models where the number of latent variables is related to sample size.

[8]For example, zero-inflated models including occupancy and capture-recapture models.

WAIC offers an attractive alternative to DIC because it uses the posterior predictive distribution as a basis for model evaluation and, as a result, is more consistent with the Bayesian philosophy than DIC is. However, the use of WAIC is limited by its inherent assumption of independence among the data. This assumption means that it is not appropriate for many time-series or spatial models. For these models, posterior predictive loss can be used.

Elegance favors a fully integrated procedure for estimating parameters and model selection. For estimating parameters and selecting among alternative models in the same analysis using within-sample data, indicator variable selection methods will be appealing. These model-based algorithms require tuning, but when properly tuned, they perform well and can be more efficient than cross-validation.

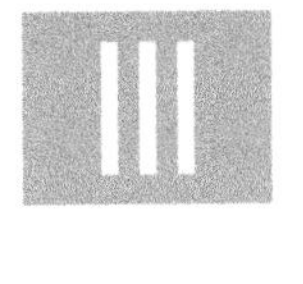

Practice in Model Building

A central aim of this book is to teach ecologists how to write Bayesian models for a broad variety of research problems. The goal of part III is to provide practice in model building. In chapter 10, we describe a framework for thinking that we routinely use to design models and bring them together with data. Next, in chapter 11 we provide a set of structured problems accompanied by solutions in chapter 12. We have found in our teaching that encouraging students to work problems is the only way to be sure they grasp the concepts we covered in parts I and II. It is all too easy to read material with a false confidence that you understand it. Confronting a problem reveals gaps in understanding that are filled by seeing the solution or by reviewing the relevant material from earlier chapters. Making a strong effort to solve problems will deepen your understanding of concepts even if your solution goes wrong somewhere, or perhaps *particularly* if it goes wrong.

10 Writing Bayesian Models

We specify a model by writing it in mathematical and statistical notation. Recall that model specification occupies a central position in the modeling process, linking design to implementation (fig. 0.0.1). The crucial skill of specifying models is often neglected in statistical texts in general and texts on Bayesian modeling in particular. The central importance of model specification motivates this chapter.

Building models is a learned craft, and no two modelers conduct their craft in exactly the same way. It can be daunting to confront a problem with no idea where to start—staring at a menacing blank page, so we offer a general set of steps for writing models that we have found useful in our own work. This approach is not immutable; the sequence of steps is not fixed, and as your craft develops you will probably develop your own way of attacking problems.

10.1 A General Approach

We first introduced a general framework for linking models of ecological processes to data in the preview of the book (chapter 1), promising that we would teach the fundamentals needed to understand how to make this linkage. We have made good on that promise if you have absorbed the material in parts I and II.

The overarching challenge in building models is to specify the components of the posterior distribution and the joint distribution and to factor the joint distribution into sensible parts. In this section we lay out a framework for doing that. We assume that you have already thought about the design of your research. This thinking has allowed you to compose an initial list of quantities that will be observed and to specify the unobserved quantities you wish to learn about from the observed ones. You will likely modify your lists of observed and unobserved quantities as you sketch the relationships between them, forcing you to think more deeply about the problem at hand. Solid progress can be made toward developing a full, properly factored expression for the posterior and joint distributions by structuring your thinking and your scribbling[1] in the following steps. The steps described here are general and somewhat abstract. We make them concrete in the next section.

1. **Start with the data.** Focus initially on a single data point to keep things simple. Think about how the data arise. Don't worry too much about subscripting the parameters and observations initially. Subscripts can be refined later as your thinking becomes more clear. Often, there is an unobservable state z that gives rise to an observation y. That state might be the size of a population, the number of species within an area of landscape, or the grams of CO_2 emitted from a unit of soil during a given interval of time. A critical idea at this point is the relationship between the data and the state you wish to understand. If you can observe the state *perfectly*, without bias or sampling error, then it is unnecessary to model the true state separately from the observation, because the true state is observable, which means that $z = y$.
2. **Think about the observing system.** Very often we observe a "surrogate" for the true state, because the surrogate is easier to observe than the state itself. For example, we use light detection and ranging (Lidar) to estimate canopy height instead of getting out big ladders and meter tapes. We use the normalized difference vegetation index (NDVI) to estimate standing crop biomass of vegetation rather than clipping plots. We might estimate the mass of individual fish from their lengths, or decomposition rate from disappearance of carbon from a buried bag. We might use distance sampling or capture-recapture models to correct for animals that are present but uncounted in a census. These are examples of observation models $h(\theta_d, z)$, models that predict what we observe from the true value producing the observations. This

[1]At the risk of sounding desperately old-fashioned, we find the best tools for supporting thinking are a sheet of paper and a pencil or a whiteboard and marker. There is no software that can combine notes, drawings, and equations as efficiently as these traditional tools.

prediction (or calibration) has its own uncertainty, which we notate as σ_o^2, realizing that σ_o^2 may be a placeholder for parameters we will derive later using moment matching. A model of the observing system represents the distribution of the observations conditional on parameters in a deterministic model relating observations to the true state and parameters for uncertainty in that relationship (fig. 10.1.1 A).

3. **Think about the sampling process.** A vector of observations $\mathbf{y}$ often corresponds to a vector of true states $\mathbf{z}$, where the correspondence is controlled by the observation model. The true states, in turn, arise as a sample from a higher-order parameter μ that we seek to understand. For example, we might want to know the average mass of fish in different streams based on a sample of lengths of fish from each stream. Our observation model would convert observed lengths to masses (with uncertainty), and a sampling model would account for uncertainty arising from taking a limited number of observations within each stream (fig. 10.1.1 B).
4. **Model the ecological process.** We are now ready to specify how the true, unobserved state μ is predicted by a deterministic model of an ecological process, which usually includes additional observations as covariates, that is, $g(\boldsymbol{\theta}_p, \mathbf{x})$. The process model might include effects resulting from imposition of treatments (e.g., sec. 6.2.3) or effects resulting from natural variation (e.g., sec. 6.2.4) or both. All the influences on μ that are not included in our model are represented stochastically, notated in this example as σ_p^2 (fig. 10.1.1 C).
5. **Think about uncertainty in the covariates.** Ecologists often pretend that covariates are measured perfectly—wishful thinking that assuredly violates a fundamental assumption of traditional linear regression. Sweeping uncertainty in covariates under the rug is unnecessary. Uncertainty arising in observing and sampling covariates can be accommodated in precisely the same way as uncertainties are included in observations of responses. To keep things simple, we assume here that covariates are measured without error (fig. 10.1.1 C). Modeling uncertainty in covariates is treated in problem 2 in the next chapter.
6. **Think about modeling variation in the parameters of your process model.** You may want to represent variation in parameters of the process model arising from differences among sites, species, time intervals, genotypes, and so on. If it is sensible to include this variation, you may wish to model it using a group-level effect (also called a random effect; fig. 10.1.1 D).
7. **Draw the full Bayesian network.** You are now equipped to sketch a full Bayesian network from component models by connecting the

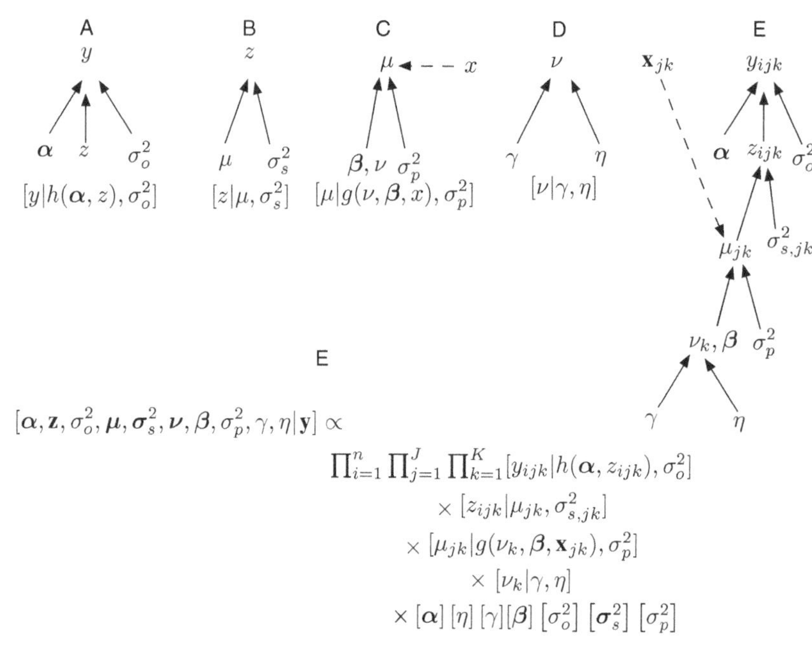

Figure 10.1.1. Bayesian networks illustrating a general approach for specifying models. Subscripts are omitted from panels **A–D** to reduce clutter. (**A**). Observation model used to correct for bias or transform the value for the true state (z) to an observation of the state (y). The parameter σ_o^2 accounts for the uncertainty in modeling the observed state based on the true state. (**B**). Model accounting for uncertainty in estimate of the true state arising because our observations are a sample of all possible instances of the true state; σ_s^2 accounts for uncertainty arising from sampling. (**C**). Process model explaining variation in the true state based on a covariate. The parameter σ_p^2 accounts for the uncertainty that arises because the process model does not include all the sources of variation in the true state. (**D**). Model of variation in the parameter ν. (**E**). Full model. Subscripts chosen for this example represent three levels in a spatial hierarchy: $k = 1, \ldots, K$ large-scale units (e.g., locations), and $j = 1, \ldots, J$ areas within large-scale units (e.g., plots) that are subsampled $i = 1, \ldots, n$ times. All unobserved quantities appear on the left-hand side of the conditioning in the posterior distribution; observed quantities appear on the right-hand side. Note that the heads of arrows appear on the left-hand side of conditioning | in the joint distribution, and the tails of arrows appear on the right-hand side. Any unobserved quantity that is at the tail of arrows and is not at the head of an arrow requires a prior with numeric arguments.

shared stochastic quantities (i.e., nodes) (fig. 10.1.1 E). At this point it is a good idea to add proper subscripts to reflect the structure of the data and latent states, particularly their nesting.

8. **Write the posterior and the joint distributions using general distribution notation.** As we discussed in section 3.3, a properly

sketched Bayesian network provides all the information you need to write an expression for the posterior distribution and a properly factored joint distribution (fig. 10.1.1 E). If you like, you can write this expression for a single data point before adding products over likelihoods and associated distributions of latent states.

9. **Choose specific probability distributions.** You must replace bracket notation with specific probability mass functions and/or probability density functions based on your knowledge of the support of each random variable. Use moment matching as needed to form proper arguments for specific distributions.
10. **Do simple things first**. It is often helpful to build complex models in stages, starting with the simplest possible model and adding complexity (and realism) incrementally. We illustrated this incremental approach in the example on spotted owls (sec. 6.2.1) and we will apply it to several problems in chapter 11.

You now have a fully specified Bayesian model that contains all the information needed to develop an MCMC algorithm from scratch or by using software. You have a drawing and an expression that you can productively discuss with your statistical colleagues.

We reemphasize that you can modify these steps to fit your own style. For example, Hooten chooses specific distributions during steps 2 through 5 rather than at the end in step 8. He dispenses with drawing the Bayesian network altogether, writing the math at each step without sketching the stochastic relationships. Hobbs generally uses the steps here but often writes specific distributions for each level of the hierarchy rather than using general bracket notation, thereby obviating step 9.

We also emphasize that all research problems will not involve all the hierarchical levels that we illustrate here. For example, the process model might make predictions at the level of the unobserved state (z) rather than the level from which states are drawn (μ). Experience and practice coupled with knowledge of your own research problem will allow you to decide which levels are needed. In the next section we make the general approach more concrete with a simple, specific example.

10.2 An Example of Model Building: Aboveground Net Primary Production in Sagebrush Steppe

We offer an example of the general approach just outlined based on Manier and Hobbs (2007), who studied effects of grazing by livestock and wild ungulates on structure and function of a sagebrush steppe ecosystem. Their analysis was not Bayesian, but the research design they used offers a nice example of the model components we described in the previous section. We have modified the design slightly to make it more accessible in this introductory example.

The design included six replications at different locations in northwestern Colorado. Each replication included three treatments on 1 ha plots established during the 1940s, providing an unusual opportunity to study long-term effects of herbivory. The first treatment used 2 m–high fences to exclude grazing by wild ungulates (mule deer, *Odocoileus hemionus*; and elk, *Cervus elaphus*) and by domestic livestock (sheep, *Ovis aries*; and cattle, *Bos taurus*). In the second treatment, three-strand barbed wire allowed grazing by wild ungulates while excluding livestock. A 1 ha unfenced control adjacent to the treatments allowed grazing by all ungulates. We will refer to each of the exclosures and the matched control as macroplots. We are interested in the way that the grazing treatments influenced aboveground net primary production (ANPP).

ANNP is estimated in our modified example as peak standing crop of live tissue at the end of the growing season. Standing crop of live tissue is notoriously tedious to measure, requiring time-consuming clipping, sorting, and weighing of plant samples. Dramatic increases in efficiency can be gained from observations more easily obtained—in this case, counts of pin hits on live plant tissue (McNaughton et al., 1966). A rectangular frame with a 10×40 grid of holes drilled on the diagonal through plywood is suspended over a plot. Steel pins are pushed through the frame until they touch vegetation or soil. The category of vegetation touched by each pin is recorded. Counts of pin touches with live biomass are regressed on clipped samples of biomass to develop an observation model.

To summarize the design, there were observations of $i = 1, \ldots, 10$ pin counts on subplots within $j = 1, \ldots, 3$ macroplots (treatments) within $k = 1, \ldots, 6$ replicates.

10.2.1 Observation Model

Pin counts ($\dot{y}_l$) and paired clip-and-weigh measurements $\dot{z}_l$ of aboveground biomass were obtained on thirty $1\,\text{m}^2$ subplots outside the macroplots indexed by l. The dot notation indicates that these data were used for calibration. The number of pin hits within a subplot was regressed on biomass in the subplot measured by clip, sort, and weigh methods. It is reasonable to assume that the clip-and-weigh samples were measured without error. Thus, we use the deterministic model (fig. 10.1.1 A):

$$h(\boldsymbol{\alpha}, \dot{z}_l) = \alpha_o + \alpha_1 \dot{z}_l.$$

We use this deterministic model to calibrate what we observe (pin hits, $\dot{y}_l$) to a measurement of what we want to understand (aboveground biomass, $\dot{z}_l$). Doing so requires estimating the coefficients $\boldsymbol{\alpha}$:

$$[\boldsymbol{\alpha}|\dot{y}_l] \propto \text{Poisson}(\dot{y}_l|h(\boldsymbol{\alpha}, \dot{z}_l))\,\text{normal}(\alpha_0|0, 1000)\,\text{normal}(\alpha_1|0, 1000). \tag{10.2.1}$$

Notice that the easily observed quantity is the response, and the quantity more difficult to observe is the predictor. The calibration equation is arranged this way because we will use it to predict what we observe from the unobserved true state in our full model.

Estimates of the coefficients ($\boldsymbol{\alpha}$) from the off-plot regression (eq. 10.2.1) are used as informative priors for the coefficients in the observation model,

$$y_{ijk} \sim \text{Poisson}(h(\boldsymbol{\alpha}, z_{ijk})), \tag{10.2.2}$$

where y_{ijk} is the pin count on subplot i of treatment j in replication k. Notice that the parameter σ_o^2 has dropped out of our model[2] (fig. 10.1.1), because we are assuming a Poisson likelihood, an assumption that will need to be checked (sec. 8.1).

10.2.2 Sampling Model

We assume that the true mass on each subplot i (fig. 10.1.1 B) is drawn from a distribution with median μ_{jk}. We can model sampling uncertainty σ_s^2 using

$$z_{ijk} \sim \text{lognormal}(\log(\mu_{jk}), \sigma_{s,jk}^2). \tag{10.2.3}$$

We choose the lognormal distribution because the true biomass is strictly nonnegative and continuous. The parameter $\sigma_{s,jk}^2$ represents the sampling variance of z_{ijk} on the log scale. We estimate a different sampling variance for each macroplot, which is the reason for the jk subscript. In our next step, we develop a model of the μ_{jk}, the median aboveground biomass on each plot of each replicate.

[2]It hasn't really left. The Poisson assumes that σ_o^2 is equal to the mean of the distribution, and the mean is predicted by $h(\boldsymbol{\alpha}, z_{ijk})$. Therefore, $\sigma_o^2 = h(\boldsymbol{\alpha}, z_{ijk})$.

10.2.3 Process Model

The process model (fig. 10.1.1) representing the effect of grazing is

$$g(\nu_k, \boldsymbol{\beta}, \mathbf{x}_{jk}) = \exp(\nu_k + \beta_1 x_{1,jk} + \beta_2 x_{2,jk}), \tag{10.2.4}$$

$$\mu_{jk} \sim \text{lognormal}(\log(g(\nu_k, \boldsymbol{\beta}, \mathbf{x}_{jk})), \sigma_p^2), \tag{10.2.5}$$

where $x_{1,jk}$ is an indicator variable equal to 1 if macroplot jk is grazed by livestock and 0 otherwise, and $x_{2,jk}$ is an indicator variable equal to 1 if the macroplot is grazed by wild ungulates and 0 otherwise. The indicator variables, of course, are treated as known, fixed quantities. The parameter σ_p^2 represents the variance of μ_{jk} on the log scale. The intercept ν_k allows for variation among locations (i.e., replicates), as described next. If we wished to understand the interaction between the two types of grazing, that is, the extent to which the effect of wild ungulate grazing depends on livestock grazing, we could add a third coefficient to the deterministic model:

$$g(\nu_k, \boldsymbol{\beta}, \mathbf{x}_{jk}) = \exp(\nu_k + \beta_1 x_{1,jk} + \beta_2 x_{2,jk} + \beta_3 x_{1,jk} x_{2,jk}). \tag{10.2.6}$$

10.2.4 Group-Level Model

We acknowledge that variation among replicates is not represented in our process model (fig. 10.1.1). We include that variation by modeling the intercept in the deterministic model using

$$\nu_k \sim \text{gamma}(\gamma, \eta). \tag{10.2.7}$$

10.2.5 Posterior and Joint Distribution

We can now write an expression for the posterior and joint distributions using the Bayesian network (fig. 10.1.1 E) as a guide and filling in specific distributions using equations 10.2.1–10.2.7. Thus,

$$\left[\boldsymbol{\alpha}, \mathbf{z}, \boldsymbol{\sigma}_s^2, \boldsymbol{\mu}, \boldsymbol{\nu}, \gamma, \eta, \boldsymbol{\beta}, \sigma_p^2 | \mathbf{y}\right] \propto \prod_{k=1}^{6} \prod_{j=1}^{3} \prod_{i=1}^{10} \text{Poisson}\left(y_{ijk} | h(\boldsymbol{\alpha}, z_{ijk})\right)$$

$$\times \text{lognormal}\left(z_{ijk} | \log(\mu_{jk}), \sigma_{s,jk}^2\right)$$

$$\times \text{lognormal}\left(\mu_{jk} | \log(g(\nu_k, \boldsymbol{\beta}, \mathbf{x}_{jk})), \sigma_p^2\right)$$

$$\times \text{ inverse gamma}(\sigma^2_{s,jk}|.001, .001)$$

$$\times \text{ gamma}(\nu_k|\gamma, \eta)\text{gamma}\left(\alpha_0 \,\middle|\, \frac{.50^2}{.03^2}, \frac{.50}{.03^2}\right)$$

$$\times \text{ gamma}\left(\alpha_1 \,\middle|\, \frac{.47^2}{.02^2}, \frac{.47}{.02^2}\right)$$

$$\times \text{ inverse gamma}(\sigma^2_p|.001, .001)\text{normal}(\beta_1|0, 1000)$$

$$\times \text{ normal}(\beta_2|0, 1000)\text{gamma}(\gamma|.001, .001)$$

$$\times \text{ gamma}(\eta|.001, .001). \tag{10.2.8}$$

Note the numeric arguments for the priors on α_0 and α_1 that were obtained from the off-site calibration of pin hits to clipped biomass (eq. 10.2.1) and converted to the parameters of the gamma distribution using moment matching. We use inverse gamma priors on the variances (σ^2_s, and σ^2_p), because the inverse gamma is a conjugate distribution for the lognormal variance.[3]

10.2.6 Model Checking and Analysis

It is particularly important to check whether the model is able to adequately represent the dispersion in the data because we initially chose a Poisson likelihood, which may not be appropriate because of the assumption that the mean and variance are equal. With a properly fit and checked model, inference will focus on the parameters ν, β_1, and β_2. Because these parameters are exponentiated in our model, we can enhance the interpretation of treatment effects by creating some derived quantities within the MCMC algorithm. The posterior distribution of $\exp(\gamma/\eta)$ gives the mean biomass in the absence of grazing.[4] Now, consider that our deterministic model can be rearranged as

$$g(\nu_k, \boldsymbol{\beta}, \mathbf{x}_{jk}) = e^{\left(\nu_k+\beta_1 x_{1,jk}+\beta_2 x_{2,jk}\right)} = e^{\nu_k} e^{\beta_1 x_{1,jk}} e^{\beta_2 x_{2,jk}}. \tag{10.2.9}$$

Thus, e^{β_2} is the *multiplicative* change in ANPP attributable to wild ungulate grazing alone. So, for example, if the mean of the posterior distribution

[3] Using a conjugate distribution obviates the need for tuning, but we would get essentially the same answer if we used a vague gamma prior on the variance.

[4] Recall that the mean of the gamma distribution is γ/η.

of e^{β_2} equals 0.90, then production is reduced by 10%, on average, in the wild ungulate grazed treatments relative to the ungrazed ones. The derived quantity $e^{(\beta_1+\beta_2)}$ is the ANPP multiplier attributable to the combination of livestock and wild ungulate grazing. The derived quantity e^{β_1} is the multiplicative effect of livestock grazing alone, and e^{β_2} is the multiplicative effect of wild ungulate grazing alone. If the posterior distributions of the exponentiated quantities strongly overlap 1, or if the posterior distributions of the coefficients (β_1, β_2, $\beta_1+\beta_2$) strongly overlap 0, then we conclude that we cannot detect effects of grazing. Note that this design allows inference about the effect of livestock grazing alone even though there was not a treatment that excluded wild ungulates and allowed grazing by livestock, because we can subtract the effect of ungulate grazing from the combined effect; that is, $e^{\beta_1} = e^{\beta_1+\beta_2-\beta_2}$.

11 Problems

Here, we offer a set of worked problems to help you actively hone your skills in model building. Each problem will require you to draw a Bayesian network and write the posterior and joint distributions. Computational syntax is deemphasized because there are many excellent sources that teach the details of writing code for MCMC software using ecological examples (e.g., McCarthy, 2007; Kéry, 2010; Kéry and Schaub, 2012) and a few that show you how to construct your own MCMC algorithms (Clark, 2007; King and Gimenez, 2009). Writing detailed code requires committing to a particular language, which would narrow what we can offer here to a single flavor of popular software.

The best way to benefit from the material is to do the problems in this chapter individually, consulting the answers in the next chapter as you proceed. Most problems will be composed of subcomponents. Because these become progressively more challenging, it might be a good idea to match your solution to the given answer for each subcomponent before proceeding to the next one. That way you will not get terribly far off the correct path to success.

Three more points are worth mentioning. We offer examples across several ecological sub-disciplines, but bear in mind that every problem can be instructive regardless of its resemblance to your own research. It might take a bit of mental effort to translate the examples here to more familiar ones, but the effort will be worthwhile. The key to success is to think about the support of random variables and imagine parallels in your field. Virtually all ecologists count things and weigh things and observe them in categories. It doesn't matter what those "things" are—grams of phosphorus in the soil, numbers of diatoms in a lake, forested area of landscape, or age and sex composition of a population. Once turned into numbers, they can usually be modeled in a similar way.

A second point is the context for model building, that is, how the process fits into a flow of research from design to analysis. (fig. 0.0.1) There might be a tendency to think the model-building process is something you do after the data have been collected. Sometimes this is necessary, but it is never preferable to building models as part of the research design. Careful thought about the model or models you will use to gain insight can be a powerful way to support your thinking on how to conduct your research, aiding in decisions on adequacy of sample, choice of locations, duration of observations, and other issues.

The final point is a bit of encouragement to forge ahead through this material. Notice that we can attack a diverse suite of problems using a single approach. Learning this approach is challenging, but rest assured that your efforts will be rewarded. Writing a Bayesian model gives you a compact expression that unequivocally describes how you will obtain insight and where uncertainties will arise. This expression can be a compelling feature of papers and proposals, showing that you know what you are doing without any bluffing. You have an accurate blueprint for writing the computer code needed to accomplish a fully Bayesian analysis; you are able to clearly communicate your analysis to your statistical colleagues in a way that will make them much more likely to collaborate with you; and you have a single tool that you can deftly apply to virtually all analysis problems in ecological research.

11.1 Fisher's Ticks

R. A. Fisher (1941) presented data on the number of ticks found on 60 sheep. For now, assume that counts were accurate and that no ticks were missed on any single count but that counts on any individual varied depending on the day they were made, creating sampling variation.

1. Diagram and write an expression for the posterior and joint distributions of the mean number of ticks per sheep assuming the counts are distributed as a Poisson random variable. Justify the Poisson as your initial choice for the data model. Write a closed-form expression for the posterior distribution of the mean number of ticks per sheep (see sect. 5.3). Describe what you would do to check your model.
2. Your model check showed that a Poisson likelihood was not capable of representing the dispersion of the observed data. Revise your model to remedy this problem. Draw a Bayesian network and write the posterior and joint distributions of the revised model.

3. Now, assume that previous research showed that ticks are undercounted. An average of 8% (standard deviation = 2%) of the true number of ticks are missed. Revise your model to include this uncertainty. Demonstrate your skill with moment matching (sec. 3.4.4) to show how you would use these data to specify values for the parameters in an informative prior distribution.
4. You discover unpublished data specifying the age, sex, and body mass of each of the 60 sheep, which you assume were measured without error. Expand on the model you composed in step 3 by including a process model using these covariates to explain individual tick burdens. Diagram and write the posterior and joint distributions for unobserved quantities conditional on the observed ones.

11.2 Light Limitation of Trees

This problem illustrates how we might model the data of Coates and Burton (1999), who studied the effects of light availability on the growth of five species of coniferous seedlings in coastal forests of British Columbia. They observed increments in heights of seedlings 5 years after they were planted in response to an index of light availability measured once at the end of the fourth growing season. There is no sampling variation because each tree has a single height at the beginning and end of the 5 years. We can safely ignore estimation error because height can be measured to the nearest centimeter for trees that are meters tall. Thus, the observed height is treated as the true height. The x_i are expressed as proportion of total sunlight, so they range from 0 to 1. The original design assumed the light index was measured perfectly.

1. Use the deterministic model

$$g(\alpha, \gamma, c, x_i) = \frac{\alpha (x_i - c)}{\frac{\alpha}{\gamma} + (x_i - c)} \tag{11.2.1}$$

   to portray light limitation for a single tree species, where α is the maximum annual growth increment at infinite light; γ is the slope of the curve at low light, and c is the light level where the growth increment is zero. Assume you have observations of height increment (y_i) paired with perfect observations of light availability (x_i) for 50 individual trees.
2. We take the liberty of modifying the design to make this problem more challenging. Assume you have three replicate observations of

light availability for each tree. Model observed light availability as a random variable arising from a distribution with a mean z_i, the unobserved, true average light availability for the ith tree.

3. Assume you have data for a second tree species similar to the data in step 2. You would like to make inference about the ability of the two tree species to become established in habitats with low light. Modify your model and create a derived quantity to examine effects of species on c.

11.3 Landscape Occupancy of Swiss Breeding Birds

This problem is a classic in landscape and wildlife ecology. We modified the original problem described by Royle and Dorazio (2008) to illustrate the utility of derived quantities in Bayesian analysis.

A fundamental challenge in landscape ecology is to understand how environmental heterogeneity shapes the spatial distribution of species. In this problem, we model landscape use by the willow tit (*Parus montanus*), a resident bird in the Swiss Alps. The Swiss Survey of Common Breeding Birds collects data on presence or absence of avian species in annual surveys of 1 km^2 quadrats distributed across Switzerland. Surveys are conducted during the breeding season on three separate days. During each survey, an observer walks a transect within each quadrat and records every visual or acoustic detection of a breeding species—in this example, the willow tit—and marks its location using a global positioning system. We reasonably assume that the true state of the quadrat (occupied or unoccupied) does not change among sample dates, because the willow tit is a resident, and we are sampling during the breeding season.

Observers recorded a 1 if a bird was observed on plot i on sampling occasion j and a 0 if it was not observed. J_i is the number of times that quadrat i was searched. Thus the data for each plot can be thought of in two ways—in binary form (a vector of zeros and ones) and as a positive integer, the number of times a bird was observed, that is, the sum of the binary data. A total of n quadrats were searched.

1. The challenge in this analysis is that birds will frequently be present but not observed. You will use the total number of times a bird was observed on quadrat i (y_i) as data for this problem. Develop a data model that uses a detection probability (p) to portray how the data arise from the true state of the quadrat. Represent the true state by the latent variable z_i, where $z_i = 1$ if the quadrat is truly occupied, and

$z_i = 0$ if it is not. Link the data model to a model of the occupancy process by modeling y_i conditional on z_i and p, and z_i conditional on ψ_i, the true probability that a bird is present on quadrat i. At least some of the J_i must be greater than 1 in this example, because repeated observations at the same location are needed to estimate detection probability. Sketch the Bayesian network and write an expression for the posterior and joint distributions.

2. You have data on the elevation of the quadrat at its centroid $x_{1,i}$ and the proportion of the quadrat in forest cover $x_{2,i}$, so that you can model ψ_i as a function of quadrat-specific covariates. Assume that you wish to estimate the posterior distribution of a derived quantity, the optimum elevation for occupancy at the average forest cover, where the optimum is defined as the elevation with the maximum probability of occupancy. Develop a model that allows you to make this inference. As a suggestion, recall the inverse logit function (sec. 2.2.1.3). Consider the kind of model you would need to find a maximum probability of occupancy as a derived quantity of the model's parameters. Draw the Bayesian network and write the posterior and joint distributions.
3. What could you learn from the coefficients predicting occupancy? Do you need to do model selection?
4. Until now we have assumed that the detection probability is constant across quadrats. You have standardized covariates that influence the detection process: length of transect $u_{1,ij}$, duration of search $u_{2,ij}$, and date $u_{3,ij}$, all measured without error. Model the detection probability using these covariates. Draw the Bayesian network and write the posterior and joint distributions. You will use the binary form of the data for this problem.

11.4 Allometry of Savanna Trees

Tredennick and colleagues (2013) tested alternative models of allometric scaling of tree size in savanna ecosystems in West Africa based on an extensive set of morphometric measurements of 286 individual branches on 25 trees of three species. Theory predicts that tree mass M scales with basal diameter D according to the power function

$$M = aD^b, \tag{11.4.1}$$

where a is a normalizing constant, and b is the scaling exponent. *Metabolic scaling theory* predicts that aboveground mass should scale as $b = 8/3$

based on optimization of resource distributions (nutrients and water) within plants. Alternatively, the *geometric stress hypothesis* predicts the scaling exponent should be $b = 5/2$ as a consequence of mechanics of supporting plant canopies. The *geometric similarity model* uses different logic based on canopy mechanics to predict $b = 3$. Inference focuses on estimating the posterior distribution of the parameter b to determine which theory has the greatest support in data.

1. Assume you have observations on diameters and masses of T individual trees of the same species. Assume that these measurements are made without error. Diagram and write a simple Bayesian model to find the posterior distributions of a and b using the log form of equation 11.4.1 as your deterministic model.
2. Assume that there are three tree species indexed by $s = 1, \ldots, 3$ and that T_s trees are measured for species s. Diagram and write out a model showing the relationship between tree mass and basal diameter where each tree species has its own parameter a_s and b_s reflecting variation among individual trees.
3. Assume a_s and b_s are drawn from a global distribution reflecting variation among tree species. Use μ_a and σ_b^2 to notate the parameters of the global distribution of a_s and μ_b and σ_b^2 for the distribution of b_s.
4. Assume that tree diameters are measured imperfectly. To account for these measurement errors, expand the model you developed in step 4 to include a Berkson "error in variables" model (Dellaportas and Stephens, 1995) assuming 5% error on at least 5% of trees. Prior data (Price et al., 2009) inform the relationship between the true diameter ρ_{st} of tree t of species s and the measured diameter D_{st},

$$\log(\rho_{st}) \sim \text{normal}(\log(D_{st}), \sigma_o^2),$$

where $\sigma_o^2 \sim \text{lognormal}(4.135, 2)$.
5. Assume you have fit and checked the model in step 4. How would you make inference about the three theories?

11.5 Movement of Seals in the North Atlantic

This problem involves the use of statistical modeling to study animal movement based on telemetry data. Suppose that an individual animal's position in two-dimensional space for a given time t can be described by a two-dimensional vector indexed by t. We could view the position vector as arising from a dynamic stochastic process of movement in either discrete or continuous time. To study this type of process, telemetry data are typically collected to provide some measure of location for the individual at a discrete set of times. Thus, for a given time during the study period, the true position of the individual is commonly unknown and hence is treated as a latent variable. Uncertainty remains about the true position because of telemetry error, even at the times for which data are available.

Jonsen et al. (2005) demonstrated an approach to using these types of data to study the movement of mammals in the North Atlantic. To simplify this problem, assume you have telemetry data on hooded seals (*Cystophora cristata*) obtained at a set of regular time intervals.

1. Construct a data model: Assume for now that the true positions of an individual seal are known for a regular set of time points. Specify a model for the telemetry observations conditioned on those true positions for all time points. Your model should have a single unknown parameter controlling the dispersion of the telemetry error. Assume the telemetry error arises from a symmetric distribution, that is, a distribution that does not have a directional bias. Assume that the telemetry error variance has a strongly informative prior. Use the notation $\mathbf{s}_t = (s_{1,t}, s_{2,t})'$ to represent the vector of telemetry observations at time t, where $s_{1,t}$ gives the latitude, and $s_{2,t}$ gives the longitude. Use $\boldsymbol{\mu}_t = (\mu_{1,t}, \mu_{2,t})'$ to describe the true position of the individual at time t for a set of times, say $t = 1, \ldots, T$.
2. Construct the process model: Assuming that a pure discrete-time random walk is a reasonable model for the movement of hooded seals, construct a dynamic process model for a given individual hooded seal position that is parameterized using no more than one unknown variable (assuming the positions at a regular set of time points are known, for now). To simplify the statistical notation, assume the time step between positions is one unit.
3. Specify prior distributions for the unknown variables in steps 1 and 2 that are flexible enough to incorporate prior information if available. At this point, there should be only two univariate prior distributions.
4. Construct the hierarchical model: Now, using the component models in steps 1–3, assume that the true position is unknown at all times, and couple the models to form a single hierarchical model in which telemetry data arise from a distribution that depends on the underlying movement process as well as the other data model parameter, and the true unknown position for any time t arises from a distribution that depends on itself at time $t - 1$ as well as an additional process

model parameter. The final component of the hierarchical model then contains the prior distributions specified in step 3.

5. Write the posterior distribution for the hierarchical model in step 4 in terms of its proportionality with respect to the joint distribution, as we have done previously.

12 Solutions

12.1 Fisher's Ticks

Diagram and write an expression for the posterior and joint distributions of the mean number of ticks per sheep assuming the counts are distributed as a Poisson random variable. Justify the Poisson as your initial choice for the data model. Write a closed-form expression for the posterior distribution of the mean number of ticks per sheep (see sec. 5.3). Describe what you would do to check your model.

There is a single stochastic relationship in this model (fig. 12.1.1 A). The data are counts. A logical place to start modeling count data is with a Poisson likelihood. The gamma distribution provides the appropriate prior for the mean number of ticks per sheep, because the mean is nonnegative and continuous and because the gamma is conjugate to the Poisson. The expression for the posterior and joint distributions is

$$[\lambda|\mathbf{y}] \propto \prod_{i=1}^{60} \text{Poisson}(y_i|\lambda)\text{gamma}(\lambda|.001, .001).$$

We specify the parameters of the gamma distribution to minimize its influence on the posterior. This simple formulation assumes all sheep have the same average number of ticks. Variation arises because the count on each sheep is a Poisson-distributed random variable drawn from a distribution with a mean (λ) shared by all individuals. Conjugacy (sec. 5.3) allows us to find the posterior distribution in closed form using

$$[\lambda|\mathbf{y}] = \text{gamma}\left(.001 + \sum_{i=1}^{60} y_i, .001 + 60\right). \tag{12.1.1}$$

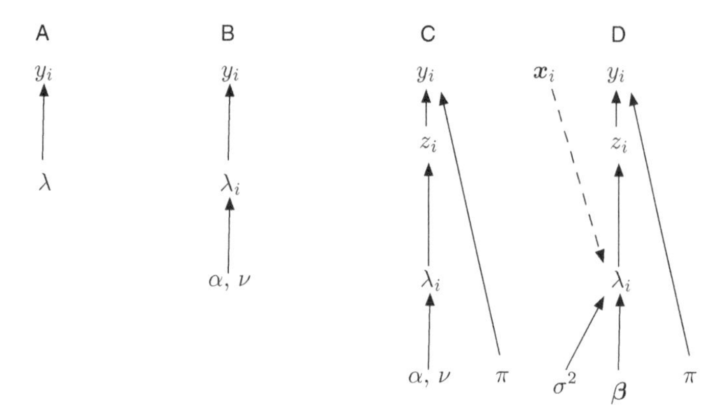

Figure 12.1.1. Bayesian networks for problem 1, Fisher's ticks. See the text for definition of symbols and corresponding models.

There are a couple of ways to check the model. The first is simply to look at the mean and the variance of the data. If the variance is not approximately equal to the mean, then the Poisson is probably not a good choice for the likelihood. The model can be checked further by estimating λ using the MCMC algorithm and simulating a new data set at each iteration $k = 1, \ldots, K$ of the MCMC output. This requires making 60 draws from the likelihood

$$y_i^{new(k)} \sim \text{Poisson}(\lambda^{(k)}) \tag{12.1.2}$$

to create the vector $\mathbf{y}^{new(k)}$. The data sets generated at each iteration are summarized using a test statistic. In this case we will use the mean and the variance:

$$\mu^{new(k)} = \frac{\sum_{i=1}^{60} y_i^{new(k)}}{60}, \tag{12.1.3}$$

$$\sigma^{2new(k)} = \frac{\sum_{i=1}^{60}\left(y_i^{new(k)} - \mu^{new(k)}\right)^2}{60}. \tag{12.1.4}$$

The same two statistics are calculated for the original data.[1] An indicator variable ($I^{(k)}$) is set to 1 if the test statistic for the simulated data is greater

[1]This need not be done at every step in this case because the statistics calculated from the data will be the same across all iterations.

than the test statistic for the observed data and set to 0 otherwise. The Bayesian p-value, P_B, is then calculated as $P_B = \sum_{k=1}^{K} I^{(k)}/K$ for each of the test statistics. If P_B is close to 1 or 0, then we conclude that the model suffers from lack of fit.

Your model check showed that your initial formulation was not capable of representing the dispersion of the observed data. Revise your model to remedy this problem. Draw a Bayesian network and write the posterior and joint distributions of the revised model.

Another way to treat the problem is to assume that each individual has its own mean tick burden, and these individual means are drawn from a distribution with its own parameters (i.e., hyperparameter α, ν) (fig. 12.1.1 B). We are not trying to explain the variation among individuals. Instead, we are simply acknowledging such variation exists using the hierarchical model

$$[\boldsymbol{\lambda}, \alpha, \nu|\mathbf{y}] \propto \prod_{i=1}^{60} \text{Poisson}(y_i|\lambda_i)\,\text{gamma}(\lambda_i|\alpha, \nu)$$
$$\times \text{gamma}(\alpha|.001, .001)\text{gamma}(\nu|.001, .001). \qquad (12.1.5)$$

The model is hierarchical because λ_i appears on both sides of the conditioning in different model components, and $\boldsymbol{\lambda}$ is now a vector containing one element (λ_i) for each individual. We can find the overall mean tick burden—that is, the mean of the distribution of the λ_i—by computing the derived quantity $\frac{\alpha}{\nu}$, the mean of a gamma distribution. The standard deviation of the mean tick burden can be calculated from the MCMC iterations for this derived quantity.

We point out that the data could also be modeled as

$$[\lambda, \kappa|\mathbf{y}] \propto \prod_{i=1}^{60} \text{negative binomial}(y_i|\lambda, \kappa)\,\text{gamma}(\lambda|.001, .001)$$
$$\times \text{gamma}(\kappa|.001, .001), \qquad (12.1.6)$$

where κ is a parameter that allows additional dispersion in the distribution created by variation among individuals in the mean number of ticks. In this case, λ in equation 12.1.6 would equal the mean of the λ_i, that is, α/ν, in equation 12.1.5. We prefer the hierarchical approach (eq. 12.1.5) because it shows more clearly the underpinning assumption of individual variation in average tick burden and because equation 12.1.5 is more easily implemented with contemporary MCMC software.

Now, assume that previous research showed that ticks are undercounted. An average of 8% (standard deviation = 2%) of the true number of ticks are missed. Revise your model to include this uncertainty. Demonstrate your skill with moment matching (sec. 3.4.4) to show how you would use these data to specify values for the parameters in an informative prior distribution.

The model can be modified to correct for bias caused by undercounting (fig. 12.1.1 C). We define z_i as the true number of ticks on sheep i. Our first thought might be to model the observations by defining parameter π as a proportion of ticks truly present z_i that are not counted, so that the average number counted is $(1-\pi)z_i$. Thus the posterior and joint distributions are

$$
\begin{aligned}
[\boldsymbol{\lambda}, z, \alpha, \nu, \pi|\mathbf{y}] \propto \prod_{i=1}^{60} & \text{Poisson}(y_i|(1-\pi)\cdot z_i) \\
& \times \text{Poisson}(z_i|\lambda_i)\,\text{gamma}(\lambda_i|\alpha,\nu) \\
& \times \text{gamma}(\alpha|.001,.001)\text{gamma}(\nu|.001,.001) \\
& \times \text{beta}(\pi|14.64, 168.36). \qquad (12.1.7)
\end{aligned}
$$

The parameters of the beta distribution for π are obtained by moment matching (sec. 3.4.4 and app. A) using

$$
\text{First parameter: } 14.64 = \frac{(.08^2 - .08^3 - .08\times .02^2)}{.02^2}, \qquad (12.1.8)
$$

$$
\text{Second parameter: } 168.36 = \frac{.08 - 2\times .08^2 + .08^3 - .02^2 + .08\times .02^2}{.02^2}. \qquad (12.1.9)
$$

However, the problem with this approach is that the y_i could be greater than the true number of ticks (z_i) on a sheep, which doesn't make sense. A better way to represent undercounting would be

$$
\begin{aligned}
[\boldsymbol{\lambda}, \mathbf{z}, \alpha, \nu, \pi|\mathbf{y}] \propto \prod_{i=1}^{60} & \text{binomial}(y_i|z_i,(1-\pi)) \\
& \times \text{Poisson}(z_i|\lambda_i)\text{gamma}(\lambda_i|\alpha,\nu) \\
& \times \text{gamma}(\alpha|.001,.001)\text{gamma}(\nu|.001,.001) \\
& \times \text{beta}(\pi|14.64, 168.36). \qquad (12.1.10)
\end{aligned}
$$

We are now treating the data as a binomial random variable where the number of trials is the true number of ticks, and the probability of a "success" (i.e., the tick is counted) is $1-\pi$. This is better than equation 12.1.7, because the support for the random variable y_i is 0 to z_i.

You discover unpublished data specifying the age, sex, and body mass of each of the 60 sheep. You assume these data are measured without error. Expand on the model you composed in step 3 by including a process model using these covariates to explain individual tick burdens. Diagram and write the posterior and joint distributions for unobserved quantities conditional on the observed ones.

The hierarchical posterior (eq. 12.1.7) represents variation among individuals but lacks any way to explain it. Covariates can be used to account for some of this variation (fig. 12.1.1 D). We choose a deterministic model,

$$g(\boldsymbol{\beta}, \mathbf{x}_i) = \exp(\beta_0 + \beta_1 x_{1,i} + \beta_2 x_{2,i} + \beta_3 x_{3,i}), \tag{12.1.11}$$

allowing us to rewrite the full posterior and joint distributions as

$$\begin{aligned}\left[\boldsymbol{\lambda}, \mathbf{z}, \pi, \boldsymbol{\beta}, \sigma^2 | \mathbf{y}\right] \propto & \prod_{i=1}^{60} \text{binomial}(y_i|z_i, (1-\pi))\text{Poisson}(z_i|\lambda_i) \quad (12.1.12)\\ & \times \text{lognormal}(\lambda_i | \log(g(\boldsymbol{\beta}, \mathbf{x}_i)), \sigma^2)\\ & \times \text{beta}(\pi | 14.64, 168.36)\\ & \times \text{inverse gamma}(\sigma^2 | .001, .001)\\ & \times \prod_{j=0}^{3} \text{normal}(\beta_j | 0, 1000),\end{aligned}$$

where σ^2 is the process variance on the log scale. The lognormal is a logical choice to represent stochasticity in the process model because the λ_i are continuous and nonnegative. The normal is a logical choice for the prior on the β_j because they are real numbers. We choose an inverse gamma distribution for the variance (σ^2) because it is continuous and nonnegative, and the inverse gamma distribution is a conjugate distribution for the lognormal variance (table A.3). The parameters chosen for all priors except the beta prior on π can be assumed to produce flat distributions with minimal influence on the posterior, at least as a starting point for analysis. This assumption should be checked by varying the values of parameters for priors and examining the posteriors to assure they are minimally sensitive to the chosen values.

12.2 Light Limitation of Trees

Use the deterministic model

$$g(\alpha, \gamma, c, x_i) = \frac{\alpha (x_i - c)}{\frac{\alpha}{\gamma} + (x_i - c)} \tag{12.2.1}$$

to portray light limitation for a single tree species, where α is the maximum growth annual increment at infinite light; γ is the slope of the curve at low light, and c is the light level where the growth increment is zero. Assume you have observations of height increment (y_i) paired with perfect observations of light availability (x_i) for 50 individual trees.

This problem can be tackled with the simple Bayesian model (fig. 12.2.1 A)

$$\begin{aligned}\left[\alpha, \gamma, c, \sigma^2 | \mathbf{y}\right] \propto & \prod_{i=1}^{50} \text{normal}(y_i | g(\alpha, \gamma, c, x_i), \sigma^2) \\ & \times \text{gamma}(\alpha | .001, .001)\, \text{gamma}(\gamma | .001, .001) \\ & \times \text{gamma}(c | .001, .001)\, \text{inverse gamma}(\sigma^2 | .001, .001).\end{aligned} \tag{12.2.2}$$

We choose a normal distribution for the likelihood because height increment is a continuous quantity that can be positive or negative. (A moose might eat the top of a seedling.) We choose gamma priors for α, γ, and c because it is biologically reasonable to assume they are nonnegative. The model would benefit from prior knowledge on these parameters, but we make them vague here. What is the interpretation of σ^2? Because we are assuming that height increment is measured perfectly, it is reasonable to view σ^2 as process variance, a parameter accounting for all the influences on growth (e.g., nutrients, herbivory, water) that are not included in our model. Note that the predictor variable does not appear in the posterior distribution, because we are treating it as known (box 6.2.2).

We take the liberty of modifying the design to make this problem more challenging. Assume you have three replicate observations of light availability for each tree. Model observed light availability as a random variable arising from a distribution with a mean z_i, the unobserved, true average light availability for the ith tree.

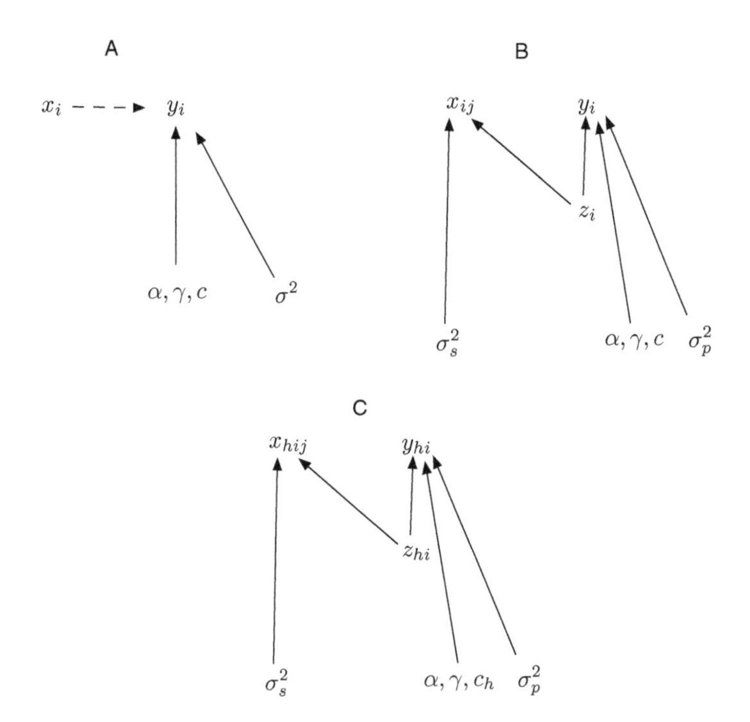

Figure 12.2.1. Bayesian network for problem 2, light limitation of trees. See the text for definition of symbols.

Sampling variation in the predictor variables can be represented using

$$\left[\alpha, \gamma, c, \mathbf{z}, \sigma_s^2, \sigma_p^2 | \mathbf{y}, \mathbf{x}\right] \propto \prod_{i=1}^{50} \prod_{j=1}^{3} \text{normal}\left(y_i | g(\alpha, \gamma, c, z_i), \sigma_p^2\right)$$
$$\times \text{beta}\left(x_{ij} | m\left(z_i, \sigma_s^2\right)\right) \text{beta}\left(z_i | 1, 1\right)$$
$$\times \text{gamma}(\alpha | .001, .001) \text{gamma}(\gamma | .001, .001)$$
$$\times \text{gamma}(c | .001, .001)$$
$$\times \text{inverse gamma}(\sigma_s^2 | .001, .001)$$
$$\times \text{inverse gamma}(\sigma_p^2 | .001, .001) \qquad (12.2.3)$$

(fig. 12.2.1 B). The function $m()$ matches the mean and variance of the beta distribution to its parameters (sec. 3.4.4 and A.2), as illustrated in the previous problem. There is a true, latent light availability z_i for the ith tree. The observations x_{ij} are drawn from the distribution beta$(x_{ij}|m(z_i, \sigma_s^2))$. Note that the z_i appears on the right-hand side of a conditioning but not on the left-hand side, which means it requires a prior. The posterior distribution now contains **x** because, like **y**, it is a random variable before it is observed. We are able to separate process variance from observation error because we have replicated observations of light availability. We estimated a single sampling variance, but we could have estimated a different variance for each tree.

Assume you have data for a second tree species similar to the data in step 2. You would like to make inference about the ability of the two tree species to become established in habitats with low light. Modify your model to examine effects of species on c.

The Bayesian network is shown in figure 12.2.1 C. The new expression for the posterior and joint distributions is

$$\begin{aligned}\left[\alpha, \gamma, \mathbf{c}, \mathbf{z}, \sigma_s^2, \sigma_p^2 | \mathbf{y}, \mathbf{x}\right] \propto & \prod_{h=1}^{2}\prod_{i=1}^{50}\prod_{j=1}^{3} \text{normal}\Big(y_{hi}|g(\alpha, \gamma, c_h, z_{hi}), \sigma_p^2\Big) \\ & \times \text{beta}\big(x_{hij}\big|m\left(z_{hi}, \sigma_s^2\right)\big)\text{beta}\left(z_{hi}|1, 1\right) \\ & \times \text{gamma}(c_h|.001, .001)\text{gamma}(\alpha|.001, .001) \\ & \times \text{gamma}(\gamma|.001, .001) \\ & \times \text{inverse gamma}(\sigma_s^2|.001, .001) \\ & \times \text{inverse gamma}(\sigma_p^2|.001, .001). \end{aligned} \tag{12.2.4}$$

The model now includes two species indexed by h. To make inference on species effects, we could create a derived quantity representing the difference between the values of c for each species; that is, $\delta = c_1 - c_2$. If the posterior distribution of δ did not strongly overlap zero, then we would conclude that there is evidence for a difference in species of tolerance for low light. Alternatively, we could examine the area of overlap between the two posterior distributions of c, from which we could infer, what is the probability that an individual of species 1 has greater tolerance for low light relative to an individual of species 2?

12.3 Landscape Occupancy of Swiss Breeding Birds

The challenge in this analysis is that birds will frequently be present but not observed. You will use the total number of times a bird was observed on quadrat i (y_i) *as data for this problem. Develop a data model that uses a detection probability* (p) *to portray how the data arise from the true state of the quadrat. Represent the true state by the latent variable* z_i*, where* $z_i = 1$ *if the quadrat is truly occupied, and* $z_i = 0$ *if it is not. Link the data model to a model of the occupancy process by modeling* y_i *conditional on* z_i *and* p*, and* z_i *conditional on* ψ_i*, the true probability that a bird is present on quadrat* i*. At least some of the* J_i *must be greater than 1 in this example because repeated observations at the same location are needed to estimate detection probability. Sketch the Bayesian network and write an expression for the posterior and the joint distribution.*

It is common in ecology to have a set of observations that contains more zeros than would be expected from a single likelihood, for example, a Poisson or a binomial. The reason for this is simple: the zeros arise from more than one mechanism. All is not lost, however, we can solve the problem by composing mixtures of distributions to represent more than one source of variation in the random variable of interest (sec. 3.4.5). These are called *zero-inflated models*.

In this example, zeros occur in the data because a site is truly unoccupied and because we fail to observe a bird on a site that is truly occupied. It is especially helpful to start by thinking about how the data arise. We define the unobserved, true state of quadrat i as $z_i = 1$ if it is occupied and $z_i = 0$ if it is not. Thus, we can model the data y_i, the number of times we observe the bird given J_i sampling occasions as

$$y_i \sim \begin{cases} 0 & z_i = 0 \\ \text{binomial}\,(J_i, p) & z_i = 1 \end{cases}, \tag{12.3.3}$$

which simply says that we will never detect a bird if the site is unoccupied, but if the site is occupied, we will detect the bird is present with probability p. Next, we must model the process governing the true state z:

$$z_i \sim \text{Bernoulli}\,(\psi_i)\,. \tag{12.3.4}$$

We use a Bernoulli distribution because the random variable z_i can take on values of 0 or 1. The frequency of these values is determined by the true probability of occupancy, (ψ_i). Combining the data model with the process

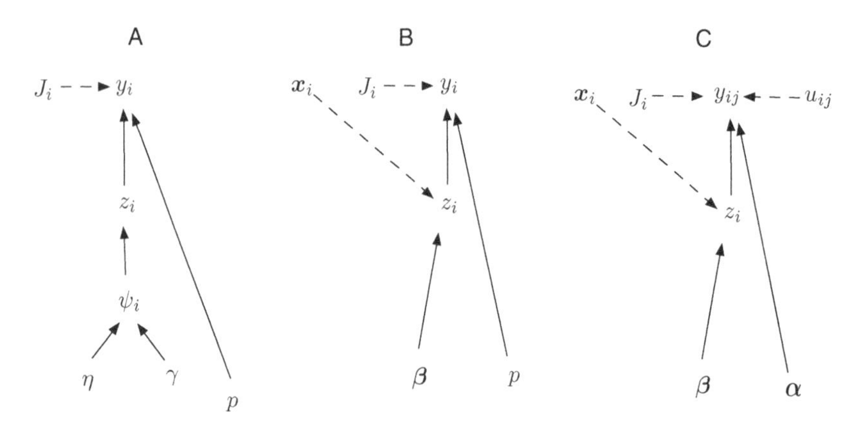

Figure 12.3.1. Bayesian networks for problems on landscape occupancy for Swiss breeding birds. Symbols are defined in the text.

model (fig. 12.3.1 A), we have

$$\begin{aligned}[\boldsymbol{\psi}, p, \mathbf{z}, \eta, \gamma | \mathbf{y}] \propto \prod_{i=1}^{n} & \text{binomial}(y_i | J_i, z_i \cdot p) \\ & \times \text{Bernoulli}(z_i | \psi_i)\text{beta}(p | 1, 1) \\ & \times \text{beta}(\psi_i | \eta, \gamma)\text{gamma}(\eta | .001, .001) \\ & \times \text{gamma}(\gamma | .001, .001). \qquad (12.3.5)\end{aligned}$$

Be sure that you understand that $z_i \cdot p$ is the product of the true state and the detection probability. Thus, when $z_i = 0$ there is no chance that the bird will be observed, and when $z_i = 1$ the bird will be observed with probability p.

You have data on the elevation of the quadrat at its centroid $x_{1,i}$ and the proportion of the quadrat in forest cover $x_{2,i}$, so that you can model ψ_i as a function of quadrat-specific covariates. Assume that you wish to estimate the posterior distribution of a derived quantity, the optimum elevation for occupancy at the average forest cover, where the optimum is defined as the elevation with the maximum probability of occupancy. Develop a model that allows you to make this inference. As a suggestion, recall the inverse logit function. Consider the kind of model you would need to find a maximum proability of occupancy as a derived quantity of the model's parameters. Draw the Bayesian network and write the posterior and joint distributions.

We choose to model the true probability of occupancy using the deterministic model

$$g(\boldsymbol{\beta}, \mathbf{x}_i) = \frac{\exp(\beta_0 + \beta_1 x_{1,i} + \beta_2 x_{1,i}^2 + \beta_3 x_{2,i})}{1 + \exp(\beta_0 + \beta_1 x_{1,i} + \beta_2 x_{1,i}^2 + \beta_3 x_{2,i})}, \tag{12.3.6}$$

which you will recognize as an inverse logit function that returns continuous values between 0 and 1. We include the term $x_{1,i}^2$ to make the linear portion quadratic, allowing for an increase and decline over the range of elevation, a feature we must have in our model to find a maximum probability of occupancy. We standardize[2] the covariates by subtracting their means and dividing by their standard deviations so that the linear model for elevation at the average forest cover is $\beta_0 + \beta_2 x_{1,i} + \beta_2 x_{1,i}^2$, because $x_{2,i} = 0$ at the mean forest cover.

The expression for the posterior and joint distributions is obtained by substituting the deterministic model for ψ_i in the Bernoulli process model and by adding the appropriate flat priors (fig. 12.3.1 B):

$$\begin{aligned}[p, \boldsymbol{\beta}, \mathbf{z}|\mathbf{y}] \propto \prod_{i=1}^{n} &\text{binomial}\,(y_i|J_i, z_i \cdot p)\,\text{Bernoulli}\,(z_i|g\,(\boldsymbol{\beta}, \mathbf{x}_i))\,\text{beta}\,(p|1, 1) \\ &\times \prod_{k=0}^{3} \text{normal}\,(\beta_k|0, 10)\,. \end{aligned} \tag{12.3.7}$$

Note also that we are treating the predictor variables as known, which is why they do not appear in the posterior distribution. This is probably fine for elevation but is certainly not reasonable for forest cover. Predictor variables derived from geographic information system (GIS) analysis are almost always treated this way, which is no excuse. They are outputs from models with error that should be included here.

This example offers a great opportunity to illustrate a mistake in model building that you are almost certain to make sometime. It is tempting to draw a Bayesian network for this problem, making ψ_i a function of the coefficients (fig. 12.3.2 A), reasoning that our process model is

$$\begin{aligned} \psi_i &= \frac{\exp\left(\beta_0 + \beta_1 x_{1,i} + \beta_2 x_{1,i}^2 + \beta_3 x_{2,i}\right)}{1 + \exp\left(\beta_0 + \beta_1 x_{1,i} + \beta_2 x_{1,i}^2 + \beta_3 x_{2,i}\right)}, \\ z_i &\sim \text{Bernoulli}\,(\psi_i)\,. \end{aligned} \tag{12.3.8}$$

[2]Standardizing is often a good idea in Bayesian regression because it speeds convergence, can make it easier to interpret the coefficients, and can facilitate specifying priors. Note that for x_1^2 we standardize first, then take the square of the standardized variable.

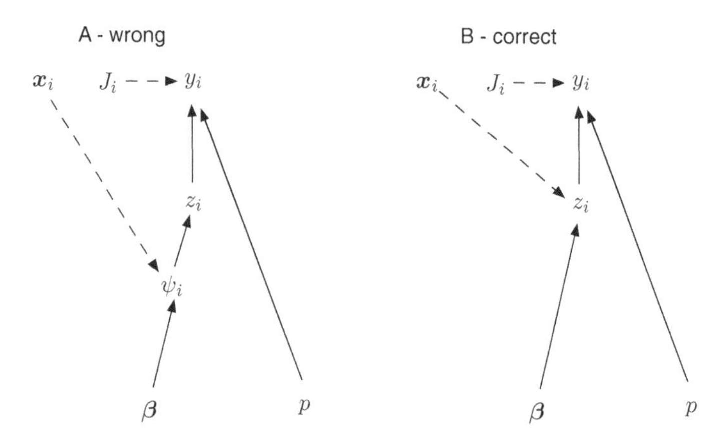

Figure 12.3.2. Alternative sketches of Bayesian networks for the landscape occupancy problem. (**A**). A drawing like this one might motivate a deterministic model formulated as $\psi_i = \exp(\beta_0 + \beta_1 x_{1,i} + \beta_2 x_{1,i}^2 + \beta_3 x_{2,i})/(1 + \exp(\beta_0 + \beta_1 x_{1,i} + \beta_2 x_{1,i}^2 + \beta_3 x_{2,i}))$. The drawing is wrong, however, because the relationship between ψ_i and $\boldsymbol{\beta}$ is deterministic not stochastic. Placing ψ_i at the head of a solid arrow implies it is a random variable, which it is not. (**B**). The correct Bayesian network represents the deterministic model $g(\boldsymbol{\beta}, \mathbf{x}_i) = \exp(\beta_0 + \beta_1 x_{1,i} + \beta_2 x_{1,i}^2 + \beta_3 x_{2,i})/(1 + \exp(\beta_0 + \beta_1 x_{1,i} + \beta_2 x_{1,i}^2 + \beta_3 x_{2,i}))$. This formulation eliminates the use of ψ_i by linking the β's directly to z_i. There is nothing wrong with calculating a derived quantity ψ_i, as defined in your computational algorithm, but including it in the Bayesian network leads to confusion about what is random and what is fixed.

There is nothing wrong with the notation, but the diagram is *wrong* because, as you recall from section 3.3, solid arrows represent stochastic relationships, that is, those indicated by the $\sim$ sign in our model, not by an $=$ sign. The heads of arrows are random variables. This means that figure 12.3.2 A is wrong because ψ_i is a deterministic quantity but is illustrated as a stochastic one. We are using ψ_i as a matter of notational convenience in equation 12.3.8 to form the parameter for the Bernoulli distribution of the true state, which *is* a random variable. We would probably want to estimate the ψ_i as a derived quantity, a deterministic function of the $\boldsymbol{\beta}$ random variables. A correct diagram omits the ψ_i (fig. 12.3.2 B) by using $g(\boldsymbol{\beta}, \mathbf{x})$ as the argument to the Bernoulli.

You also may be wondering, what happened to the process variance and the observation error? In this example the stochastic relationships were sufficiently clear that we progressed directly to the choice of distributions (binomial, Bernoulli) for the likelihood and the process model, forgoing the preliminary step of writing things in general bracket notation containing stand-ins for variance and central tendency. This doesn't mean that we have lost the stochasticity in the observations or the process. Remember, there is an observation variance for each y_i given by $J_i \psi_i z_i (1 - \psi_i z_i)$, the variance of the binomial likelihood. There is a process variance for each of the z_i given by $g(\mathbf{x}_i, \boldsymbol{\beta})(1 - g(\mathbf{x}_i, \boldsymbol{\beta}))$, the variance of the Bernoulli process model. If it is not evident where these expressions come from, review the fundamentals on these distributions (sec. 3.4.3.1).

What could you learn from the coefficients predicting occupancy? Do you need to do model selection?

The posterior distributions of the model coefficients can tell us about the role of forest cover and elevation in controlling the distribution of willow tits. If the distribution of β_1 or β_2 strongly overlaps zero then we can conclude that the predictor variable is just as likely to be related to an increase in use as it is to a decline in use—we have no evidence for an effect. If the posteriors do not overlap zero, then we can conclude that occupancy is associated with the predictor variable. The magnitude of the coefficients allows us to infer the sensitivity of the occupancy to changes in the predictor variables, because the predictors have been standardized such that they are on the same scale. Dusting off our calculus allows us to use the linear and quadratic terms (β_1, β_2) to discover that the elevation associated with a maximum probability of occupancy as a derived quantity $x_1^* = -\frac{1}{2}\beta_1/\beta_2$. Embedding the expression for the optimum in our MCMC code $\left(\text{e.g., } x_1^{*(k)} = -\frac{1}{2}\beta_1^{(k)}/\beta_2^{(k)}\right)$ would provide the samples needed to approximate the posterior distribution of x^*, but to make this interpretable, it would be good to include the transformation equation $x_{1,\text{meters}}^{*(k)} = x_1^{*(k)} \times \text{sd}(\boldsymbol{x}_1) + \text{mean}(\boldsymbol{x}_1)$.

We could use a model selection method to compare alternative models, but if our objective was to "estimate the optimum elevation at the average forest cover," then the parameters in the model are set by that objective—we must have coefficients for elevation, elevation squared, and forest cover. In this case, model selection is probably not necessary to meet our objective.

Until now we have assumed that the detection probability is constant across quadrats. Assume you have standardized covariates that influence the detection process: length of transect $u_{1,ij}$, duration of search $u_{2,ij}$, and date $u_{3,ij}$, all measured without error. Model the detection probability using these covariates. Draw the Bayesian network and write the posterior and joint distributions. You will use the binary form of the data for this Problem.

We can model the detection probability as follows. We define $h(\boldsymbol{\alpha}, \mathbf{u}_{ij}) = \frac{\exp(\alpha_0 + \alpha_1 u_{1,ij} + \alpha_2 u_{2,ij} + \alpha_3 u_{3,ij})}{1 + \exp(\alpha_0 + \alpha_1 u_{1,ij} + \alpha_2 u_{2,ij} + \alpha_3 u_{3,ij})}$. The posterior and joint distributions are (fig. 12.3.1 C)

$$[\boldsymbol{\alpha}, \boldsymbol{\beta}, \mathbf{z}|\mathbf{y}] \propto \prod_{i=1}^{n}\prod_{j=1}^{J_i} \text{Bernoulli}\big(y_{ij}|z_i \cdot h(\boldsymbol{\alpha}, \mathbf{u}_{ij})\big)\, \text{Bernoulli}(z_i|g\,(\boldsymbol{\beta}, \mathbf{x}_i))$$
$$\times \prod_{k=0}^{3} \text{normal}(\beta_k|0, 100)\text{normal}(\alpha_k|0, 100). \tag{12.3.9}$$

It is critical to see in this example that the detection covariates are specific to each search on each day. This means we need to model whether a bird was observed on day j (0 = not observed; 1 = observed); which requires the j subscript on y and u. It also changes our likelihood from binomial to Bernoulli to accommodate the binary observations.

12.4 Allometry of Savanna Trees

Assume you have observations on diameters and masses of T individual trees of the same species. Assume that these measurements are made without error. Diagram and write a simple Bayesian model to find the posterior distributions of a and d using the log form of equation 11.4.1 as your deterministic model.

We use the log form of equation 11.4.1 for the deterministic model[3]:

$$g(a, b, D_t) = \log(a) + b\log(D_t). \tag{12.4.1}$$

The likelihood is

$$\log(M_t) \sim \text{normal}(g(a, b, D_t), \sigma_p^2), \tag{12.4.2}$$

which is the same as saying the random variable M_t is lognormally distributed. The lognormal is a good choice, because tree mass is a continuous random variable that cannot be negative. The posterior and joint

[3]We have made a small departure (M_t, D_t) from our usual use of lowercase for scalars to conform with the notation used by the authors. So, $\boldsymbol{M}$ is a vector, and M is a scalar.

distribution (fig. 12.4.1 A) is

$$\begin{aligned}\left[a, b, \sigma_p^2|\mathbf{M}\right] &\propto \prod_{t=1}^{T} \text{normal}\left(\log(M_t)|g(a, b, D_t), \sigma_p^2\right)\\ &\times \text{gamma}(.001, .001)\\ &\times \text{normal}(b|0, 100)\text{inverse gamma}(\sigma_p^2|.001, .001).\end{aligned} \quad (12.4.3)$$

The response $\mathbf{M}$ appears in the posterior distribution because we model it as a random variable to reflect uncertainty arising from the process variance. The predictor variable $\mathbf{D}$ is not found in the posterior because we are assuming that it is known. Priors on a are gamma distributed because a must be nonnegative. We could also use a uniform distribution or a vague lognormal distribution here. It is biologically sensible for the b to be nonnegative, but we choose a normal prior to allow the data to determine its sign. We choose an inverse gamma prior for σ_p^2 because it must be nonnegative and because the inverse gamma is a conjugate for the lognormal variance. Note that this is a simple Bayesian model. There is no hierarchical structure.

Assume that there are three tree species indexed by $s = 1, \ldots, 3$ *and that* T_s *trees of species* s *are measured. Diagram and write out a model showing the relationship between tree mass and basal diameter where each tree species has its own parameters* a_s *and* b_s *reflecting variation among individual trees within species* s.

The Bayesian network is virtually the same as before, changed only by new subscripts reflecting multiple species (Figure 12.4.1 B). The deterministic model is now

$$g\left(a_s, b_s, D_{st}\right) = \log(a_s) + b_s \log\left(D_{st}\right). \quad (12.4.4)$$

We need to add subscripts for species to the expression for the joint and posterior,

$$\begin{aligned}\left[\mathbf{a}, \mathbf{b}, \sigma_p^2|\mathbf{M}\right] &\propto \prod_{s=1}^{3}\prod_{t=1}^{T_s} \text{normal}\Big(\log\left(M_{st}\right)|g\left(a_s, b_s, D_{st}\right), \sigma_p^2\Big) \quad (12.4.5)\\ &\times \text{gamma}(a_{s|}.001, .001)\text{normal}(b_s|0, 100)\\ &\times \text{inverse gamma}\Big(\sigma_p^2|.001, .001\Big).\end{aligned}$$

Assume a_s *and* b_s *are drawn from a global distribution reflecting variation among tree species. Use* μ_a *and* σ_a^2 *to notate the parameters of the distribution of* a_s *and* μ_b *and* σ_b^2 *for the distribution of* b_s.

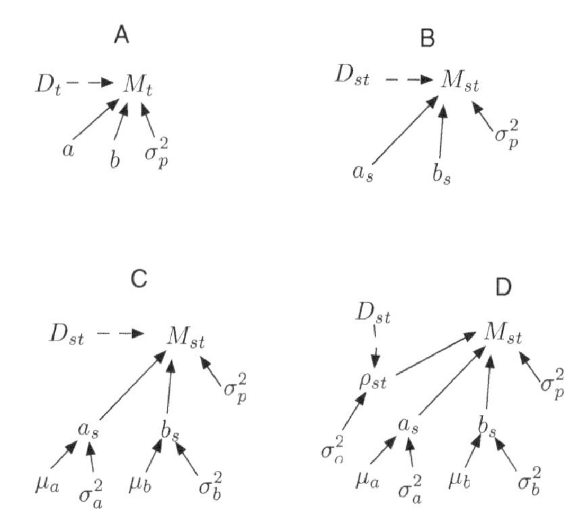

Figure 12.4.1. Bayesian networks for allometric scaling of trees problem. Symbols are defined in the text.

We add a level to the hierarchy such that the parameters for each species are now drawn from a distribution reflecting interspecific variation (Figure 12.4.1 C)

$$
\begin{aligned}
&\left[\mathbf{a}, \mathbf{b}, \mu, \boldsymbol{\sigma}_a^2, \boldsymbol{\mu}_b, \boldsymbol{\sigma}_b^2, \sigma_p^2 | \mathbf{M}\right] \\
&\quad \propto \prod_{s=1}^{3}\prod_{t=1}^{T_s} \text{normal}\left(\log\left(M_{st}\right) | g\left(a_s, b_s, D_{st}\right), \sigma_p^2\right) \\
&\quad \times \text{gamma}\left(a_s | \frac{\mu_a^2}{\sigma_a^2}, \frac{\mu_a}{\sigma_a^2}\right) \text{normal}\left(b_s | \mu_b, \sigma_b^2\right) \\
&\quad \times \text{inverse gamma}\left(\sigma_a^2 | .001, .001\right) \\
&\quad \times \text{inverse gamma}\left(\sigma_b^2 | .001, .001\right) \\
&\quad \times \text{gamma}(\mu_a | .001, .001)\text{normal}(\mu_b | 0, 1000) \\
&\quad \times \text{inverse gamma}\left(\sigma_p^2 | .001, .001\right).
\end{aligned}
\tag{12.4.6}
$$

Assume that tree diameters are measured imperfectly. To account for these measurement errors, expand the model you developed in step (4) to include *a Berkson "error in variables" model (Dellaportas and Stephens, 1995) assuming 5% error on at least 5% of trees. Prior data (Price et al., 2009) inform the relationship between the true diameter ρ_{st} of tree t of species s and the measured diameter D_{st}, so that* $\log(\rho_{st}) \sim \text{normal}(\log(D_{st}), \sigma_o^2)$, *where* $\sigma_o^2 \sim \text{lognormal}(4.135, 2)$.

We have prior information that allows us to include errors in measurements of tree diameters. There is a true, unobserved diameter ρ_{st} of tree t of species s that is related to the measured diameter, $\log(\rho_{st}) \sim \text{normal}(\log(D_{st}), \sigma_o^2)$. We modify the likelihood in equation 12.4.7 to include the errors in observations using

$$\prod_{s=1}^{3}\prod_{t=1}^{T_s} \text{normal}\big(\log(M_{st})|g(a_s, b_s, \rho_{st}), \sigma_p^2\big)\text{normal}\big(\rho_{st}|D_{st}, \sigma_o^2\big) \quad (12.4.7)$$

and add an informative prior for the described σ_o^2 (fig. 12.4.1 E).

Assume you have fit and checked the model in step 4. How would you make inference about the three theories?

We could calculate a derived quantity for each theory as the difference between the exponent predicted by the $j = 1, \ldots, 3$ theories (b_j^*) and the exponent estimated from the data. Thus, at each iteration in the MCMC algorithm we would calculate

$$\Delta_j^{(k)} = b_j^* - \mu_b^{(k)} \quad (12.4.8)$$

to evaluate the alternative theories across species and

$$\delta_{js}^{(k)} = b_j^* - b_s^k \quad (12.4.9)$$

to evaluate the alternative theories within species. These derived quantities (sec. 8.3) allow us to sample the posterior distributions of the difference between the theoretical prediction and the observations.[4] The theories with the greatest support in the data would have the smallest differences between the predicted and the estimated exponents. Distributions of Δ_j and δ_j that failed to probabilistically exclude zero would provide evidence that theory j is inconsistent with the observations.

[4]We could also calculate ratios of the prediction to the expectation.

12.5 Movement of Seals in the North Atlantic

Construct a data model: Assume for now that the true positions of the individual seal are known for a regular set of time points. Specify a model for the telemetry observations conditioned on those true positions for all time points. Your model should have a single unknown parameter controlling the dispersion of the telemetry error. Assume the telemetry error arises from a symmetric distribution; that is, a distribution that does not have a directional bias. Assume that the telemetry error variance has a strongly informative prior. Use the notation $\mathbf{s}_t = (s_{1,t}, s_{2,t})'$ *to represent the vector of telemetry observations at time* t*, where the* $s_{1,t}$ *gives the latitude, and* $s_{2,t}$ *gives the longitude.*[5] *Use* $\boldsymbol{\mu}_t = (\mu_{1,t}, \mu_{2,t})'$ *to describe the true position of the individual at time* t *for a set of times, say* $t = 1, \ldots, T$.

Perhaps the simplest telemetry data distribution is the multivariate normal, where

$$\mathbf{s}_t \sim \text{multivariate normal}(\boldsymbol{\mu}_t, \sigma_s^2 \mathbf{I}), \tag{12.5.1}$$

where, σ_s^2 controls the dispersion (or amount of telemetry error), and $\mathbf{I}$ is a 2×2 matrix with 1s on the diagonal.[6] In essence, the term $\sigma_s^2 \mathbf{I}$ is written as a proper covariance matrix, which is a necessary element of the multivariate normal distribution. It is important to note that this distribution could easily be generalized by allowing the diagonal elements of this covariance matrix to be different, implying more dispersion in longitude versus latitude for example, and/or by allowing for nonzero off-diagonal elements, that allow for various elliptical shapes for the error distribution in any direction. Also, importantly, because our chosen covariance matrix is, in fact, diagonal, we could write the data model componentwise, rather than in matrix notation;

$$s_{1,t} \sim \text{normal}(\mu_{1,t}, \sigma_s^2), \tag{12.5.2}$$

$$s_{2,t} \sim \text{normal}(\mu_{2,t}, \sigma_s^2). \tag{12.5.3}$$

[5]The $'$ symbol is called a "prime" and denotes a vector transpose.

[6]Many papers and texts use $\mathbf{s}_t \sim \text{N}(\boldsymbol{\mu}_t, \sigma_s^2 \mathbf{I})$, abbreviating, normal with N. Because the parameters are well described as a mean vector and a covariance matrix by the fact that they are boldface clearly identifies this as a multivariate normal. Others abbreviate using $\text{MVN}(\boldsymbol{\mu}_t, \sigma_s^2 \mathbf{I})$. We write out the full distribution name, as we have done throughout the book.

Notice that the boldface variables disappear, indicating that we are now dealing with one-dimensional probability distributions for univariate quantities.

Construct the process model: Assuming that a pure discrete-time random walk is a reasonable model for the movement of hooded seals, construct a dynamic process model for a given individual hooded seal position that is parameterized using no more than one unknown variable (assuming the positions at a regular set of time points are known, for now). To simplify the statistical notation, assume that the time step between positions is one unit.

Now that we have the notation set up for the position process, assuming that the true positions are known at each time $t = 1, \ldots, T$, we could specify a simple discrete-time multivariate random walk as

$$\boldsymbol{\mu}_t \sim \text{multivariate normal}(\boldsymbol{\mu}_{t-1}, \sigma_\mu^2 \mathbf{I}), \tag{12.5.4}$$

where the temporal process depends on itself at the previous time step, and the parameter σ_μ^2 controls the distance an individual moves during one unit of time. As with the preceding data model, we could generalize this stochastic model for the position to accommodate much more mechanistic dynamics simply by premultiplying the mean vector by a matrix, say $\mathbf{M}$, so that the mean would become a translation of the last location in space. Also, higher-order temporal dependence could be induced by letting $\boldsymbol{\mu}_t$ depend on both $\boldsymbol{\mu}_{t-1}$ and $\boldsymbol{\mu}_{t-2}$ (perhaps through separate translations), as described in Johnson et al. (2005). However, in our simple random walk with a single variance component, we could write the process model componentwise, as we did in step (1):

$$\mu_{1,t} \sim \text{normal}(\mu_{1,t-1}, \sigma_\mu^2), \tag{12.5.5}$$

$$\mu_{2,t} \sim \text{normal}(\mu_{2,t-1}, \sigma_\mu^2). \tag{12.5.6}$$

Specify prior distributions for the unknown variables in steps 1 and 2 that are flexible enough to incorporate prior information if available. At this point, there should be only two univariate prior distributions.

There is only a single unknown parameter in each of the two preceding models, a variance component. Thus a conjugate choice for a prior distribution for each would be the inverse gamma,

$$\sigma_s^2 \sim \text{inverse gamma}(r_s, q_s), \tag{12.5.7}$$

$$\sigma_\mu^2 \sim \text{inverse gamma}(r_\mu, q_\mu), \tag{12.5.8}$$

though, like always, other choices are available. Remember that r_s, q_s, r_μ, and q_μ should be numeric arguments and that r_s and q_s are strongly informative.

Construct the hierarchical model: Now, using the component models in steps 1–3, assume that the true position is unknown at all times, and couple the models to form a single hierarchical model in which the telemetry data arise from a distribution that depends on the underlying movement process as well as the other data model parameter, and the true unknown position for any time t arises from a distribution that depends on itself at time t − 1 as well as an additional process model parameter. The final component of the hierarchical model then contains the prior distributions specified in step 3.

To construct our full hierarchical model, we simply need to assume that the true position process ($\boldsymbol{\mu}_t$) is unknown and group all the models together starting with the data model, such that it is clear that each is conditioned on the next. If we assume the true process is known at the initial time ($\boldsymbol{\mu}_0$), then the hierarchical model is

$$\mathbf{s}_t \sim \text{multivariate normal}(\boldsymbol{\mu}_t, \sigma_s^2\mathbf{I}), \quad \text{for } t = 1, \ldots, T, \tag{12.5.9}$$

$$\boldsymbol{\mu}_t \sim \text{multivariate normal}(\boldsymbol{\mu}_{t-1}, \sigma_\mu^2\mathbf{I}) \text{ , for } t = 1, \ldots, T, \tag{12.5.10}$$

$$\sigma_s^2 \sim \text{inverse gamma}(r_s, q_s), \tag{12.5.11}$$

$$\sigma_\mu^2 \sim \text{inverse gamma}(r_\mu, q_\mu). \tag{12.5.12}$$

Of course, the true position will not be known initially, and a more formal way to handle the issues is with a prior. Thus, we could specify $\boldsymbol{\mu}_0 \sim$ multivariate normal($\boldsymbol{\alpha}, \sigma_0^2\mathbf{I}$) as the final model component (where $\boldsymbol{\alpha}$ and σ_0^2 are assumed to be known a priori).

Write the posterior distribution for the hierarchical model in step 4 in terms of its proportionality with respect to the joint distribution, as we have done previously.

Now that we have a hierarchical model fully specified in step 4, we can write the posterior as proportional to the properly factored joint distribution:

$$[\boldsymbol{\mu}_0, \ldots, \boldsymbol{\mu}_T, \sigma_s^2, \sigma_\mu^2 | \mathbf{s}_1, \ldots, \mathbf{s}_T] \propto \prod_{t=1}^{T} [\mathbf{s}_t | \boldsymbol{\mu}_t, \sigma_s^2][\boldsymbol{\mu}_t | \boldsymbol{\mu}_{t-1}, \sigma_\mu^2][\sigma_s^2][\sigma_\mu^2][\boldsymbol{\mu}_0] \,. \tag{12.5.13}$$

AFTERWORD

Reviewing Bayesian Models in Papers and Proposals

We recently submitted a paper reporting a Bayesian analysis of population dynamics to a well-known ecological journal. We discussed the paper with the editor, who remarked that it was challenging to find reviewers qualified to evaluate this type of work. We hope this book will expand the number of ecologists who can constructively review papers and proposals using Bayesian methods. To that end, we offer a few suggestions for peer review. A discussion of peer review provides a fitting conclusion to the book, because it allows us to touch on many of the topics we have covered.

Is the Model Clearly Specified?

It should come as no surprise that we believe that all papers and proposals[1] that use Bayesian analyses should contain a clearly written expression for the posterior and joint distributions (chapters 6, 10, and 12). It is fine that authors describe the individual components of the joint distribution separately, as we have often done (e.g., eqs. 12.5.9–12.5.12); however, the job is not complete until these have been assembled into a properly factored expression for the joint distribution and its proportionality to the posterior distribution (e.g., eq. 12.5.13). These expressions might be usefully placed in an appendix, perhaps, but they should be accessible for review. There is no justification for omitting them entirely.

One of the reasons that the expression for the posterior and joint distributions is so important is that it provides a blueprint for analysis that can be evaluated by the reviewer. All observed and unobserved quantities found in the posterior distribution must be present in the joint distribution. All unobserved quantities that appear on the left-hand side of a conditioning and not on the right-hand side must have priors with numeric arguments. Latent variables that appear on both sides of a conditioning symbol in the joint distribution should not have priors. Distributions must have support appropriate for the random variables being modeled. Assumptions about conditional independence needed to simplify the joint distribution (as described in sec. 3.3) must be sensible. Product terms must make sense in terms of the sampling or experimental design, and the independence of random variables in products should be justified. Choices of prior distributions and their parameters should be discussed in the body of the text or in a table. As a reviewer, you should consider whether these choices of priors have been made thoughtfully, as discussed in section 5.4.

[1]Of course, the level of detail needed in a scientific paper will exceed what you should expect in a proposal. However, we have reviewed proposals in which authors stated, "We will use a Bayesian hierarchical model...," a statement that is only slightly more informative than, "We will use statistics...." Use your judgment about how much detail is needed, but don't allow bluffing.

Is the Algorithm for Implementing the Model Adequately Described?

MCMC is currently the most frequently used algorithm for analysis of Bayesian hierarchical models, but faster alternatives are emerging (INLA Development Team, 2014; Stan Development Team, 2014). Authors should describe the algorithm they used for analysis and the software used to implement it. If they wrote their own MCMC algorithm, then they should write out the full-conditionals in an appendix. Journals often require code submitted in an appendix if authors relied on software to implement MCMC. As a reviewer, you should be able to see the correspondence between the code and the statistical model specified in the paper.

Has Convergence Been Assured and the Model Checked?

We discussed the importance of convergence of the MCMC algorithm. Authors should communicate how they evaluated convergence (sec. 7.3.4). They must also describe what they did to check the fit of the model. Posterior predictive checks should be a standard component of papers reporting Bayesian analyses (sec. 8.1).

Are There Appropriate Inferences from a Single Model?

There are no rigid requirements for reporting results from Bayesian analyses, but there are things you should look for. We think it is useful to plot posterior distributions overlaid on priors because these overlays allow readers to see how much was learned from the data (e.g., figure 8.5.2). If all the priors and posteriors overlap strongly with no shrinkage in the posterior, then it reasonable to question what was learned from the study that was not already known. However, overlap between some priors and posteriors is not evidence of a problem. We would expect this in complex models. Strongly overlapping prior and posterior distributions can confirm theory in a useful way. Strong priors on some parameters may be needed to learn about other parameters and states for which very little was known at the outset of an investigation (sec. 5.4).

Authors should provide enough information to allow the posterior to be used as a prior in future studies. It is good practice to tabulate means, medians, variances (or standard deviations), and quantiles of posterior distributions of parameters and other quantities of interest (sec. 8.2). Highest posterior density intervals should be reported if posterior distributions are strongly skewed or multimodal.

Are There Appropriate Inferences from More Than One Model?

We reiterate our view and the view of others (Gelman and Rubin, 1995; Gelman and Shalizi, 2013; Ver Hoef, 2015) that much can be learned from analysis of a single properly checked model. However, there will also be cases in which inference from more than one model is desired. Chapter 9 showed that model selection is a deep, nuanced topic. As a reviewer, you should be satisfied that model selection was based on thoughtful consideration of alternatives rather than mere convenience of computation. The gold standard for model selection based on predictive ability is out-of-sample validation. If the success of proposed research depends on model comparison, then proposals will be enhanced by plans for obtaining out-of-sample data needed to support those comparisons. We have seen that model averaging is inherently Bayesian, and there are many approaches for arriving at Bayesian averaged quantities, but the simple approach that is most commonly used (i.e., AIC weights; Burnham and Anderson, 2002) is not inherently Bayesian nor necessarily best suited for Bayesian model averaging. Furthermore, many other approaches for multimodel inference, including regularization and model-based approaches, wrap up the fitting, selection, and averaging procedure into one tidy package (chapter 9).

Are There Derived Quantities That might Enhance the Analysis?

Much of the power of the methods we have taught comes from the ability to make inferences from quantities derived from parameters and states. An alert and informed reviewer may be able to identify quantities that would be particularly informative but were included in the analysis. Your review could enhance the research by identifying the opportunity to compute these quantities.

Continuing Your Learning

We have all heard it said that some particular ecologist or another is a "modeler," as if the rest of us were something else. An important idea in this book is that strong, creative science starts with deep thinking about processes. The ability to evaluate our thinking with observations requires writing models that are mathematically and statistically coherent. It follows that we are all modelers. Developing skill in modeling ecological systems needs to be a lifelong commitment. It is central to what we do every day in our research.

The best way to benefit from your investment in this book is to find a problem relevant to your research and implement what you have learned here to solve it. Every problem we attack in our research teaches us something new. Allowing problems to guide your learning is a very efficient way to master Bayesian modeling.

Not all ecologists have access to Bayesian statisticians, but the effort to develop collaborations with them is worth the effort, even if they are not next door. There is much that you can do on your own, but you will probably be able to tackle problems of greater complexity and with greater confidence if you work with a statistical colleague. These collaborations will yield many mutual benefits. Developing them is one of the best ways for ecologists to continue learning statistics and for statisticians to learn ecology.

We have not emphasized coding in this book, believing that we had enough to cover. A logical next step, particularly if you are going to attack a specific problem in your research, is to obtain one of the many books (McCarthy, 2007; Royle and Dorazio, 2008; King and Gimenez, 2009; Gelman and Hill, 2009; Link and Barker, 2010; Kéry, 2010; Kéry and Schaub, 2012) that teach coding using ecological examples. McCarthy (2007) is probably the most accessible and with the broadest range of examples; the others tend to emphasize population ecology. We also admire Gelman and Hill (2009) for its wisdom and clarity, but you will need to do a bit of mental translation of the problems, because most are drawn from social science. Bear in mind that the various versions of software for implementing MCMC (Lunn et al., 2000; Plummer, 2002; Stan Development Team, 2014) are just languages for encoding expressions for the joint distribution and quantities derived from its elements. So, if you can write that expression, you have gone a long way toward writing the proper code.[2] You will need to learn a bit of syntax and will suffer some frustration fixing inevitable coding errors, but learning to write code is a straightforward task if you have a solid understanding of your model.

Packages in R (R Core Team, 2013) provide interfaces to all the popular flavors of MCMC software, and R tools can help you write your own MCMC algorithms. These change almost monthly, so we urge you to research what's most current and attractive. However, you would be wise to avoid the point-and-click approach provided by some software (e.g., WinBUGS, OpenBUGS). We have found that most people who start out using these graphical user interfaces soon switch to writing scripts in R to implement their models because of the greater flexibility and ease of documentation that it provides.

Perhaps the best way to continue your learning is to actively seek out scientific papers that use Bayesian methods and read them with a new, trained eye. You would not be alone if you confessed to skipping over equations in earlier reading, a habit that is no longer necessary or productive. You can learn a great deal from considering the authors' models, perhaps diagramming and writing their posterior and joint distributions if they failed to do so. As we learned in developing the examples in chapters 6 and 12, this can be more challenging and instructive than you might think.

Many problems in ecological research involve sampling over space and time, raising the issue of spatial and temporal dependence in data and how to model it properly. We felt this topic was beyond the scope of a primer. A logical next step is to learn about spatio-temporal modeling in the Bayesian context. Cressie and Wikle (2011) have written a superb source for self-teaching this material. Our book gives you the technical grounding you need to make progress in their more advanced text.

Our last suggestion is this: if your office lacks a whiteboard, get one.

[2]As we have said before, the converse is not true.

ACKNOWLEDGMENTS

Hobbs and Hooten received support to prepare this book from the National Science Foundation (Awards EF-0914489, DEB-1145200, EF-1241856). Hobbs was also supported by a Sabbatical Fellowship from the National Socio-Environmental Synthesis Center (Award DBI-1052875 to the University of Maryland). We thank Henrik Andrén, Ben Bolker, Christian Che-Castaldo, Matt Low, Nadja Rüger, Kate Searle, Christina Staudhammer, Andrew Tredennick, Ryan Wilson, and Amy Yackel Adams for comments on early drafts. We are especially grateful for the relentless efforts of our students Kristin Broms, Brian Brost, Frances Buderman, Alison Ketz, Henry Scharf, John Tipton, and Perry Williams, who tracked down errors and made helpful suggestions to enhance clarity. The book is stronger for their scrutiny. We are grateful to Anthony Sinclair, Ray Hilborn, and Grant Hopcraft for sharing the Serengeti wildebeest data. We also would like to thank Chris Wikle, Noel Cressie, Jim Clark, Ray Hilborn, Jay Ver Hoef, Jennifer Hoeting, Devin Johnson, Simon Tavener, and many other colleagues who contributed to our understanding of statistics and ecological modeling.

Appendix A

Probability Distributions and Conjugate Priors

TABLE A.1
Probability Distributions Used in Ecological Modeling to Represent Stochasticity in Discrete Random Variables (z)

Distribution	*Random variable*	*Parameters*	*Moments*
Poisson $[z\|\lambda] = \frac{\lambda^z e^{-\lambda}}{z}$	Counts of things that occur randomly over time or space, e.g., the number of birds in a forest stand, the number of fish in a kilometer of river, the number of prey captured per minute	λ, the mean number of occurrences per time or space $\lambda = \mu$	$\mu = \lambda$ $\sigma^2 = \lambda$
Binomial $[z \mid \eta, \phi] = \binom{\eta}{z}\phi^z(1-\phi)^{\eta - z}$ $\binom{\eta}{z} = \frac{\eta!}{z!(\eta - z)!}$	Number of "successes" on a given number of trials, e.g., number of survivors in a sample of individuals, number of plots containing an exotic species from a sample, number of terrestrial pixels that are vegetated in an image	η, the number of trials ϕ, the probability of a success $\phi = 1 - \sigma^2/\mu$ $\eta = \mu^2/(\mu - \sigma^2)$	$\mu = \eta\phi$ $\sigma^2 = \eta\phi(1-\phi)$
Bernoulli $[z\|\phi] = \phi^z(1-\phi)^{1-z}$	A special case of the binomial where the number of trials= 1 and the random variable can take on values 0 or 1; widely used in survival analysis, occupancy models	ϕ, the probability that the random variable= 1 $\phi = \mu$ $\phi = 1/2 + 1/2\sqrt{1 - 4\sigma^2}$	$\mu = \phi$ $\sigma^2 = \phi(1-\phi)$
Negative binomial $[z\|\lambda, \kappa] = \frac{\Gamma(z+\kappa)}{\Gamma(\kappa)z!}\left(\frac{\kappa}{\kappa+\lambda}\right)^{\kappa} \times \left(\frac{\lambda}{\kappa+\lambda}\right)^{z}$	Counts of things occurring randomly over time or space, as with the Poisson; includes dispersion parameter κ allowing the variance to exceed the mean	λ, the mean number of occurrences per time or space κ, the dispersion parameter $\lambda = \mu$ $\kappa = \mu^2/(\sigma^2 - \mu)$	$\mu = \lambda$ $\sigma^2 = \lambda + \lambda^2/\kappa$

TABLE A.1
(continued)

Distribution	*Random variable*	*Parameters*	*Moments*
Multinomial $[\mathbf{z} \mid \eta, \phi] = \eta! \prod_{i=1}^{k} \frac{\phi_i^{z_i}}{z_i}$	Counts that fall into $k > 2$ categories, e.g., number of individuals in age classes, number of pixels in different landscape categories, number of species in trophic categories in a sample from a food web	$\mathbf{z}$, a vector giving the number of counts in each category ϕ, a vector of the probabilities of occurrence in each category $\sum_{i=1}^{k} \phi_i = 1$ $\sum_{i=1}^{k} z_i = \eta$	$\mu_i = \eta\phi_i$ $\sigma_i^2 = \eta\phi_i(1-\phi_i)$

Note: We use μ to symbolize the first moment of the distribution, $\mu = E(z)$, and σ^2 to symbolize the second central moment, $\sigma^2 = E\left((z-\mu)^2\right)$.

TABLE A.2
Probability distributions Used in Ecological Modeling to Represent Stochasticity in Continuous Random Variables (z)

Continuous Distributions	*Random variable* (z)	*Parameters*	*Moments*
Normal $\left[z \mid \mu, \sigma^2\right] = \frac{1}{\sigma\sqrt{2\pi}} e^{-\frac{(z-\mu)^2}{2\sigma^2}}$	Continuously distributed quantities that can take on positive or negative values; sums of things.	μ, σ^2	μ, σ^2
Lognormal $[z \mid \alpha, \beta]$ $\frac{1}{z\sqrt{2\pi\beta^2}} e^{-\frac{(\log(z)-\alpha)^2}{2\beta^2}}$	Continuously distributed quantities with nonnegative values. Random variables with the property that their logs are normally distributed. Thus, if z is normally distributed, then exp(z) is lognormally distributed. Represents products of things. The variance increases with the mean squared.	α, the mean of z on the log scale β, the standard deviation of z on the log scale $\alpha = \log(\text{median}(z))$ $\alpha = \log(\mu) - 1/2 \log\left(\frac{\sigma^2+\mu^2}{\mu^2}\right)$ $\beta = \sqrt{\log\left(\frac{\sigma^2+\mu^2}{\mu^2}\right)}$	$\mu = e^{\alpha+\frac{\beta^2}{2}}$ $\text{median}(z_i) = e^{\alpha}$ $\sigma^2 = (e^{\beta^2}-1)e^{2\alpha+\beta^2}$

TABLE A.2
(continued)

Continuous Distributions	*Random variable* (z)	*Parameters*	*Moments*
Gamma $[z\|\alpha,\beta] = \frac{\beta^\alpha}{\Gamma(\alpha)} z^{\alpha-1} e^{-\beta z}$ $\Gamma(\alpha) = \int_0^\infty t^{\alpha-1} e^{-t}\, dt$.	The time required for a specified number of events to occur in a Poisson process; any continuous quantity that is nonnegative.	α = shape β = rate $\alpha = \frac{\mu^2}{\sigma^2}$ $\beta = \frac{\mu}{\sigma^2}$ Note–be very careful about rate, defined as above, and scale = $\frac{1}{\beta}$.	$\mu = \frac{\alpha}{\beta}$ $\sigma^2 = \frac{\alpha}{\beta^2}$
Inverse gamma $[z\|\alpha,\beta] = \frac{\beta^\alpha}{\Gamma(\alpha)} z^{-\alpha-1} \exp\left(\frac{-\beta}{z}\right)$	The reciprocal of a gamma-distributed random variable.	α = shape β = scale $\alpha = \frac{\mu^2}{\sigma^2} + 2$ $\beta = \mu\left(\frac{\mu^2}{\sigma^2} + 1\right)$	$\mu = \frac{\beta}{\alpha-1}$ for $\alpha > 1$ $\sigma^2 = \frac{\beta^2}{(\alpha-1)^2(\alpha-2)}$ for $\alpha>2$
Exponential $[z\|\alpha,\beta] = \lambda e^{-\lambda z}$	Intervals of time between sequential events that occur randomly over time or space. If the number of events is Poisson distributed, then the times between events are exponentially distributed.	λ, the mean number of occurrences per time or space $\lambda = \frac{1}{\mu}$	$\mu = \frac{1}{\lambda}$ $\sigma^2 = \left(\frac{1}{\lambda}\right)^2$
Beta $[z\|\alpha,\beta] = B\, z^{\alpha-1}(1-z)^{\beta-1}$ $B = \frac{\Gamma(\alpha+\beta)}{\Gamma(\alpha)\Gamma(\beta)}$ Because B is a normalizing constant, $[z \mid \alpha,\beta] \propto z^{\alpha-1}(1-z)^{\beta-1}$	Continuous random variables that can take on values between 0; and 1, any random variable that can be expressed as a proportion; survival; proportion of landscape invaded by exotic; probabilities of transition from one state to another.	$\alpha = \frac{(\mu^2 - \mu^3 - \mu\sigma^2)}{\sigma^2}$ $\beta = \frac{\mu - 2\mu^2 + \mu^3 - \sigma^2 + \mu\sigma^2}{\sigma^2}$	$\mu = \frac{\alpha}{\alpha+\beta}$ $\sigma^2 = \frac{\alpha\beta}{(\alpha+\beta)^2(\alpha+\beta+1)}$
Dirichlet $[z\|\boldsymbol{\alpha}] = \Gamma\left(\sum_{i=1}^{k} \alpha_i\right) \times \frac{\prod_{j=1}^{k} z_j^{\alpha_j - 1}}{\Gamma(\alpha_j)}$	Vectors of more than two elements of continuous random variables that can take on values between 0 and 1 and that sum to 1.	$\alpha_i = \mu_i \alpha_0$ $\alpha_0 = \sum_{i=1}^{k} \alpha_i$	$\mu_i = \frac{\alpha_i}{\sum_{i=1}^{k} \alpha_i}$ $\sigma_i^2 = \frac{\alpha_i(\alpha_0 - \alpha_i)}{\alpha_0^2(\alpha_0+1)}$,

TABLE A.2
(continued)

Continuous Distributions	*Random variable* (z)	*Parameters*	*Moments*
Uniform $[z\|\alpha,\beta] =$ $\frac{1}{\beta-\alpha}$ for $\alpha \le z \le \beta$, 0 for $z < \alpha$ or $z > \beta$	Any real number.	$\alpha =$ lower limit $\beta =$ upper limit $\alpha = \mu - \sigma\sqrt{3}$ $\beta = \mu + \sigma\sqrt{3}$	$\mu = \frac{\alpha+\beta}{2}$ $\sigma^2 = \frac{(\beta-\alpha)^2}{12}$

Note: We use μ to symbolize the first moment of the distribution, $\mu = E(z)$, and σ^2 to symbolize the second central moment, $\sigma^2 = E\left((z-\mu)^2\right)$.

TABLE A.3
Table of Conjugate Distributions

Likelihood	*Prior distribution*	*Posterior distribution*
$y_i \sim \text{binomial}(n, \phi)$	$\phi \sim \text{beta}(\alpha, \beta)$	$\phi \sim$ beta $\left(\sum y_i + \alpha, n - \sum y_i + \beta\right)$
$y_i \sim \text{Bernoulli}(\phi)$	$\phi \sim \text{beta}(\alpha, \beta)$	$\phi \sim$ beta $\left(\sum_{i=1}^n y_i + \alpha, \sum_{i=1}^n (1-y_i) + \beta\right)$
$y_i \sim \text{Poisson}(\lambda)$	$\lambda \sim \text{gamma}(\alpha, \beta)$	$\lambda \sim$ gamma $\left(\alpha + \sum_{i=1}^n y_i,\ \beta + n\right)$
$y_i \sim \text{normal}\left(\mu, \sigma^2\right)$; σ^2 is known	$\mu \sim \text{normal}(\mu_0, \sigma_0^2)$	$\mu \sim$ normal $\left(\frac{\left(\frac{\mu_0}{\sigma_0^2} + \frac{\sum_{i=1}^n y_i}{\sigma^2}\right)}{\left(\frac{1}{\sigma_0^2} + \frac{n}{\sigma^2}\right)}, \left(\frac{1}{\sigma_0^2} + \frac{n}{\sigma^2}\right)^{-1}\right)$
$y_i \sim \text{normal}\left(\mu, \sigma^2\right)$; μ is known.	$\sigma^2 \sim$ inverse gamma(α, β)	$\sigma^2 \sim$ inverse gamma $\left(\alpha + \frac{n}{2}, \beta + \frac{\sum_{i=1}^n (y_i-\mu)^2}{2}\right)$
$y_i \sim \text{lognormal}(\mu, \sigma^2)$; μ is known.	$\sigma^2 \sim$ inverse gamma(α, β),	$\sigma^2 \sim$ inverse gamma $\left(n/2 + \alpha, \frac{(\sum_{i=1}^n \log(y_i) - \mu)^2}{2} + \beta\right)$
$y_i \sim \text{lognormal}\left(\mu, \sigma^2\right)$; σ^2 is known.	$\mu \sim \text{normal}\left(\mu_0, \sigma_0^2\right)$	$\mu \sim$ normal $\left(\frac{\left(\frac{\mu_0}{\sigma_0^2} + \frac{\sum_{i=1}^n \log y_i}{\sigma^2}\right)}{\left(\frac{1}{\sigma_0^2} + \frac{n}{\sigma^2}\right)}, \left(\frac{1}{\sigma_0^2} + \frac{n}{\sigma^2}\right)^{-1}\right)$

BIBLIOGRAPHY

Ahrestani, F. S., M. Hebblewhite, and E. Post, 2013. The importance of observation versus process error in analyses of global ungulate populations. *Scientific Reports* **3**:31250.

Anderson, R. M., and R. M. May, 1979. Population biology of infectious diseases: Part 1. *Nature* **280**:361–367.

Ando, T., and R. Tsay, 2010. Predictive likelihood for Bayesian model selection and averaging. *International Journal of Forecasting* **26**:744–763.

Araujo, H. A., C. Holt, J.M.R. Curtis, R. I. Perry, J. R. Irvine, and C.G.J. Michielsens, 2013. Building an ecosystem model using mismatched and fragmented data: A probabilistic network of early marine survival for coho salmon *Oncorhynchus kisutch* in the Strait of Georgia. *Progress in Oceanography* **115**:41–52.

Azzalini, A., 1996. Statistical inference based on the likelihood. Chapman and Hall, London.

Barker, R. J., and W. A. Link, 2013. Bayesian multimodel inference by RJMCMC: A Gibbs sampling approach. *American Statistician* **67**:150–156.

Berliner, L. M., 1996. Hierarchical Bayesian time-series models. In *Maximum entropy and Bayesian methods*, 15–22. Dordrecht: Kluwer.

Bolker, B., 2008. *Ecological models and data in R*. Princeton, NJ: Princeton University Press.

———, 2013. emdbook: Ecological models and data (book support). R package vernon 1.3.4. http://ms.mcmaster.ca/bolker/emdbook/.

Boone, R. B., S. J. Thirgood, and J.G.C. Hopcraft, 2006. Serengeti wildebeest migratory patterns modeled from rainfall and new vegetation growth. *Ecology* **87**:1987–1994.

Braakhekke, M. C., T. Wutzler, C. Beer, J. Kattge, M. Schrumpf, B. Ahrens, I. Schoening, M. R. Hoosbeek, B. Kruijt, P. Kabat, and M. Reichstein, 2013. Modeling the vertical soil organic matter profile using Bayesian parameter estimation. *Biogeosciences* **10**:399–420.

Broquet, G., F. Chevallier, F. M. Breon, N. Kadygrov, M. Alemanno, F. Apadula, S. Hammer, L. Haszpra, F. Meinhardt, J. A. Morgui, J. Necki, S. Piacentino, M. Ramonet, M. Schmidt, R. L. Thompson, A. T. Vermeulen, C. Yver, and P. Ciais, 2013. Regional inversion of CO_2 ecosystem fluxes from atmospheric measurements: Reliability of the uncertainty estimates. *Atmospheric Chemistry and Physics* **13**:9039–9056.

Brown, J. H., V. K. Gupta, B. L. Li, B. T. Milne, C. Restrepo, and G. B. West, 2002. The fractal nature of nature: Power laws, ecological complexity and biodiversity. *Philosophical Transactions of the Royal Society B: Biological Sciences* **357**:619–626.

Burnham, K. P., and D. R. Anderson, 2002. *Model selection and multimodel inference*, 2nd ed. New York: Springer-Verlag.

Campbell, E. W., III, A. A. Adams, S. J. Converse, T. H. Fritts, and G. H. Rodda, 2012. Do predators control prey species abundance? An experimental test with brown treesnakes on Guam. *Ecology* **93**:1194–1203.

Carey, R. K., 2007. Modeling NO_2 emissions from agricultural soils using a multi-level linear regression. PhD thesis, Duke University, Durham, NC.

Carlin, B., and S. Chib, 1995. Bayesian model choice via Markov chain Monte Carlo methods. *Journal of the Royal Statistical Society, Series B* **57**:473–484.

Casella, G., and R. L. Berger, 2002. *Statistical inference*. Pacific Grove, CA: Duxbury.

Caswell, H., 1988. *Matrix population models*. Sunderland, MA: Sinauer.

Celeux, G., F. Forbes, C. P. Robert, and D. M. Titterington, 2006. Deviance information criteria for missing data models. *Bayesian Analysis* **1**:651–673.

Chave, J., R. Condit, S. Aguilar, A. Hernandez, S. Lao, and R. Perez, 2004. Error propagation and scaling for tropical forest biomass estimates. *Philosophical Transactions of the Royal Society B: Biological Sciences* **359**:409–420.

Clark, J. M., 2007. *Models for ecological data*. Princeton, NJ: Princeton University Press.

Clark, J. S., 2003a. Uncertainty and variability in demography and population growth: A hierarchical approach. *Ecology* **84**:1370–1381.

———, 2003b. Uncertainty in ecological inference and forecasting. *Ecology* **84**:1349–1350.

———, 2005. Why environmental scientists are becoming Bayesians. *Ecology Letters* **8**:2–14.

Clark, J. S., P. Agarwal, D. M. Bell, P. G. Flikkema, A. Gelfand, X. L. Nguyen, E. Ward, and J. Yang, 2011. Inferential ecosystem models, from network data to prediction. *Ecological Applications* **21**:1523–1536.

Coates, K. D., and P. J. Burton, 1999. Growth of planted tree seedlings in response to ambient light levels in northwestern interior cedar-hemlock forests of British Columbia. *Canadian Journal of Forest ResearchRevue Canadienne de Recherche Forestière* **29**:1374–1382.

Congdon, P., 2006. Bayesian model choice based on Monte Carlo estimates of posterior model probabilities. *Computational Statistics and Data Analysis* **50**:346–357.

Cressie, N., C. A. Calder, J. S. Clark, J. M. Ver Hoef, and C. K. Wikle, 2009. Accounting for uncertainty in ecological analysis: The strengths and limitations of hierarchical statistical modeling. *Ecological Applications* **19**:553–570.

Cressie, N., and C. K. Wikle, 2011. *Statistics for spatio-temporal data*. New York: Wiley.

Crome, F.H.J., M. R. Thomas, and L. A. Moore, 1996. A novel Bayesian approach to assessing impacts of rain forest logging. *Ecological Applications* **6**:1104–1123.

Dellaportas, P., and D. A. Stephens, 1995. Bayesian analysis of error-in-variables regression models. *Biometrics* **51**:1085–1095.

Dennis, B., J. M. Ponciano, S. R. Lele, M. L. Taper, and D. F. Staples, 2006. Estimating density dependence, process noise, and observation error. *Ecological Monographs* **76**:323–341.

Diekman, O., H. Heesterbeek, and T. Britton, 2012. *Mathematical tools for understanding infectious disease dynamics*. Princeton, NJ: Princeton University Press.

Dobson, F. S., B. Zinner, and M. Silva, 2003. Testing models of biological scaling with mammalian population densities. *Canadian Journal of ZoologyRevue Canadienne de Zoologie* **81**:844–851.

Eaton, M. J., and W. A. Link, 2011. Estimating age from recapture data: Integrating incremental growth measures with ancillary data to infer age-at-length. *Ecological Applications* **21**:2487–2497.

Edelstein-Keshet, L., 1988. *Mathematical models in biology*. New York: McGrawh Hill.

Edwards, E.W.F., 1992. *Likelihood*. Baltimore, MD: Johns Hopkins University Press.

Elbroch, L. M., and H. U. Wittmer, 2012. Puma spatial ecology in open habitats with aggregate prey. *Mammalian Biology* **77**:377–384.

Elderd, B. D., V. M. Dukic, and G. Dwyer, 2006. Uncertainty in predictions of disease spread and public health responses to bioterrorism and emerging diseases. *Proceedings of the National Academy of Sciences* **103**:15693–15697.

Farnsworth, M. L., J. A. Hoeting, N. T. Hobbs, and M. W. Miller, 2006. Linking chronic wasting disease to mule deer movement scales: A hierarchical Bayesian approach. *Ecological Applications* **16**:1026–1036.

Fiechter, J., R. Herbei, W. Leeds, J. Brown, R. Milliff, C. Wikle, A. Moore, and T. Powell, 2013. A Bayesian parameter estimation method applied to a marine ecosystem model for the coastal Gulf of Alaska. *Ecological Modelling* **258**:122–133.

Fisher, R. A., 1941. The negative binomial distribution. *Annals of Eugenics* **11**:182–187.

Gelfand, A. E., and S. K. Ghosh, 1998. Model choice: A minimum posterior predictive loss approach. *Biometrika* **85**:1–11.

Gelfand, A. E., and A.F.M. Smith, 1990. Sampling-based approaches to calculating marginal densities. *Journal of the American Statistical Association* **85**:398–409.

Gelman, A., 2006. Prior distributions for variance parameters in hierarchical models. *Bayesian Analysis* **1**:1–19.

Gelman, A., J. B. Carlin, H. S. Stern, and D. B. Rubin, 2004. *Bayesian data analysis*. London: Chapman and Hall/CRC.

Gelman, A., and J. Hill, 2009. *Data analysis using regression and multilevel/hierarchical models*. Cambridge: Cambridge University Press.

Gelman, A., J. Hwang, and A. Vehtari, 2014. Understanding predictive information criteria for Bayesian models. *Statistics and computing* **24**: 997–1016.

Gelman, A., G. O. Roberts, and W. R. Gilks, 1996. Efficient Metropolis jumping rules. *Bayesian Statistics* **5**:599–607.

Gelman, A., and D. B. Rubin, 1992. Inference from iterative simulation using multiple sequences. *Statistical Science* **7**:457–511.

———, 1995. Avoiding model selection in Bayesian social research. *Sociological Methodology* **25**:165–173.

Gelman, A., and C. R. Shalizi, 2013. Philosophy and the practice of Bayesian statistics. *British Journal of Mathematical and Statistical Psychology* **66**:8–38.

George, E. I., and R. E. McCulloch, 1993. Variable selection via Gibbs sampling. *Journal of the American Statistical Association* **88**:881–889.

Geweke, J., 1992. Evaluating the accuracy of sampling-based approaches to calculating posterior moments. Oxford: Clarendon Press.

Gimenez, O., J.-D. Lebreton, J.-M. Gaillard, R. Choquet, and R. Pradel, 2012. Estimating demographic parameters using hidden process dynamic models. *Theoretical Population Biology* **82**:307–316.

Gneiting, T., 2011. Making and evaluating point forecasts. *Journal of the American Statistical Association* **106**:746–762.

Golley, F. B., 1993. *A history of the ecosystem concept in ecology: More than the sum of the parts*. New Haven, CT: Yale University Press.

Good, I. J., 1983. Explicativity, corroboration, and the relative odds of hypotheses. Minneapolis: University of Minnesota Press.

Gotelli, N. J., and A. M. Ellison, 2004. A primer of ecological statistics. Sunderland, MA: Sinauer.

Gross, K., B. A. Craig, and W. D. Hutchinson, 2002. Bayesian estimation of a demographic matrix model from stage-frequency data. *Ecology* **83**:3285–3298.

Gross, K., A. R. Ives, and E. V. Nordheim, 2005. Estimating fluctuating vital rates from time-series data: A case study of aphid biocontrol. *Ecology* **86**:740–752.

Gudimov, A., E. O'Connor, M. Dittrich, H. Jarjanazi, M. E. Palmer, E. Stainsby, J. G. Winter, J. D. Young, and G. B. Arhonditsis, 2012. Continuous Bayesian network for studying the causal links between phosphorus loading and plankton patterns in Lake Simcoe, Ontario, Canada. *Environmental Science and Technology* **46**:7283–7292.

Hamel, S., J. M. Craine, and E. G. Towne, 2012. Maternal allocation in bison: Co-occurrence of senescence, cost of reproduction, and individual quality. *Ecological Applications* **22**:1628–1639.

Haskell, J. P., M. E. Ritchie, and H. Olff, 2002. Fractal geometry predicts varying body size scaling relationships for mammal and bird home ranges. *Nature* **418**:527–530.

Hastie, D. I., and P. J. Green, 2012. Model choice using reversible jump Markov chain Monte Carlo. *Statistica Neerlandica* **66**:309–338.

Hastie, T., R. Tibshivani, and J. Friedman, 2009. *The elements of statistical learning: Data mining, inference, and prediction*, 2nd ed. New York: Springer.

Hefley, T. J., A. J. Tyre, and E. E. Blankenship, 2013. Fitting population growth models in the presence of measurement and detection error. *Ecological Modelling* **263**:244–250.

Heidelberger, P., and P. Welch, 1983. Simulation run length control in the presence of an initial transient. *Operations Research* **31**:1109–1044.

Hilborn, R., and M. Mangel, 1997. *The ecological detective: Confronting models with data*. Princeton, NJ: Princeton University Press.

Hill, A. V., 1910. The possible effects of the aggregation of the molecules of hæmoglobin on its dissociation curves. *Journal of Physiology* **40**:i–vii.

Hoeting, J. A., D. Madigan, A. E. Raftery, and C. T. Volinsky, 1999. Bayesian model averaging: A tutorial. *Statistical Science* **14**:382–401.

Holling, C. S., 1959. Some characteristics of simple types of predation and parasitism. *Canadian Entomologist* **91**:385–398.

Hooten, M., and N. T. Hobbs, 2015. A guide to Bayesian model selection for ecologists. *Ecological Monographs* **85**:3–28.

INLA Development Team, 2014. www.r-inla.org.

Innis, G., 1978. *Grassland simulation model*. New York: Springer-Verlag.

Jeffreys, H., 1961. *Theory of probability*, 3rd ed. Oxford: Oxford University Press.

Johnson, D. S., and J. A. Hoeting, 2011. Bayesian multimodel inference for geostatistical regression models. *PLOS ONE* **6**: e25677.

Jonsen, I. D., J. M. Flemming, and R. A. Myers., 2005. Robust state-space modeling of animal movement data. *Ecology* **86**:2874–2880.

Judson, O. P., 1994. The rise of the individual-based model in ecology. *Trends in Ecology & Evolution* **9**:9–14.

Kass, R. E., and A. E. Raftery, 1995. Bayes factors. *Journal of the American Statistical Association* **90**:773–795.

Kéry, M., 2010. Introduction to WinBUGS for ecologists: *A Bayesian approach to regression, ANOVA, mixed models and related analyses*. Walthem, MA: Academic Press.

Kéry, M., and M. Schaub, 2012. *Bayesian population analysis using WinBUGS: A hierarchical perspective*. Waltham, MA: Academic Press.

King, R., and O. Gimenez, 2009. *Bayesian analysis for population ecology*. Cambridge: Cambridge University Press.

Kingsland, S. E., 1985. *Modeling nature: Episodes in the history of population ecology*. Chicago: University of Chicago Press.

Latimer, A. M., S. S. Wu, A. E. Gelfand, and J. A. Silander, 2006. Building statistical models to analyze species distributions. *Ecological Applications* **16**:33–50.

Levins, R., 1966. The strategy of model building in population biology. *American Scientist* **54**:421–431.

———, 1969. Some demographic and genetic consequences of environmental heterogeneity for biological control. *Bulletin of the Entomological Society of America* **15**:237–240.

Liberg, O., G. Chapron, P. Wabakken, H. C. Pedersen, N. T. Hobbs, and H. Sand, 2012. *Shoot, shovel and shut up: Cryptic poaching slows restoration of a large carnivore in Europe. Proceedings of the Royal Society B: Biological Sciences* **279**:910–915.

Link, W. A., and R. J. Barker, 2006. Model weights and the foundations of multimodel inference. *Ecology* **87**:2626–2635.

———, 2010. *Bayesian inference with ecological applications*. Waltham, MA: Academic Press.

Link, W. A., and M. J. Eaton, 2012. On thinning of chains in MCMC. *Methods in Ecology and Evolution* **3**:112–115.

Lunn, D., T. A., N. Best, and D. Spiegelhalter, 2000. WinBUGS: A Bayesian modelling framework, Concepts, structure, and extensibility. *Statistics and Computing* **10**:325–337.

Luo, Y., K. Ogle, C. Tucker, S. Fei, C. Gao, S. LaDeau, J. S. Clark, and D. S. Schimel, 2011. Ecological forecasting and data assimilation in a data-rich era. *Ecological Applications* **21**:1429–1442.

MacArthur, R., and E. Wilson, 2001. *The theory of island biogeography*. Princeton, NJ: Princeton University Press.

Madigan, D., and A. E. Raftery, 1994. Model selection and accounting for model uncertainty in graphical models using Occam's window. *Journal of the American Statistical Association* **89**:1535–1546.

Manier, D. J., and N. T. Hobbs, 2007. Large herbivores in sagebrush steppe ecosystems: Livestock and wild ungulates influence structure and function. *Oecologia* **152**:739–750.

Marquet, P. A., R. A. Quiñones, S. Abades, F. Labra, M. Tognelli, M. Arim, and M. Rivadeneira, 2005. Scaling and power-laws in ecological systems. *Journal of Experimental Biology* **208**:1749–1769.

May, R. M., 1977. Thresholds and breakpoints in ecosystems with a multiplicity of stable states. *Nature* **269**:471–477.

McCarthy, M. A., 2007. *Bayesian methods for ecology*. Cambridge: Cambridge University Press.

McCarthy, M. A., R. Citroen, and S. C. McCall, 2008. Allometric scaling and Bayesian priors for annual survival of birds and mammals. *American Naturalist* **172**:216–222.

McCarthy, M. A., and P. Masters, 2005. Profiting from prior information in Bayesian analyses of ecological data. *Journal of Applied Ecology* **42**:1012–1019.

McNaughton, S. J., D. G. Milchunas, and D. A. Frank, 1966. How can net primary productivity be measured in grazing ecosystems? *Ecology* **77**:974–997.

Mduma, S.A.R., A.R.E. Sinclair, and R. Hilborn, 1999. Food regulates the Serengeti wildebeest: A 40-year record. *Journal of Animal Ecology* **68**:1101–1122.

Miller, A., 2002. *Subset selection in regression*, 2nd ed. Boca Raton, FL: Chapman and Hill/CRC.

Newman, K. B., S. T. Buckland, S. T. Lindley, L. Thomas, and C. Fernandez, 2006. Hidden process models for animal population dynamics. *Ecological Applications* **16**:74–86.

Norton-Griffiths, M., 1973. Counting the Serengeti migratory wildebeest using two-stage sampling. *African Journal of Ecology* **11**:135–149.

O'Hara, R. B., and M. J. Sillanpaa, 2009. A review of Bayesian variable selection methods: What, how and which. *Bayesian Analysis* **4**:85–117.

Otto, S. P., and T. Day, 2007. *A biologist's guide to mathematical modeling in ecology and evolution*. Princeton, NJ: Princeton University Press.

Pacella, S. R., B. Lebreton, P. Richard, D. Phillips, T. H. DeWitt, and N. Niquil, 2013. Incorporation of diet information derived from Bayesian stable isotope mixing models into mass-balanced marine ecosystem models: A case study from the Marennes-Oleron estuary, France. *Ecological Modelling* **267**:127–137.

Part, T., and P. Forslund, 1996. Age and breeding performance in monogamous birds: The influence of mate experience. *Trends in Ecology and Evolution* **11**:220–220.

Pawitan, Y., 2001. *In all likelihood: Statistical modeling and inference using likelihood*. Oxford: Oxford Scientific.

Pennycuick, C. J., 1992. *Newton rules biology*. Oxford: Oxford University Press.

Peters, 1983. *The ecological implications of body size*. Cambridge: Cambridge University Press.

Peterson, E. E., J. M. Ver Hoef, D. J. Isaak, J. A. Falke, M. J. Fortin, C. E. Jordan, K. McNyset, P. Monestiez, A. S. Ruesch, A. Sengupta, N. Som, E. A. Steel, D. M. Theobald, C. E. Torgersen, and S. J. Wenger, 2013. Modelling dendritic ecological networks in space: An integrated network perspective. *Ecology Letters* **16**:707–719.

Plowright, W., 1982. The effects of rinderpest and rinderpest control on wildlife in Africa. *Symposium of Zoological Society of London* **50**:1:28.

Plummer, M., 2002. Discussion of the paper by Spiegelhalter et al. *Journal of the Royal Statistical Society B* **64**:620.

Plummer, M., 2003. JAGS: A program for analysis of Bayesian graphical models using Gibbs sampling. *DSC Working Papers, Proceedings of the 3rd International Workshop on Distributed Statistical Computing,* March 20–22, Technische Universität Wien, Vienna, Austria, `www.ci.tuwien.ac.at/Conferences/DSC-2003/`.

Plummer, M., 2008. Penalized loss functions for Bayesian model comparison. *Biostatistics* **9**:523–539.

Plummer, M., N. Best, K. Cowles, and K. Vines, 2010. coda: Output analysis and diagnostics for MCMC. R package version 0.14-4. `http://CRAN.R-project.org/package=coda`.

Polishchuk, L. V., and V. B. Tseitlin, 1999. Scaling of population density on body mass and a number-size trade-off. *Oikos* **86**:544–556.

Price, C. A., K. Ogle, E. P. White, and J. S. Weitz, 2009. Evaluating scaling models in biology using hierarchical Bayesian approaches. *Ecology Letters* **12**: 641–651.

Qian, S. S., T. F. Cuffney, I. Alameddine, G. McMahon, and K. H. Reckhow, 2010. On the application of multilevel modeling in environmental and ecological studies. *Ecology* **91**:355–361.

R Core Team, 2013. R: A Language and Environment for Statistical Computing. R Foundation for Statistical Computing, Vienna, Austria.

Raftery, A., and S. Lewis, 1995. *The number of iterations, convergence diagnostics and generic Metropolis algorithms*. London: Chapman and Hall.

Railsback, S. F., and V. Grimm, 2012. *Agent-based and indvidual-based modeling: A practical introduction*. Princeton, NJ: Princeton University Press.

Ramsey, F. and D. Schafer, 2012. *The statistical sleuth: A course in methods of data analysis*, 3rd ed. Boston, MA: Cengage Learning.

Richardson, S., 2002. Discussion of the paper by Spiegelhalter et al. *Journal of the Royal Statistical Society B* **64**:626–627.

Ritchie, M. E., 1998. Scale-dependent foraging and patch choice in fractal environments. *Evolutionary Ecology* **12**:309–330.

Rotella, J. J., W. A. Link, J. D. Nichols, G. L. Hadley, R. A. Garrott, and K. M. Proffitt, 2009. An evaluation of density-dependent and density-independent influences on population growth rates in Weddell seals. *Ecology* **90**:975–984.

Royall, R., 1997. *Statistical evidence: A likelihood paradigm*. Boca Raton, FL: Chapman and Hall/CRC.

Royle, J. A., 2004. *N*-mixture models for estimating population size from spatially replicated counts. *Biometrics* **60**:108–115.

Royle, J. A., and R. M. Dorazio, 2008. *Hierarchical modeling and inference in ecology: The analysis of data from populations, metapopulations, and communities*. London: Academic Press.

Rüger, N., U. Berger, S. P. Hubbell, G. Vieilledent, and R. Condit, 2011. Growth strategies of tropical tree species: Disentangling light and size effects. *PLOS ONE* **6**.

Rüger, N., C. Wirth, S. J. Wright, and R. Condit, 2012. Functional traits explain light and size response of growth rates in tropical tree species. *Ecology* **93**:2626–2636.

Schneider, D. C., 1997. *Quantitative ecology: Spatial and temporal scaling*. New York: Academic Press.

Schwartzman, G. L., and S. P. Kaluzny, 1987. *Ecological simulation primer. Series in biological resource management*. New York: Macmillan.

Schwarz, G., 1978. Estimating the dimension of a model. *Annals of Statistics* **6**:461–464.

Seaman J. W., III, J. W. Seaman Jr., and J. D. Stamey, 2012. Hidden dangers of specifying noninformative priors. *American Statistician* **66**.

Silva, M., and J. A. Downing, 1995. The allometric scaling of density and body-mass: A nonlinear relationship for terrestrial mammals. *American Naturalist* **145**:704–727.

Sinclair, A.R.E., 2003. Mammal population regulation, keystone processes and ecosystem dynamics. *Philosophical Transactions of the Royal Society of London B: Biological Sciences* **358**:1729–1740.

Sokal, R. R., and F. J. Rohlf, 1995. *Biometry: The principles and practices of statistics in biological research*. New York: W. H. Freeman.

Spalinger, D. E., and N. T. Hobbs, 1992. Mechanisms of foraging in mammalian herbivores: New models of functional response. *American Naturalist* **140**:325–348.

Spiegelhalter, D. J., N. G. Best, B. P. Carlin, and A. van der Line, 2002. Bayesian measures of model complexity and fit. *Journal of the Royal Statistical Society B* **64**:583–639.

Stan Development Team, 2014. Stan: A C++ library for probability and sampling, version 2.2. mc-stan.org/.

Stauffer, H. B., 2008. *Contemporary Bayesian and frequentist statistical research methods for natural resource scientists*. Hoboken, NJ: Wiley Interscience.

Tanner, M. A., 1996. *Tools for statistical inference: Methods for the exploration of posterior distributions and likelihood functions*. New York: Springer.

Tavecchia, G., P. Besbeas, T. Coulson, B.J.T. Morgan, and T. H. Clutton-Brock, 2009. Estimating population size and hidden demographic parameters with state-space modeling. *American Naturalist* **173**:722–733.

Tredennick, A. T., L. P. Bentley, and N. P. Hanan, 2013. Allometric convergence in savanna trees and implications for the use of plant scaling models in variable ecosystems. *PLOS ONE* **8**.

van Kleunen, M., E. Weber, and M. Fischer, 2010. A meta-analysis of trait differences between invasive and non-invasive plant species. *Ecology Letters* **13**:235–245.

Ver Hoef, J. M., 2015. The hidden costs of multimodel inference. *Journal of Wildlife Management*. In press.

Watanabe, S., 2010. Asymptotic equivalence of Bayes cross validation and widely applicable information criterion in singular learning theory. *Journal of Machine Learning Research* **11**:3571–3594.

———, 2013. A widely applicable Bayesian information criterion. *Journal of Machine Learning Research* **14**:867–897.

Webb, C. T., J. A. Hoeting, G. M. Ames, M. I. Pyne, and N. L. Poff, 2010. A structured and dynamic framework to advance traits-based theory and prediction in ecology. *Ecology Letters* **13**:267–283.

West, G. B., V. M. Savage, J. Gillooly, B. J. Enquist, W. H. Woodruff, and J. H. Brown, 2003. Why does metabolic rate scale with body size? *Nature* **421**:713–713.

Westoby, M., D. S. Falster, A. T. Moles, P. A. Vesk, and I. J. Wright, 2002. Plant ecological strategies: Some leading dimensions of variation between species. *Annual Review of Ecology and Systematics* **33**:125–159.

Westoby, M., and I. J. Wright, 2006. Land-plant ecology on the basis of functional traits. *Trends in Ecology & Evolution* **21**:261–268.

Wikle, C. K., 2003. Hierarchical models in environmental science. *International Statistical Review* **71**:181–199.

Wikle, C. K., R. F. Milliff, R. Herbei, and W. B. Leeds, 2013. Modern statistical methods in oceanography: A hierarchical perspective. *Statistical Science* **28**:466–486.

Wilson, A. M., A. M. Latimer, J. A. Silander, A. E. Gelfand, and H. de Klerk, 2011. A hierarchical Bayesian model of wildfire in a Mediterranean biodiversity hotspot: Implications of weather variability and global circulation. *Ecological Modelling* **221**:106–112. Corrigendum.

Xiao, X., E. P. White, M. B. Hooten, and S. L. Durham, 2011. On the use of log-transformation vs. nonlinear regression for analyzing biological power laws. *Ecology* **92**:1887–1894.

INDEX

Milton Keynes UK
Ingram Content Group UK Ltd.
UKHW011845110624
443863UK00003B/28